Bonobos

Bonobos

Unique in Mind, Brain and Behavior

EDITED BY

Brian Hare
Associate Professor of the Department of Evolutionary Anthropology, Duke University, USA

Shinya Yamamoto
Associate Professor of Kyoto University Institute for Advanced Study, Japan

OXFORD
UNIVERSITY PRESS

OXFORD
UNIVERSITY PRESS

Great Clarendon Street, Oxford, OX2 6DP,
United Kingdom

Oxford University Press is a department of the University of Oxford.
It furthers the University's objective of excellence in research, scholarship,
and education by publishing worldwide. Oxford is a registered trade mark of
Oxford University Press in the UK and in certain other countries

© Oxford University Press 2017

The moral rights of the authors have been asserted

First Edition published in 2017

Impression: 1

All rights reserved. No part of this publication may be reproduced, stored in
a retrieval system, or transmitted, in any form or by any means, without the
prior permission in writing of Oxford University Press, or as expressly permitted
by law, by licence or under terms agreed with the appropriate reprographics
rights organization. Enquiries concerning reproduction outside the scope of the
above should be sent to the Rights Department, Oxford University Press, at the
address above

You must not circulate this work in any other form
and you must impose this same condition on any acquirer

Published in the United States of America by Oxford University Press
198 Madison Avenue, New York, NY 10016, United States of America

British Library Cataloguing in Publication Data
Data available

Library of Congress Control Number: 2017943719

ISBN 978–0–19–872851–1 (hbk.)
ISBN 978–0–19–872852–8 (pbk.)

DOI: 10.1093/oso/ 9780198728511.003

Printed and bound by
CPI Litho (UK) Ltd, Croydon, CR0 4YY

Links to third party websites are provided by Oxford in good faith and
for information only. Oxford disclaims any responsibility for the materials
contained in any third party website referenced in this work.

To all bonobos and those who protect them

Foreword

It is hard to imagine now, but a hundred years ago bonobos and chimpanzees were still viewed as a single species by Western science. There were rumours of differences between them by a perceptive Dutch zoologist, Frits Portielje, who in 1916 speculated that a popular young male ape at the Amsterdam Zoo named Mafuca might represent a new species. A photograph shows that Mafuca was a bonobo. A few years later, Robert Yerkes, the famous American primatologist, was impressed by 'Prince Chim'. He considered this ape more sensitive, altruistic and intelligent than any other he knew (Yerkes, 1925). Prince Chim, too, turned out to have been a bonobo. Harold Coolidge, the anatomist who gave *Pan paniscus* its eventual taxonomic status as a separate species, provocatively concluded from a post-mortem on Prince Chim that the bonobo 'may approach more closely to the common ancestor of chimpanzees and man than does any living chimpanzee' (Coolidge, 1933, p. 56).

The first study of substance comparing bonobo and chimpanzee behavior was carried out in the 1930s at a zoo in Munich. Eduard Tratz and Heinz Heck published their findings after the Second World War in a succinct German-language paper that is still very much apropos (Tratz and Heck, 1954). Modern research started when Takayoshi Kano set up a field site at Wamba in the Democratic Republic of the Congo, in 1973, when Kanzi's training on symbolic language began at the Yerkes Primate Center, in 1980, and when I began watching bonobos at the San Diego Zoo, in 1983. In the meantime, research at the Lomako Forest site in the DRC was gearing up. These early studies left little doubt that bonobos are behaviorally quite distinct. Compared to chimpanzees, they are more peaceful, more sex- than power-oriented, and the females occupy a more central, even dominant position in society (de Waal, 1997; Kano, 1992).

Until late last century, these apes were known as 'pygmy chimpanzees'. This was the only accepted common name in science. It had the great drawback, however, of obscuring species differences by representing the bonobo as merely a diminutive version of the 'real' chimpanzee. Since the size difference is not nearly as great as the term 'pygmy' suggests, and is surely one of the least interesting aspects of the species comparison, my own writing always followed the advice of Tratz and Heck, who promoted the bonobo name as a way to grant these apes the special attention they deserve.

We are now benefiting from a second wave of studies on this fascinating species, both in the field and in captivity, dealing with more specific problems, including for the first time also those regarding cognition. The main theme remains a direct comparison between both *Pan* species, as also attested by the chapters in the present volume. One characteristic that has received much attention is female social cohesiveness and status in bonobo society. Given that females are the migratory sex in the *Pan* genus, one would expect males to be more cooperative in both ape species. Males are related to each other, whereas the females—arriving from diverse origins—are mostly unrelated. The bonobo poses a huge challenge to the common dismissal of non-human cooperation as mostly kinship-based, because female bonobos band together and collectively dominate the males despite a lack of shared genes. Along the same lines, traditional views of prosociality have been up-ended by the experiments reported by Tan and Hare (this volume), who conclude that bonobos do not even need to live in the same group to help each other. Reporting remarkable cooperation and food sharing without tangible returns in bonobos, they conclude that 'future models of human prosociality will need to incorporate findings from both *Pan* species'.

There has been an unfortunate tendency in the literature on human evolution to ignore or marginalize the bonobo. That the species arrived much later on the scene than the chimpanzee is hardly an excuse. Perhaps they are too peaceful or too female-dominated to appeal to those trying to explain our own origins. Evolutionary scenarios tend to favour the chimpanzee, with its male dominance, group hunting and territorial aggression, as model of the last common ancestor. There is no need, however, to choose between bonobo and chimpanzee as ancestral models, nor does it seem wise to focus exclusively on male behavior. The earlier citation from Coolidge was not offered to make the point that the bonobo makes for a *better* ancestral model, but at the very least both *Pan* species should be put on an equal footing. Genome comparisons place chimpanzees and bonobos genetically equidistant from our species (Prüfer et al., 2012).

Without going into detail about the chapters in this volume, which treat not only behavior and cognition but also the dire conservation situation of this primate, it is obvious that the bonobo is attracting increasing attention. We are finally making up for years of relative neglect. This is made possible by the decrease in major warfare in the Democratic Republic of Congo, and also by an ever more sophisticated testing arsenal for cognition, cooperation, prosociality, tool use and so on. We have an enthusiastic cadre of field-workers who work under the most difficult conditions, as well as observers and experimenters focused on captive apes, who test a wide range of ideas to see how the two *Pan* species compare. There exist many fundamental similarities, but also striking differences related to playfulness, empathy, conflict resolution, communication and cortical organization. There is also the puzzle of why bonobos, known as great tool users in captivity, rarely use tools in the wild. The answer to some of these questions needs to come from detailed comparisons between the ecologies of both *Pan* species, which may help us understand the selection pressures responsible for the differences. Other questions will be answered by non-invasive neuroscience, eye-tracking, physiological measures and other novel techniques possible only under controlled conditions or with the help of dedicated laboratories. Broad collaboration will be absolutely essential for progress, and it is great to see how lab and field approaches merge in this volume.

It is no exaggeration to speak of an explosion of new knowledge, which is very well captured in *Bonobos: Unique in Mind, Brain and Behavior.* This landmark volume will be required reading not only for students of primate behavior but also for those interested in human evolution in the broadest sense. Because in the end, if we manage to understand how the behavior of two sister species came to diverge to such a degree, we will also learn more about our own species. Not only are we each others' closest living relatives but in our own lineage, too, a relatively small number of genetic modifications has given rise to a great many special traits.

Frans B. M. de Waal

Contents

Foreword   vii
List of Contributors   xiii

1 Minding the bonobo mind   1
Brian Hare and Shinya Yamamoto

PART I Society   15

2 Female contributions to the peaceful nature of bonobo society   17
Takeshi Furuichi

3 Affiliations, aggressions and an adoption: Male–male relationships in wild bonobos   35
Martin Surbeck and Gottfried Hohmann

PART II Social Development   47

4 Bonobo baby dominance: Did female defense of offspring lead to reduced male aggression?   49
Kara Walker and Brian Hare

5 *Pan paniscus* or *Pan ludens*? Bonobos, playful attitude and social tolerance   65
Elisabetta Palagi and Elisa Demuru

PART III Mind and Communication   79

6 Does the bonobo have a (chimpanzee-like) theory of mind?   81
Christopher Krupenye, Evan L. MacLean and Brian Hare

7 What did we learn from the ape language studies?   95
Michael Tomasello

8 Natural communication in bonobos: Insights into social awareness and the evolution of language   105
Zanna Clay and Emilie Genty

PART IV Cooperation — 123

9 Courtesy food sharing characterized by begging for social bonds in wild bonobos — 125
Shinya Yamamoto and Takeshi Furuichi

10 Prosociality among non-kin in bonobos and chimpanzees compared — 140
Jingzhi Tan and Brian Hare

PART V Foraging Strategies — 155

11 Ecological variation in cognition: Insights from bonobos and chimpanzees — 157
Alexandra G. Rosati

12 Bonobos, chimpanzees and tools: Integrating species-specific psychological biases and socio-ecology — 171
Josep Call

PART VI Mind and Brains Compared — 181

13 Bonobo personality: Age and sex effects and links with behavior and dominance — 183
Nicky Staes, Marcel Eens, Alexander Weiss and Jeroen M.G. Stevens

14 Social cognition and brain organization in chimpanzees (*Pan troglodytes*) and bonobos (*Pan paniscus*) — 199
William D. Hopkins, Cheryl D. Stimpson and Chet C. Sherwood

15 Cognitive comparisons of genus *Pan* support bonobo self-domestication — 214
Brian Hare and Vanessa Woods

PART VII Evolution — 233

16 The formation of Congo River and the origin of bonobos: A new hypothesis — 235
Hiroyuki Takemoto, Yoshi Kawamoto and Takeshi Furuichi

PART VIII Conservation and Captive Care — 249

17 Geospatial information informs bonobo conservation efforts — 251
Janet Nackoney, Jena Hickey, David Williams, Charly Facheux, Takeshi Furuichi and Jef Dupain

**18 Bonobo population dynamics: Past patterns and future predictions
for the Lola ya Bonobo population using demographic modelling** **266**
Lisa J. Faust, Claudine André, Raphaël Belais, Fanny Minesi, Zjef Pereboom,
Kerri Rodriguez and Brian Hare

Afterword 275
  Richard Wrangham

Index 279

List of Contributors

Claudine André Lola ya Bonobo, Kinshasa, Democratic Republic of Congo

Raphaël Belais Lola ya Bonobo, Kinshasa, Democratic Republic of Congo

Josep Call School of Psychology and Neuroscience, University of St Andrews, United Kingdom, and Max Planck Institute for Evolutionary Anthropology, Leipzig, Germany

Zanna Clay Department of Psychology, Durham University, Durham, United Kingdom

Elisa Demuru Centro di Ateneo, Museo di Storia Naturale, Università di Pisa, Calci, Pisa, Italy

Jef Dupain African Wildlife Foundation, Nairobi, Kenya

Marcel Eens Department of Biology, Behavioral Ecology and Ecophysiology Group, University of Antwerp, Belgium

Charly Facheux African Wildlife Foundation, Nairobi, Kenya

Lisa J. Faust Alexander Center for Applied Population Biology, Department of Conservation and Science, Lincoln Park Zoo, Chicago, Illinois, USA

Takeshi Furuichi Primate Research Institute, Kyoto University, Inuyama, Japan

Emilie Genty Department of Comparative Cognition, Institute of Biology, University of Neuchatel, Neuchatel, Switzerland

Brian Hare Department of Evolutionary Anthropology, Duke University, and Center for Cognitive Neuroscience, Duke University, Durham, North Carolina, USA

Jena Hickey International Gorilla Conservation Programme, Kigali, Rwanda

Gottfried Hohmann Max Planck Institute for Evolutionary Anthropology, Department of Primatology, Leipzig, Germany

William D. Hopkins Neuroscience Institute, Georgia State University, Atlanta, Georgia 30302, and Division of Developmental and Cognitive Neuroscience, Yerkes National Primate Research Center, Atlanta, Georgia, and Ape Cognition and Conservation Initiative, Des Moines, Iowa, USA

Yoshi Kawamoto Primate Research Institute, Kyoto University, Inuyama, Japan

Christopher Krupenye Department of Evolutionary Anthropology, Duke University, Durham, North Carolina, USA

Evan L. MacLean School of Anthropology, University of Arizona, Tucson, Arizona, USA

Fanny Minesi Lola ya Bonobo, Kinshasa, Democratic Republic of Congo

Janet Nackoney Department of Geographical Sciences, University of Maryland, College Park, Maryland, USA

Elisabetta Palagi Centro di Ateneo, Museo di Storia Naturale, Università di Pisa, Calci, Pisa and Istituto di Scienze e Tecnologie della Cognizione, Unità di Primatologia Cognitiva, CNR, Rome, Italy

Zjef Pereboom Centre for Research and Conservation, Royal Zoological Society of Antwerp, Belgium

Kerri Rodriguez Department of Evolutionary Anthropology and Center for Cognitive Neuroscience, Duke University, North Carolina, USA

Alexandra G. Rosati Department of Psychology, University of Michigan, Ann Arbor, Michigan, USA

Chet C. Sherwood Department of Anthropology and Center for the Advanced Study of Human Paleobiology, George Washington University, Washington, DC, USA

Nicky Staes Department of Biology, Behavioral Ecology and Ecophysiology Group, University of Antwerp, Belgium, Centre for Research and Conservation, Royal Zoological Society of Antwerp, Belgium, and Department of Anthropology and Center for the Advanced Study of Human Paleobiology, George Washington University, Washington, DC, USA

Jeroen M.G. Stevens Centre for Research and Conservation, Royal Zoological Society of Antwerp, Belgium, and Department of Biology, Behavioral Ecology and Ecophysiology Group, University of Antwerp, Belgium

Cheryl D. Stimpson Department of Anthropology and Center for the Advanced Study of Human Paleobiology, George Washington University, Washington, DC, USA

Martin Surbeck Max Planck Institute for Evolutionary Anthropology, Department of Primatology, Leipzig, Germany

Hiroyuki Takemoto Primate Research Institute, Kyoto University, Inuyama, Japan

Jingzhi Tan Department of Evolutionary Anthropology, Duke University, Durham, North Carolina, USA, and Zoo Atlanta, Atlanta, Georgia, USA

Michael Tomasello Max Planck Institute for Evolutionary Anthropology and Department of Psychology and Neuroscience, Duke University, North Carolina, USA

Frans B.M. de Waal Psychology Department and Yerkes National Primate Research Center, Emory University, Georgia, USA

Kara Walker Department of Evolutionary Anthropology, Duke University, North Carolina, USA

Alexander Weiss Department of Psychology, School of Philosophy, Psychology and Language Sciences, University of Edinburgh, United Kingdom, and the Scottish Primate Research Group, United Kingdom

David Williams African Wildlife Foundation, Washington, DC, USA

Vanessa Woods Department of Evolutionary Anthropology, Duke University, North Carolina, USA

Richard Wrangham Department of Human Evolutionary Biology, Harvard University, Cambridge, Massachusetts, USA

Shinya Yamamoto Kyoto University Institute for Advanced Study, and Wildlife Research Center, Kyoto University, Kyoto, Japan

CHAPTER 1

Minding the bonobo mind

Brian Hare and Shinya Yamamoto

Abstract In this chapter we introduce the central role the bonobo plays in testing evolutionary hypotheses regarding ape minds (including our own). The importance of bonobos has become apparent only recently with sustained fieldwork at multiple sites in the Congo Basin as well as the first direct quantitative comparisons between bonobos, chimpanzees and humans. This recent work has revealed a number of traits in which bonobos and chimpanzees are more similar to humans than they are to each other. This means that bonobos are crucial to determining the evolutionary processes by which cognitive traits evolved in our own lineage. Based on the evidence within, it becomes clear that one can no longer know chimpanzees or humans without also knowing bonobos. We argue this makes investing in bonobo research and improved protection for bonobos in captivity and the wild an even higher priority.

Nous illustrons le rôle central joué par le bonobo pour tester les hypothèses relatives à l'évolution de l'esprit des grands singes (y compris le nôtre). L'importance des bonobos n'est apparue que récemment grâce à un travail de terrain soutenu sur de multiples sites dans le bassin du Congo ainsi qu'aux premières comparaisons quantitatives directes entre les bonobos, les chimpanzés et les humains. Ces récents travaux ont révélé un certain nombre de caractéristiques pour lesquelles les bonobos et les chimpanzés présentent plus de similarités avec les humains que l'un envers l'autre. Cela signifie que les bonobos sont essentiels pour déterminer les processus d'évolution par lesquels les caractéristiques cognitives ont évolué dans notre propre lignée. Sur la base des preuves contenues dans ce document, il devient clair que l'on ne peut plus connaître les chimpanzés ou les humains sans connaître les bonobos. Cela rend donc d'autant plus primordiaux l'investissement dans la recherche sur les bonobos et l'amélioration de la protection des bonobos en captivité comme à l'état sauvage.

This is an exciting moment for research into bonobos. The contributors to this book have helped demonstrate the growing momentum behind this work that will help us get to know our closest relative better. In particular, there has been remarkable progress in studies examining the mind, brain and behavior of bonobos. Much of this work has involved an experimental approach that complements the tradition of studying wild bonobos initiated by Professor Takayoshi Kano in 1973. These experiments and large-scale studies have corroborated many pioneering observational studies (de Waal, 1987; Idani, 1991; Kano, 1992; Kuroda, 1989), revealed many new phenomena (e.g. episodic memory: Mulcahy and Call, 2006; heightened eye contact: Kano et al., 2015; xenophilia: Tan and Hare, 2013; differential hormonal reactivity: Wobber et al., 2010; intolerance: Jaeggi et al., 2010; Cronin et al., 2015, etc.; risk aversion: Rosati and Hare, 2012, 2013; maternal influence on sons: Surbeck et al., 2011; Surbeck and Hohmann, 2013; female–female cooperation: Furuichi, 2011; Tokuyama and Furuichi, 2015; Yamamoto, 2015; communicative complexity: Clay et al., 2015; Demuru et al., 2015, etc.), and suggested at least two novel hypothesis for bonobo evolution (Hare et al., 2012; Takemoto et al., 2015). Thus, this book acts as a venue displaying the latest treasures uncovered during the bonobo renaissance now underway. *Pan paniscus* has gone from neglect to an increasingly popular species for scientific study. Figure 1.1 suggests that the gap between publications on bonobos compared to chimpanzees is closing. Hopefully in the future, the research and publication effort will be divided equally between each species, as phylogeny would suggest.

Nowhere is the scientific value of bonobos more obvious than when trying to answer questions about how ape and human cognition evolves.

Hare, B. and Yamamoto, S., *Minding the bonobo mind*. In: *Bonobos: Unique in Mind, Brain, and Behavior*. Edited by Brian Hare and Shinya Yamamoto: Oxford University Press (2017). © Oxford University Press. DOI: 10.1093/oso/ 9780198728511.003.0001

2 BONOBOS: UNIQUE IN MIND, BRAIN AND BEHAVIOR

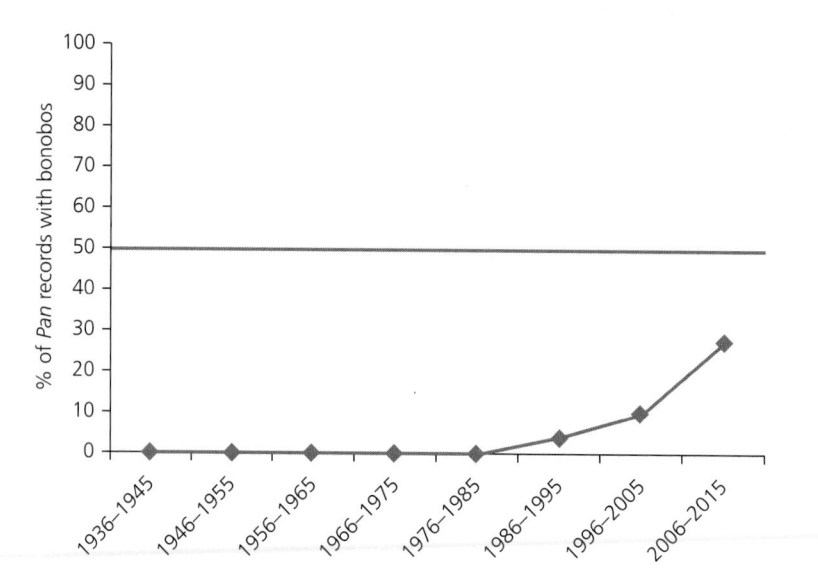

Figure 1.1 The percentage of records in Google Scholar for the search term 'bonobo' out of the total records for both species since discovery of bonobos in ten-year increments. Highly similar results are produced when using each species taxanomic name. However, simply examining total number of publication records in ISI Web of Science and Google Scholar for 'bonobo' and 'chimpanzee' reveals that the bonobo makes up only 3 per cent and 9 per cent of the total citations indexed for both species collectively. This is far from the 50 per cent that phylogeny would predict. (Le pourcentage des enregistrements de recherche dans 'Google Scholar' pour le mot 'bonobo' parmis les enregistrements pour les deux espèces depuis la découverte du bonobo en incréments de dix ans. On voit de résultats très similairs quand on utilise le nom taxonomique de chaque espèce. Cependant, simplement examiner le nombre total de publications dans le ISI Web of Science et Google Scholar pour 'bonobo' et 'chimpanzee' montre que le bonobo constitue 3 per cent seulement et 9 per cent du total des citations pour les deux espèces collectivement. Ces nombres sont loins du 50 per cent que la phylogénie aura prévu.)

Phylogenetic comparisons with both living and extinct apes are crucial for reconstructing human evolution. Unfortunately cognition, temperament, personality and all other psychological features neither fossilize nor have typically been linked to traits that can be measured in fossils. This means that for the study of psychological evolution we must rely on the remaining living apes. Although each species has its own unique evolutionary path, inferences about human evolution can be made through careful comparisons of bonobos, chimpanzees and humans. For example, all three species co-orient in the direction of another individual's gaze, but only humans have an explicit understanding of false beliefs in others (e.g. Hare, 2011, but see Tomasello et al., 2007; Krupenye et al., 2016; Herrmann et al., 2010). This means our last common ancestor (LCA) followed gaze to predict and manipulate others' behavior but did not have the false belief understanding that is tightly linked to language development in humans. Therefore, false belief but not gaze following is a derived feature of human evolution. Fossils simply cannot provide this level of resolution regarding the internal cognitive processes that evolved in the human lineage (Hare, 2011).

Given the central role that bonobos can play in testing ideas about human evolution, why then have they not been centre-stage in comparative studies? Figure 1.2 presents a short history of major events in bonobo research. Examining history seems to suggest that the answer is largely socio-political since bonobos were the last ape to be recognized, are relatively rare in captivity compared to the more ubiquitous chimpanzee and are only endemic to the Democratic Republic of Congo (DRC) where few scientists have traditionally ventured, given that work with chimpanzees in East and West Africa has been far more feasible (i.e. there has been relatively little work on any species of great ape in the DRC even though it is home to the largest populations). Scientifically, the lack of attention given to bonobos may simply be a result of the overwhelming

1925	R. Yerkes describes Prince Chim in *Almost Human* (Yerkes, 1925)
1929	Bonobo skull recognized (Schwartz, 1929; Coolidge, 1933)
1973	First long-term study of wild bonobos begins (Kano, 1992)
1978	First *paniscus-troglodytes* behavioral comparison (Savage-Rumbaugh & Wilkerson, 1978)
1980	Language training of Kanzi begins (Savage-Rumbaugh & Lewin, 1994)
1988	First ethogram comparing bonobo-chimpanzee behavior (de Waal, 1988)
1997	Publication of *Bonobos: The Forgotten Ape* (de Waal & Lanting, 1997)
2001	Self-domestication proposed for bonobo evolution (Wrangham & Pilbeam, 2001; Hare et al, 2012)
2012	Bonobo genome published (Prufer et al, 2012)

Figure 1.2 Timeline of major events in history of research with bonobos (*Pan paniscus*).
(Chronologie des événements majeurs dans l'histoire des recherches avec les bonobos (Pan paniscus)).

phenotypic and genetic similarity between bonobos and chimpanzees. The two species are so similar that Robert Yerkes, the father of American primatology, did not realize that the infant bonobo and chimpanzee living with him for a year were different species (i.e. Chim and Panzee: de Waal and Lanting, 1997). Decades of research have subsequently shown that both species have highly similar life history, feeding adaptations, morphology, social systems and cognition. Both bonobos and chimpanzees are highly social, knuckle-walking, fruit-eating, large-brained apes. They both live in fission–fusion societies that are multi-female and multi-male and that can include over 100 individuals. They form friendships, groom, fight and reconcile within their groups. Males of both species are larger than females and females have sexual swellings and mate promiscuously. In both species females tend to emigrate and males are philopatric (Boesch et al., 2002; Kano, 1992; Muller and Mitani, 2005). Genomic comparisons also reveal the two species only differ by a mere 0.4 per cent and have a divergence time around 1 to 2 million years (de Manuel et al., 2016; Prufer et al., 2012). Researchers looking for a way to invest limited resources have traditionally favoured chimpanzees simply because they provided the fastest, safest, most affordable and familiar way to answer most questions. Chimpanzees became the model species for the genus *Pan* and by default the LCA (Figure 1.2). This book is designed to challenge the wisdom of continuing this approach in the future.

Evolution is descent with modification. Darwinian evolution rejects the very notion that extreme similarity rules out meaningful differences. The two species of *Pan* are no exception. The two species evolved in part due to allopatric speciation. The Congo River provides an ancient and significant geographical barrier since apes cannot swim (the formation of this river was previously considered 1.5–2.5 million years ago (Thompson, 2003), but new evidence suggests a more ancient origin, 34 million years ago (Takemoto et al., 2015)). Only limited genetic admixture between bonobos south of the Congo River and chimpanzees to the north of the river occurred since their separation (de Manuel et al., 2016). This effect can be considered small and most likely occurred during extreme climatic periods when the Congo River's flow was reduced. Yet, bonobos still contributed less than 1 per cent of the total central chimpanzee genome, and far less to other chimpanzee subspecies.

In comparison to other primates with graded hybridization zones, this signals relative reproductive isolation of the two populations. The relative lack of gene flow played a central role in speciation as well as allowing each species to evolve unique genetic and phenotypic features to exist amongst their overwhelming similarities. The consequence of this isolation is nowhere more apparent than in comparisons to humans. Genomic comparisons to humans suggest that about 3 per cent of the human genome is more similar to either the bonobo or chimpanzee genome than the two sister species are to each other. A quarter of human genes have some regions more similar to one species of *Pan* over the other, but again, in approximately equal proportions (Prufer et al., 2012). This pattern is echoed in behavioral and cognitive comparisons with bonobos and chimpanzees, both showing traits more similar to that observed in humans than to each other (Hare and Yamamoto, 2015; Table 1.1). Many of these are phenotypic traits thought to be central in explaining the evolution of our own species. This means that understanding how bonobos and chimpanzees diverged from one another can allow for inference about cognitive evolution in related traits in our own species (Hare, 2007, 2009, 2011). These significant phenotypic differences are particularly exciting given how genetically similar the two species are. Comparisons between bonobos and chimpanzees raise the spectre of identifying the genetic basis and evolutionary origin of traits that otherwise would be too technically challenging to tackle, given the relative gulf between *Homo* and *Pan* (Prufer et al., 2012). Moreover, a careful comparison of traits like those presented in Table 1.1 shows the danger of only considering chimpanzees when determining what behavioral or cognitive traits are derived in humans. Focusing exclusively on human comparisons to chimpanzees would lead us to conclude, erroneously, that humans are unique among apes for non-conceptive sex, a reliance on mothers in adulthood, for showing adult play, sharing with strangers or having female alliances. In each of these traits bonobos are more similar to humans than to chimpanzees, meaning these traits are either shared between bonobos and humans through common descent or convergent evolution. If shared, it suggests the cognitive state

Table 1.1 Behaviors in bonobos and chimpanzees more similar to humans than each other
(Les comportements des bonobos et des chimpanzés sont plus similaires aux comportements humains qu'ils sont l'un à l'autre).

	Bonobo	Chimpanzee	Human foragers
Extractive foraging[1]	Only captivity	Frequent	Frequent
Non-conceptive sexual behavior[2]	Frequent	Absent	Frequent
Lethal aggression between groups[3]	Absent	Present	Present
Mother's importance to adult offspring[4]	High	Low	High
Infanticide/Female coercion[5]	Absent	Present	Present
Levels of adult play[6]	High	Low	High
Cooperative hunting[7]	Absent	Present	Present
Sharing between strangers[8]	Present	Absent	Present
Male–male alliances[9]	Absent	Frequent	Frequent
Female gregariousness[10]	High	Low	High

[1] Hohmann and Fruth, 2003a; Gruber et al., 2010; Furuichi et al., 2015; Hopkins et al., 2015;
[2] Kano, 1992; Hashimoto and Furuichi, 2006; Hohmann and Fruth, 2000; Hare et al., 2007; Hare and Woods, 2011; Ryu et al., 2015; Clay and deWaal, 2015;
[3] Wrangham, 1999; Wilson et al., 2014;
[4] De Lathouwers and Van Elsacker, 2006; Surbeck et al., 2011; Schubert et al., 2013; [5] Hohmann and Fruth, 2002; Surbeck et al., 2011;
[6] Palagi and Paoli, 2007; Wobber et al., 2010;
[7] Ihobe, 1992; Mitani and Watts, 2001 Surbeck and Hohmann, 2008;
[8] Yamamoto et al., in prep.; Tan and Hare, 2013; Tan et al., 2015;
[9] Kano, 1992; Wrangham, 1999;
[10] Furuichi, 2011; Stevens et al., 2015.

of our last common ancestor, while if convergent it can help uncover the process that may have shaped both human and bonobo psychology in similar ways (MacLean et al., 2012). Either evolutionary result would heavily influence how we think about human evolution. Descent with modification flies in the face of any suggestion that the species that is the best overall model of our common ancestor is somehow more valuable scientifically. Bonobos are anything but redundant in the study of human evolution. Bonobos are as valuable as chimpanzees to research on humans.

It is not just in comparison to humans that meaningful differences appear between the two species. Bonobos and chimpanzees differ significantly from one another across a host of phenotypic traits. Ethologists first began to notice these differences in the earliest studies of their social behavior, particularly involving the expression of aggression. Chimpanzees are characterized by clear dominance hierarchies within-group and hostile between-group relationships. Males are dominant, coerce females, commit infanticide, strive for alpha status, are highly territorial, xenophobic and display coalitionary aggression that can be lethal (Feldblum et al., 2014; Muller and Mitani, 2005; Muller et al., 2011; Wilson et al., 2014; Wrangham et al., 2006). Female chimpanzees, especially in East Africa, support their offspring but tend not to form female coalitions while being aggressive towards immigrants (Kahlenberg et al., 2008; Pusey et al., 2008; Pusey and Schroepfer-Walker, 2013; Riedel et al., 2011). In contrast, bonobos show much less intense forms of aggression within and between their groups. Bonobo males do strive for status but do not form coalitions with other males or attain alpha status within their own groups (Kano, 1992). Mothers are typically a male's most important social partner, with mothers potentially impacting male social status (Furuichi, 2011; Schubert et al., 2013; Surbeck et al., 2011). Bonobo females do form coalitions with each other and are intolerant of male coercion (Hohmann and Fruth, 2003b; Kano, 1992; Tokuyama and Furuichi 2016; Wrangham, 2002). Bonobos do not show the pattern of lethal aggression observed in chimpanzees and have even been seen to interact socially and travel with neighbouring groups (Furuichi, 2011). Female immigrants are prized social partners for resident female bonobos and they receive intense attention but attract relatively little physical aggression (Ryu et al., 2015; Sakamaki et al., 2015). Differences between bonobos and chimpanzees go well beyond those seen in aggressive behavior as well. In their natural interactions, bonobos are more playful as adults than chimpanzees (Palagi, 2006, 2007) and begin using sexual behavior during social interactions as infants (Woods and Hare, 2011). In developmental comparisons, bonobos show delay relative to chimpanzees across behavioral, cognitive, physiological and morphological traits (see respectively Wobber et al., 2010, 2014; Zollikofer and de Leon, 2010; Behringer et al., 2014). When compared quantitatively in a comprehensive, cognitive battery chimpanzees are better with causal reasoning while bonobos are more skilful in social tasks (Herrmann et al., 2010). Their foraging preferences also differ, with bonobos being averse and chimpanzees being prone to risk in simulated foraging tasks (Rosati and Hare, 2013). Bonobos and chimpanzees have opposite reactions to strangers and males of each species have the opposite physiological response to social stress (Tan and Hare, 2013; Wobber et al., 2010). Taken together, while bonobos and chimpanzees are remarkably similar, where they differ, the differences are extreme. It is hard to avoid wondering how such modifications have occurred and what they might tell us about evolution between closely related species more generally. What was it about the ecology, social world and epigenetic influences that led to the robust differences we see within a sea of similarity? Many chapters in the book aim to answer questions like these.

The renewed interest in bonobos clearly shows the central role they should play in testing ideas about both human and ape evolution. Given the quantity and quality of research being produced, we predict a very exciting future for bonobo research. In addition, more opportunities to study bonobos are presenting themselves. With relative political stability over the past decade within bonobo habitat, field researchers are observing wild bonobos more frequently than ever. This includes the Wamba field site in the Luo Scientific Reserve that was the location of the first and longest study of wild bonobos, established in 1973 by Takayoshi Kano (Kano, 1992) and Lui Katole in Salonga National Park that has been what is probably the most productive study site for wild bonobos, supported by the Max Planck Society for over a decade now (Hohmann and Fruth, 2003c). In addition, Lola ya Bonobo, the bushmeat orphanage outside the capital of Kinshasa, has also become a major captive research site hosting scores of researchers from over a dozen institutions over the same period (Wobber and Hare, 2011; Figure 1.3). Moreover, researchers have increasingly looked to zoo populations as the source of studies on bonobo behavior, particularly bonobo colonies in Europe (as an aside, it is unclear why zoos in the United States lag so far behind in research productivity; only a handful of studies have been produced in the same period). In addition,

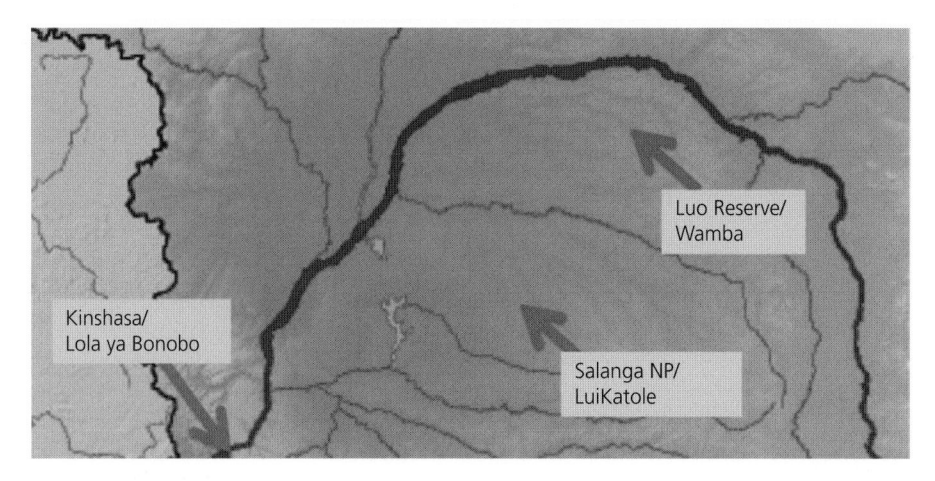

Figure 1.3 Bonobos are only endemic to tropical forest south of the Congo River in the Democratic Republic of Congo. They are listed as endangered with decreasing populations according to the IUCN red book (http://www.iucnredlist.org). Their suspected historical range (~500,000 km²) is nearly the size of France and greater than California (http://www.iucnredlist.org/details/full/15932/0). The map shows the Congo River and the location of the three research sites described in the text as the most productive over the last decade for bonobo behavior and cognition. Map from Wikicommons: CongoLualaba_watershed_topo.png.
(Les bonobos sont endémiques à la forêt tropicale au Sud du Fleuve Congo dans la République Démocratique du Congo. Ils sont considérés en voie de disparition avec décroissance de population selon le livre rouge du IUCN http://www.iucnredlist.org≥). Leur portée (~500,000 km2) est presque la même taille que la France et est plus grande que la Californie http://www.iucnredlist.org/details/full/15932/0≥). La carte montre le Fleuve Congo et l'emplacement des trois sites de recherche décrits dans le texte comme les sites plus productives durant la dernière décennie pour le comportement et la cognition des bonobos. Carte de wikicommons: File:CongoLualaba_watershed_topo.png).

the Kumamoto Sanctuary of Kyoto University has recently established a colony of captive bonobos ($N = 6$) for the purpose of studying their cognition and behavior (the first population of bonobos ever in Japan). Researchers of captive bonobo have also finally begun to break the shackles of the small sample size from which most bonobo research has long suffered. Lola ya Bonobo, in particular, has allowed for a series of large-scale experimental comparisons of development and cognition between the two *Pan* species where at least 20–30 individuals of each species have participated (e.g. Hare et al., 2007; Herrmann et al., 2010, 2011; Maclean and Hare, 2012, 2013; Rosati and Hare, 2012, 2013; Wobber et al., 2010a,b, 2014). Researchers in European zoos have also been able to collaborate and boost sample sizes. In some studies several zoo populations were sampled to allow for powerful analyses (Behringer et al., 2014a,b; Jaeggi et al., 2010; Stevens et al., 2007).

Despite growing momentum, there are still major pieces of infrastructure that are missing that would support sustainable growth of the field. The first and most obvious is the simple fact that no dedicated bonobo field researcher that we are aware of has received a tenure-tracked assistant professorship at a major research university in the past decade or more in the United States, and only one in Europe. Over the same period, a host of researchers focused primarily on chimpanzee behavior have—deservedly—taken up positions at top universities. Figure 1.4 also shows that in the United States the National Science Foundation (NSF) has historically funded ten times more research projects with chimpanzees than bonobos. At the time of writing, only two small NSF graduate student grants are funded for bonobo work, and never in the history of NSF has a grant over US$500,000 been awarded for a project on bonobos (14 such projects have been funded that included chimpanzees). We assume this imbalance is representative of the funding internationally as well. Unless addressed with an increase in funding overall for work with great apes or a redistribution of existing funding, the disparity in research effort will continue. Things should become more balanced soon as a new crop of young bonobo researchers—many of whose work is published in

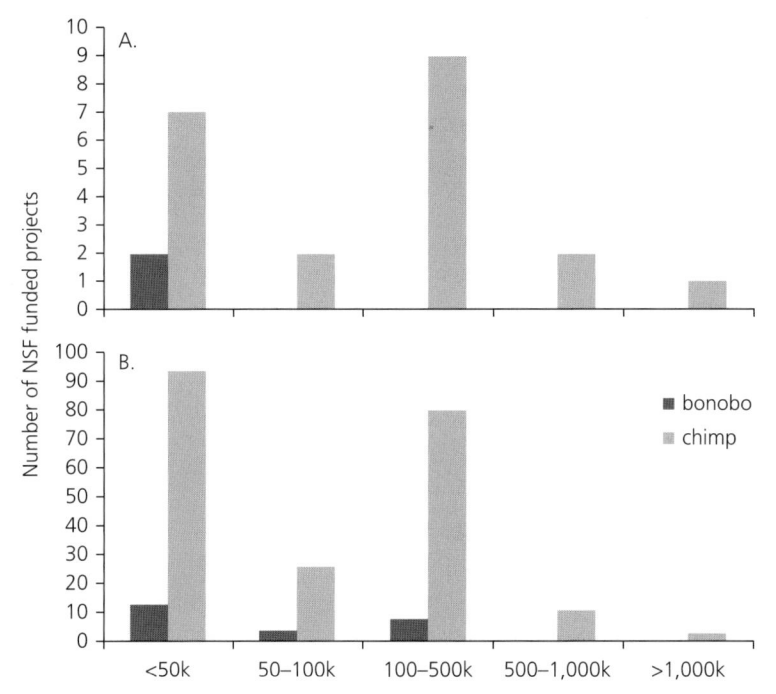

Figure 1.4 The number of National Science Foundation (NSF) funded projects in A) the current funding period of 2015 and B) during the entire history of NSF for bonobo projects compared to chimpanzee projects. The horizontal axis represents the size of the grants made in thousands of US dollars. The NSF grants archive was searched using the terms 'bonobo' and 'chimpanzee': http://www.nsf.gov/awardsearch/. The same results were obtained using scientific names for each species.
(Le nombre de projets financés par la Fondation de Sciences Nationale (NSF) dans A) la periode de financement courante de 2015 et B) durant l'histoire du NSF pour les projets bonobo en comparaison avec les projets chimpanzés. L'axe horizontal représente la quantité de bourses en U.S. dollars. L'archive des bourses du NSF a été cherché en utilisant les mots 'bonobo' et 'chimpanzee' http://www.nsf.gov/awardsearch/. Les mêmes résultats ont été obtenu en utilisant les noms scientifiques de chaque espèce).

this book—are reaching the stage where they will successfully compete for future positions. Similarly, as more high-impact research is produced, departments and funding agencies will seek to lead in this exciting area of research. Hopefully, more zoos, and especially those in the United States, will also begin to recognize the benefits of participating in this type of research.

Another issue that must be addressed is the shortage of Congolese scientists who focus primarily on bonobo behavior or conservation (i.e. only one paper in this book includes a Congolese collaborator). It will be these scientists that will be most successful at promoting the value of the bonobo, the only ape that is wholly Congolese, with the Congolese public, politicians and students. Finally, primatologists across the board desperately need to turn their focus increasingly to Asia. China's influence is growing in all areas where endangered primates live, the Congo Basin being no exception. China is now Africa's primary economic partner. The DRC has quickly become dependent on China; it is one of only three countries in Central Africa which solds China over 30 per cent of its total exports in 2014 (Republic of the Congo and Central African Republic being the other two; see *Economist*, Taking a Tumble, 29 August 2015). Students need to be recruited from China to work with apes in Africa, conferences on conservation need to be held in China and Chinese academics and zoological societies must be engaged to build a vibrant conservation community that can respond to future challenges. Ultimately the conservation of bonobos will depend on a vibrant international community of primatologists focused on their survival in the wild and their welfare in captivity. We are hopeful that rapid progress will be made in these areas and we hope that this book will play a small role in moving bonobo research centre stage where it belongs. We also hope this book will both publicize the most exciting new findings and also help those who want to promote more bonobo research in the future.

The organization of this book

This is the first book to bring together scholars to report on our new understanding of the bonobo's unusual cognitive profile. Therefore, much of the book will focus on studies of bonobo cognition. In this introduction we have described the recent interest in bonobo cognition while briefly reviewing the history of research with bonobos. Here we again emphasize the importance of aggregating fieldwork and laboratory work to help build up an entire picture of the bonobo both with observation in the animals' natural environment and with controlled experiments. The following 17 chapters are organized into eight subsections and have been contributed by experts representing a diverse set of disciplines studying bonobos living in a range of settings.

In order to place this new work in its evolutionary contexts, researchers from the two most active bonobo field sites start the book by reporting on recent discoveries regarding the social behavior of bonobos (Chapters 2 and 3). Furuichi describes the bonobo's peaceful society compared to the chimpanzees' more agonistic one. He searches for the reason behind the difference between these closely related species and concludes that the critical difference is seen in the contribution of bonobo females who maintain high social status by relying on strong social bonds among females, together with prolonged pseudo-estrus (Chapter 2). Surbeck and Hohmann agree that bonobos form strong female–female bonds and they also investigate the nature of male relationships in bonobo society. Their findings suggests the benefit of male–male bonding is limited. They then provide the exception to this rule in reporting the adoption of an orphan by an adult male (Chapter 3). These natural observations provide a solid ethological foundation inspiring further investigation of their social cognition.

The second section of the book explores how the social behavior of bonobos develops (Chapters 4 and 5). Walker and Hare tackle the mystery of female dominance in bonobo society by proposing and testing the offspring dominance hypothesis. This hypothesis predicts that infant and juvenile bonobos play a central role in motivating conflicts of interests that lead to female aggression, and also that males defer to infants and juveniles with their mothers in a variety of situations. These are all supported by observation in a bonobo sanctuary (Chapter 4). Palagi and Demuru focus on development of another social aspect: play, an affiliative rather than aggressive interaction. Summarizing bonobos' uniquely playful traits, their work sheds light on the importance of social play in scaffolding the socio-emotional and communicative competence of individuals. They suggest that early infancy is a critical time window in which the degree of trust in others grows and gives the infants the opportunity to develop their emotional competence (Chapter 5).

The third section includes chapters exploring the limit and extent of the bonobo mind as it communicates and, potentially, reads the minds of others (Chapters 6, 7 and 8). Krupenye and colleagues deal with one of the most controversial topics in ape social cognition, theory of mind. There have been few studies investigating theory of mind in bonobos in comparison to chimpanzees. The comparison between the two species with this sophisticated skill set will lead directly to our understanding of the evolution of human-like social cognition. Krupenye and colleagues lay the theoretical base for future studies (Chapter 6). Tomasello explores the evolution of human language, reviewing previous ape language projects including the famous language-trained bonobo, Kanzi. He concludes the lack of human forms of language in non-human apes comes predominantly from their inability to recognize shared intentionality in others, despite many requisite cognitive skills being present in non-human apes, including aspects such as basic symbol learning, categorization and sequential learning (Chapter 7). Clay and Genty provide a review of the first set of studies exploring the natural communication of bonobos. They report on a growing literature of their vocal and gestural communication that points to remarkable complexity, flexibility and intentionality. They show where these findings are relevant to thinking about the evolution of human language in particular (Chapter 8).

The forth section consists of chapters exploring the cooperative nature of bonobos as they share food with others (Chapters 9 and 10). Yamamoto and Furuichi describe details of food sharing among wild bonobos in Wamba and make some comparisons between species and between field sites. Unlike chimpanzees, where most sharing occurs when males obtain meat, in bonobos the majority of sharing occurs when fruit is shared between adult females. The highlight of this chapter is their suggestion of bonobos' 'courtesy' food sharing, characterized by begging for a social bond rather than the food itself. They discuss the importance of this exchange as it relates to female–female bonding, especially between a new immigrant female and other resident females (Chapter 9). Tan and Hare examined the bonobos' prosocial propensity not only in sharing but also in various prosocial choice tests and instrumental helping tasks. Reviewing previous experimental works including their own, they show that, compared to chimpanzees, bonobos are more tolerant and proactively prosocial towards non-relatives, and even more so towards strangers. They review findings of xenophilia (a preference for strangers) in bonobos that possibly has important implications for explaining the origin of human co-operation (Chapter 10).

The fifth section considers the cognitive abilities deployed by bonobos as they forage for and process food. The chapters explore species-specific psychological biases in bonobos as they relate to socio-ecology (Chapters 11 and 12). Rosati points out differences between chimpanzees and bonobos in some cognitive skills which are related to their foraging strategies. Bonobos seem to exhibit less accurate spatial memory, reduced levels of patience and greater risk aversion. These traits seem to map onto their wild foraging behavior that depends on relatively stable food availability (Chapter 11). Call tackles the puzzle of why bonobos are expert extractive foragers in captivity but have never been seen using tools to obtain food in the wild. Neither socio-ecological factors nor cognitive processes underlying tool use can explain the difference. However, he suggests that the combination of three species-specific psychological biases in cognition, temperament and emotion could contribute to explaining the difference in tool use between bonobos and chimpanzees observed in the wild (Chapter 12).

The sixth section will focus on large-scale comparisons of bonobos to both chimpanzees and humans in their cognitive abilities and brain anatomy (Chapters 13, 14 and 15). Staes and colleagues examine individual differences among 154 captive bonobos using the Hominoid Personality Questionnaire. They also integrate this with behavioral observation and experiments to validate their personality coding. This study shows that personality dimensions in bonobos are correlated with sex, age and behaviors; some are similar to humans and chimpanzees but others are different. They discuss these differences as they relate to bonobo socio-ecology (Chapter 13). Hopkins and colleagues explore bonobo–chimpanzee differences in social cognition by investigating their brain organization. Although chimpanzees and bonobos are quite similar in terms of total brain size, surface area and gyrification, their voxel-based morphometry analyses show that there is some differentiation in grey matter concentration and location in certain cortical regions between chimpanzees and bonobos. Hopkins and colleagues emphasize the importance of further research to identify the role of social and ecological factors on the evolution of cognition and the brain (Chapter 14). Hare and Woods put the bonobo–chimpanzee cognitive comparison in a broader evolutionary context. They sum up previous quantitative cognitive studies, highlighting the most distinctive aspects of bonobo social cognition such as cooperation and xenophilia. These traits are differentiated from chimpanzees' but are similar to other domestic animals. Based on these facts, Hare and Woods explain theories of both human and animal cognitive evolution with their self-domestication hypothesis (Chapter 15).

Finally, the last two sections include chapters exploring the past and future of the bonobo, providing novel perspectives on how to promote the survival of this highly endangered species (Chapters 16, 17 and 18). Takemoto and colleagues re-examine the history of the Congo River and how it functions as a strong geographical barrier between bonobo and chimpanzee habitats. Their investigation of the recent information of the seafloor sediments near the mouth of the river and the geophysical survey on the continent suggests that the river formed 34 million years ago and is therefore much older than it has been considered. If this is true, the origin of the bonobo most likely resulted as a small group crossed the Congo River at shallow points when the water level was at its lowest during Pleistocene dry periods. This finding will further stimulate tests of exactly when and how bonobo and chimpanzee speciation occurred (Chapter 16). Nackoney and colleagues look to the future and report on powerful tools for bonobo conservation: geospatial information analysis and models. They demonstrate how spatial data and maps are used for monitoring threats and prioritizing locations to safeguard bonobo habitat, including identifying areas of highest conservation value to bonobos and collaboratively mapping community-based natural resource management zones for reducing deforestation in key corridor areas. This will be central to the development of dynamic conservation strategies that will help strengthen bonobo protection (Chapter 17). Faust and colleagues explore the past and future of captive bonobos by investigating population dynamics in the world largest bonobo sanctuary beyond Kinshasa. Several model scenarios clarify potential future growth trajectories for the sanctuary. This research illustrates how data on historic dynamics can be modelled to inform future sanctuary capacity and management needs, and it allows careful planning for this essential captive population (Chapter 18).

Because of the important hypotheses that are detailed throughout this book, scholars and students alike will find it an invaluable reference on bonobos. Crucially, each chapter will include a French translations of the abstract and figure legends. This will allow the book to inspire an entirely new generation of Congolese scientists to work for the protection of this species that is 100 per cent Congolese. It is these scientists that will undoubtedly play the most important role in assuring the future of the bonobo.

We would like to give our sincere thanks to many contributors and collaborators who committed large amounts of time and effort to this project. First of all, the research included in this book could not be conducted without tremendous collaboration from field assistants, villagers, the government of the Democratic Republic of Congo, laboratory colleagues, zoo and sanctuary keepers and many other staff. Frans de Waal and Richard Wrangham kindly read all the chapters and have contributed a brilliant Foreword and an Afterword. It is an honour to have them start and end this book since they inspired so many people to pursue research with bonobos and they are ideally placed to put the current work in its historical context. We also acknowledge many specialists who gave constructive comments for earlier versions of the chapters: Josep Call, Hillary Fouts, Ian Gilby, William Hopkins, Tatyana Humle, Fumihiro Kano, Kelsey Lucca, Evan MacLean, Gema Martin-Ordas, John Mitani, Masato Nakatsukasa, Sergio Pellis, Alexandra Rosati, Steve Ross, Aaron Sandel, Amanda Seed, Katie Slocombe, Martin Surbeck, Jingzhi Tan and Felix Warneken. We thank them all for their tremendous contribution that improved the quality of the work we present here. Finally, we would like to express our gratitude to Oxford University Press. Our special thanks go to Ian Sherman, Lucy Nash and Bethany Kershaw for their continual patience and support. Without their encouragement, this book would never have been possible. We are thrilled to publish this book with Oxford University Press, and hail its responsible sourcing of paper used to produce its books. The Press asks every printer that it employs to provide it with the name and contact information of the paper mill that the printer sources paper from, and the mill's PREPS (Publishers Database for Responsible Environmental Paper Sourcing) rating for the specific paper being used in the book. For this book, the publisher has used a 3* PREPS-rated paper supplied from non-African countries. Because this book is the result of a collaboration among such large, diverse and international groups of people, we hope that it will provide stimulating reading to all those interested in the most amazing of all apes, the bonobo.

References

Andre, C., Kamate, C., Mabonzo, P., Morel, D., and Hare, B. (2008). The conservation value of Lola ya Bonobo Sanctuary. In: Furuichi, T. and Thompson, J. (eds). *The Bonobos: Behavior, Ecology and Conservation*. New York, NY: Springer, pp. 303–22.

Beaune, D., Bretagnolle, F., Bollache, L., Hohmann, G., and Fruth, B. (2015). Can fruiting plants control animal behavior and seed dispersal distance? *Behaviour*. DOI:10.1163/1568539X-00003205.

Behringer, V., Deschner, T., Deimel, C., Stevens, J., and Hohmann, G. (2014a). Age related changes in urinary testosterone levels suggest differences in puberty onset and divergent life history strategies in bonobos and chimpanzees. *Hormones and Behavior*, 66, 525–33.

Behringer, V., Deschner, T., Murtagh, R., Stevens, J., and Hohmann, G. (2014b). Age-related changes in Thyroid hormone levels of bonobos and chimpanzees indicate heterochrony in development. *Journal of Human Evolution*, 66, 83–8.

Boesch, C., Hohmann, G. and Marchant, L. (2002). *Behavioural diversity in chimpanzees and bonobos*, Cambridge, Cambridge University Press.

Clay, Z. and de Waal, F.B.M. (2015). Sex and strife: post conflict sexual contacts in bonobos. *Behaviour*. DOI:10.1163/1568539X-00003155.

Cronin, K. A., de Groot, E., and Stevens, J. M. (2015). Bonobos show limited social tolerance in a group setting: A comparison with chimpanzees and a test of the relational model. *Folia Primatologica*, 86, 164–177.

de Manuel, M., Kuhlwilm, M., Frandsen, P., Sousa, V. C., Desai, T., Prado-Martinez, J., Hernandez-Rodriguez, J., Dupanloup, I., Lao, O., and Hallast, P. (2016). Chimpanzee genomic diversity reveals ancient admixture with bonobos. *Science*, 354, 477–481.

de Lathouwers, M. and Van Elsacker, L. (2006). Comparing infant and juvenile behavior in bonobos (*Pan paniscus*) and chimpanzees (*Pan troglodytes*): a preliminary study. *Primates*, 47, 287–93.

Demuru, E., Ferrari, P.F., and Palagi, E. (2015). Emotionality and intentionality in bonobo playful communication. *Animal cognition*,18, 333–344.

de Waal, F.B.M. and Lanting, F. (1997). *Bonobo: The Forgotten Ape*. Los Angeles: California University Press.

de Waal, F.B.M. (1987). Tension regulation and non-reproductive functions of sex among captive bonobos (*Pan paniscus*). *National Geographic Research*, 3, 318–35.

Feldblum, J. T., Wroblewski, E. E., Rudicell, R. S., Hahn, B. H., Paiva, T., Cetinkaya-Rundel, M., Pusey, A. E., and Gilby, I. C. (2014). Sexually coercive male chimpanzees sire more offspring. *Current Biology*, 24, 2855–2860.

Furuichi, T. (2011). Female contributions to the peaceful nature of bonobo society. *Evolutionary Anthropology*, 20, 131–42.

Furuichi,T., Sanz, C., Koops, K., Sakamaki, T., Ryu, H., Tokuyama, N., and Morgan, D. (2015). Why do wild bonobos not use tools like chimpanzees do? *Behaviour*. DOI:10.1163/1568539X-00003226.

Gruber, T., Clay, Z., and Zuberbühler, K. (2010). A comparison of bonobo and chimpanzee tool use: evidence for a female bias in the Pan lineage. *Animal Behaviour*, 80, 1023–33.

Hare, B. (2007). From nonhuman to human mind: what changed and why. *Current Directions in Psychological Science*, 16, 60–4.

Hare, B. (2009). What is the effect of affect on bonobo and chimpanzee problem solving? In: Berthoz, A. and Christen, Y. (eds). *The Neurobiology of the Umwelt: How Living Beings Perceive the World*. Springer Press, Berlin, pp. 89–102.

Hare, B. (2011). From hominoid to hominid mind: what changed and why? *Annual Review of Anthropology*, 40. 293–309.

Hare, B.,Melis, A., Woods, V., Hastings, S., and Wrangham, R. (2007). Tolerance allows bonobos to outperform chimpanzees in a cooperative task. *Current Biology*, 17, 619–23.

Hare, B., Wobber, T., and Wrangham, R. (2012). The self-domestication hypothesis: bonobo psychology evolved due to selection against male aggression. *Animal Behaviour*, 83, 573–85.

Hare, B. and Yamamoto, S. (2015). Moving bonobos off the scientifically endangered list. *Behaviour*, 152, 247–58.

Hashimoto, C. and Furuich, T. (2006). Comparison of behavioral sequence of copulation between chimpanzees and bonobos. *Primates*, 47, 51–5.

Herrmann, E., Hare, B. Call, J., and Tomasello, M. (2010). Differences in the cognitive skills of Bonobos and Chimpanzees. *PLoS One*, 5, e12438.

Herrmann, E., Hare, B. Cisseski, J., and Tomasello, M. (2011). The origins of human temperament: children avoid novelty more than other apes. *Developmental Science*, 14, 1393–405.

Herrmann, E., Keupp, S., Hare, B., Vaish, A., and Tomasello, M. (2013). Direct and Indirect reputation formation in great apes and human children. *Journal of Comparative Psychology*, 127, 63–75.

Hohmann, G. (2001). Association and social interactions between strangers and residents in bonobos (*Pan paniscus*). *Primates*, 42, 91–9.

Hohmann, G. andFruth, B. (2000). Use and function of genital contacts among female bonobos. *Animal Behaviour*, 60, 107e120.

Hohmann, G. and Fruth, B. (2002). Dynamics in social organization of bonobos (*Pan paniscus*). In: Boesch, C., Marchant, L., and Hohmann, G. (eds). *Behavioural Diversity in Chimpanzee*. Cambridge: Cambridge University Press, pp. 138–50.

Hohmann, G. and Fruth, B. (2003a). Culture in bonobos? Between-species and within-species variation in behavior. *Current Anthropology*, 44, 563–609.

Hohmann, G. and Fruth, B. (2003b). Intra- and inter-sexual aggression by bonobos in the context of mating. *Behaviour*, 140, 1389–413.

Hohmann, G. and Fruth, B. (2003c). Lui Kotal—A new site for field research on bonobos in Salonga National Park. *Pan African News*, 10(2), 1–10. http://mahale.main.jp/PAN/10_2/10(2)_05.html.

Hopkins, W.,Schaeffer, J., Russell, J., Bogart, S., Meguerditchian, A., and Coulon, O. (2015). A comparative assessment of handedness and its potential neuroanatomical correlates in chimpanzees and bonobos. *Behaviour*. DOI:10.1163/1568539X–00003204.

Idani, G. (1991). Social relationships between immigrant and resident bonobo (*Pan paniscus*) females at Wamba. *Folia Primatologica*, 57, 83–95.

Ihobe, H. (1992). Observations on the meat-eating behavior of wild bonobos (Pan paniscus) at Wamba, Republic of Zaire. *Primates*, 33, 247–250.

Jaeggi, A., Stevens, J., and van Schaik, C. (2010). Tolerant food sharing and reciprocity is precluded by despotism among bonobos but not chimpanzees. *American Journal of Physical Anthropology*, 143, 41–51.

Kano, T. (1992). *The Last Ape: Pygmy Chimpanzee Behavior and Ecology*. Stanford, CA: Stanford University Press.

Kano, F., Hirata, S., and Call, J. (2015). Social attention in the two species of pan: Bonobos make more eye contact than chimpanzees. *PloS one*, 10, e0129684.

Kahlenberg, S. M., Thompson, M. E., and Wrangham, R. W. (2008). Female competition over core areas in Pan troglodytes schweinfurthii, Kibale National Park, Uganda. *International Journal of Primatology*, 29, 931.

Krupenye, C., Kano, F., Hirata, S., Call, J., and Tomasello M. (2016). Great apes anticipate that other individuals will act according to false beliefs. *Science*, 354, 110–14.

Kuroda, S. (1989). Developmental retardation and behavioral characteristics of pygmy chimpanzees. In: Heltne, P. and Marquardt, L. (eds). *Understanding Chimpanzees*. Cambridge, MA: Harvard University Press, pp. 184–93.

Maclean, E. and Hare, B. (2012). Bonobos and chimpanzees infer the target of an actor's attention. *Animal Behaviour*, 83, 345–53.

Maclean, E. and Hare, B. (2013). Spontaneous triadic play in bonobos and chimpanzees. *Journal of Comparative Psychology*, 127, 245–55.

Maclean, E. and Hare, B. (2015). Bonobos and chimpanzees read helpful gestures better than prohibitive gestures. *Behaviour*. DOI:10.1163/1568539X–00003203.

Maclean, E. L., Matthews, L. J., Hare, B. A., Nunn, C. L., Anderson, R. C., Aureli, F., Brannon, E. M., Call, J., Drea, C. M., and Emery, N. J. (2012). How does cognition evolve? Phylogenetic comparative psychology. *Animal cognition*, 15, 223–238.

Mitani, J. C. and Watts, D. P. (2001). Why do chimpanzees hunt and share meat? *Animal Behaviour*, 61, 915–924.

Mulcahy, N. J. and Call, J. (2006). Apes save tools for future use. *Science*, 312, 1038–1040.

Muller, M. N. and Mitani, J. C. (2005). Conflict and cooperation in wild chimpanzees. *Advances in the Study of Behavior*, 35, 275–331.

Muller, M. N., Thompson, M. E., Kahlenberg, S. M., and Wrangham, R. W. (2011). Sexual coercion by male chimpanzees shows that female choice may be more apparent than real. *Behavioral Ecology and Sociobiology*, 65, 921–933.

Palagi, E. (2006). Social play in bonobos (Pan paniscus) and chimpanzees (Pan troglodytes): Implicationsfor natural social systems and interindividual relationships. *American Journal of Physical Anthropology*, 129, 418–426.

Palagi, E. and Paoli, T. (2007). Play in adult bonobos (*Pan paniscus*): modality and potential meaning. *American Journal of Physical Anthropology*, 134, 219–25.

Prufer, K., Munch, K., Hellmann, I., Akagi, K. Miller, J. R., Walenz, B., Koren, S., Sutton, G., Kodira, C., and Winer, R. (2012). The bonobo genome compared to the chimpanzee and human genome. *Nature*, 486, 527–31.

Pusey, A. E. and Schroepfer-Walker, K. (2013). Female competition in chimpanzees. *Philosophical Transactions of the Royal Society B*, 368, 20130077.

Pusey, A., Murray, C., Wallauer, W., Wilson, M., Wroblewski, E., and Goodall, J. (2008). Severe aggression among female *Pan troglodytes schweinfurthii* at Gombe National Park, Tanzania. *International Journal of Primatology*, 29, 949–73.

Riedel, J., Franz, M., and Boesch, C. (2011). How feeding competition determines female chimpanzees gregariousness and ranging in Tai National Park, Cote d'Ivoire. *American Journal of Primatology*, 73, 305–13.

Rosati, A. (2015). Context influences spatial frames of reference in bonobos (*Pan paniscus*). *Behaviour*, 152, 375–406.

Rosati, A. G. and Hare, B. (2012). Decision making across social contexts: competition increases preferences for risk in chimpanzees and bonobos. *Animal Behaviour*, 84, 869–879.

Rosati, A. and Hare, B. (2013). Chimpanzees and bonobos exhibit emotional responses to decision outcomes. *PLoS One*, 8 (5), e63058.

Ryu, H., Hill, D., and Furuichi, T. (2015). Prolonged maximal sexual swelling in wild bonobos facilitates affiliative interactions between females. *Behaviour*. DOI:10.1163/1568539X–00003212.

Sakamaki, T., Behncke, I., Laporte, M., Mulavwa, M., Ryu, H., Takemoto, H., Tokuyama, N., Yamamoto, S., and Furuichi, T. (2015). Intergroup transfer of females and social relationships between immigrants and residents in bonobo (Pan paniscus) societies. *Dispersing primate females*. New York, Springer.

Schroepfer-Walker, K., Wobber, T., and Hare, B. (2015). Experimental evidence that grooming and play are social currency in bonobos and chimpanzees. *Behaviour*, 152, 545–562.

Schubert, G., Vigilant, L., Boesch, C., Klenke, R., Langergraber, K., Mundry, R., Surbeck, M., and Hohmann, G. (2013). Co-residence between males and their mothers and grandmothers is more frequent in bonobos than chimpanzees. *PLoS One*. DOI:10.1371/journal.pone.0083870.

Stevens, J., De Groot, E., and Staes, N. (2015). Relationship quality in captive bonobo groups. *Behaviour*, 152, 259–283.

Stevens, J. M., Vervaecke, H., de Vries, H., and van Elsacker, L. (2007). Sex differences in the steepness of dominance hierarchies in captive bonobo groups. *International Journal of Primatology*, 28, 1417–1430.

Surbeck, M. and Hohmann, G. (2008). Primate hunting by bonobos at LuiKotale, Salonga National Park. *Current Biology*, 18, R906–R907.

Surbeck, M., Mundry, R., and Hohmann, G. (2011). Mothers matter! Maternal support, dominance status and mating success in male bonobos (*Pan paniscus*). *Proceedings of the Royal Society B*, 278, 590.

Surbeck, M. and Hohmann, G. (2013). Intersexual dominance relationships and the influence of leverage on the outcome of conflicts in wild bonobos (Pan paniscus). *Behavioural Ecology and Sociobiology*, 67, 1767–1780.

Takemoto, H., Kawamoto, Y. and Furuichi, T. (2015). How did bonobos come to range south of the Congo River? reconsideration of the divergence of *Pan paniscus* from other pan populations. *Evolutionary Anthropology*, 24, 170–84.

Tan, J. and Hare, B. (2013). Bonobos share with strangers. *PLoS One*, 8(1), e51922.

Tan, J., Kwetuenda, S., and Hare, B. (2015). Preference or Paradigm? Bonobos do not share in 'the' prosocial choice task. *Behaviour*. 152, 521–544.

Thompson, J. (2003). A model of the biogeographical journey from Proto-pan to *Pan paniscus*. *Primates*, 44, 191–7.

Tokuyama, N. and Furuichi, T. (2016). Do friends help each other? Patterns of female coalition formation in wild bonobos at Wamba. *Animal Behaviour*, 119, 27–35.

Tomasello, M., Hare, B., Lehmann, H., and Call, J. (2007). Reliance on head versus eyes in the gaze following of great apes and human infants: the cooperative eye hypothesis. *Journal of Human Evolution*, 52, 314–320.

Watts, D. and Mitani, J. (2002). Hunting behaviour of chimpanzees at Ngogo, Kibale National Park, Uganda. In: Boesch, C., Hohmann, G., and Marchant, L. (eds). *Behavioral Diversity in Chimpanzees and Bonobos*. Cambridge: Cambridge University Press, pp. 244–57.

Wilson, M., Boesch, C., Fruth, B., Furuichi, T., Gilby, I. C., Hashimoto, C., Hobaiter, C. L., Hohmann, G., Itoh, N., and Koops, K. (2014). Lethal aggression better explained by adaptive strategies than by human impact. *Nature*, 513, 414–17.

Wobber, V., Wrangham, R., and Hare, B. (2010). Bonobos exhibit delayed development of social behavior and cognition relative to chimpanzees. *Current Biology*, 20, 226–230.

Wobber, V. and Herrmann, E. (2015). The influence of testosterone of cognitive performance in bonobos and chimpanzees. *Behaviour*. DOI:10.1163/1568539X–00003202.

Wobber, T. and Hare, B. (2011). Psychological health of orphan bonobos and chimpanzees in African sanctuaries. *PLoS One*, 6, e17147.

Wobber, V., Hare, B., Maboto, J., Lipson, S. Wrangham, R., and Ellison, P. (2010). Differential reactivity of steroid hormones in chimpanzees and bonobos when anticipating food competition. *Proceedings of the National Academy of Sciences*, 107, 12457–62.

Wobber, T., Herrmann, E., Hare, B., Wrangham, R., and Tomasello, M. (2014). The evolution of cognitive development in *Pan* and *Homo*. *Developmental Psychobiology*, 5, 547–73.

Wobber, V., Wrangham, R., and Hare, B. (2010). Evidence for delayed development of social behavior and cognition in bonobos relative to chimpanzees. *Current Biology*, 20, 226–30.

Woods, V. and Hare, B. (2011). Bonobo but not chimpanzee infants use socio-sexual contact with peers. *Primates*, 52, 111–16.

Wrangham, R. (1999). Evolution of coalitionary killing. *Yearbook of Physical Anthropology*, 42, 1–30.

Wrangham, R. (2002). The cost of sexual attraction: is there a trade-off in female Pan between sex appeal and received coercion. *Behavioural diversity in chimpanzees and bonobos*, 204–215.

Wrangham, R. W., Wilson, M L., and Muller, M. N. (2006). Comparative rates of violence in chimpanzees and humans. *Primates*, 47, 14–26.

Yamamoto, S. (2015). Non-reciprocal but peaceful fruit sharing in the wild bonobos of Wamba. *Behaviour*, 152, 335–357.

PART I

Society

CHAPTER 2

Female contributions to the peaceful nature of bonobo society

Takeshi Furuichi

Abstract Although chimpanzees (*Pan troglodytes*) and bonobos (*Pan paniscus*) are closely related, their societies show surprisingly large differences in many aspects. Bonobo females, in comparison with chimpanzee females, are characterized by their gregariousness, close social association with one another, leadership and prolonged pseudo-estrus. Males are also different in their aggression between the two species; bonobo males are less aggressive than chimpanzee males and do not commit infanticide nor any other conspecific killing, which was reported in chimpanzees. This chapter investigates these differences between such closely related species and explains how bonobos achieve the peaceful nature of their society. The behavioral characteristics of female bonobos seem to contribute to the peaceful nature of bonobo society.

Bien que les chimpanzés (*Pan troglodytes*) et les bonobos (*Pan paniscus*) soient deux espèces particulièrement proches et étroitement liées, de grandes différences peuvent être observées au sein de leur société. Les femelles bonobos, comparées aux femelles chimpanzés, se montrent plus grégaires, entretiennent des relations plus étroites les unes avec les autres, font preuve d'un certain leadership et présentent des périodes d'œstrus prolongées. Des différences comportementales sont également observées entre les mâles de ces deux espèces: les mâles bonobos apparaissent moins agressifs, ne commettant, à l'inverse des mâles chimpanzés, ni infanticide, ni meurtre intra-spécifique. La présente étude, consacrée donc aux différences entre ces deux espèces pourtant si proches, a pour but d'expliquer la nature pacifique des groupes de bonobos. Il semble ainsi que les femelles bonobos, par certains comportements caractéristiques, contribuent fortement à la structure non-violente de leur société.

Introduction

Although chimpanzees (*Pan troglodytes*) and bonobos (*Pan paniscus*) are closely related, females of the two species show surprisingly large differences in many behavioral aspects.[1] While female chimpanzees tend to range alone or in small parties during non-estrous periods, female bonobos aggregate even more often than do males. Female chimpanzees do not have frequent social interactions with other females whereas female bonobos maintain close social associations with one another. Although the ranging patterns of chimpanzee parties are generally led by males, female bonobos often take the initiative in ranging behavior. While female chimpanzees usually do not exhibit estrus during post-partum amenorrhea or pregnancy, female bonobos exhibit a prolonged pseudo-estrus during such non-conceptive periods.

Studies of these two species have also shown great differences in agonistic behaviors performed by males. Male chimpanzees frequently fight with other males to compete for estrous females but male bonobos seldom do. While there are many records of infanticide by male chimpanzees there is no confirmed record of such an event among bonobos. Several cases of within-group killing among adult male chimpanzees have been reported but there is no such record for bonobos. While intergroup conflicts among chimpanzees sometimes involve killing members of the other group, intergroup conflicts

[1] Original version of this chapter was published in Evolutionary Anthropology 20:131–142 (2011). Minor correction were made on the original version.

Furuichi, T., *Female contributions to the peaceful nature of bonobo society*. In: *Bonobos: Unique in Mind, Brain, and Behavior*.
Edited by Brian Hare and Shinya Yamamoto: Oxford University Press (2017).
© Oxford University Press. DOI: 10.1093/oso/9780198728511.003.0002

among bonobos are considerably more moderate. In some cases, bonobos from two different groups may even range together for several days while engaging in various peaceful interactions.

I will address two important questions that arise from these comparisons, exploring why females of such closely related species show clear differences in behavior and whether or not the behavioral characteristics of female bonobos contribute to the peaceful nature of bonobo society.

Grouping patterns and female initiative in determining ranging

The social systems of both chimpanzees and bonobos are characterized by fission and fusion. Their social groups, called unit groups or communities, are semi-closed and have fairly stable membership, except for the transfer of females between unit groups (Goodall, 1983; Kano, 1982; Nishida, 1979; Van Elsacker et al., 1995). Within these groups, however, animals characteristically split into smaller subgroups or parties in which memberships change over time (Furuichi, 1987; Goodall, 1986; Kuroda, 1979; Nishida, 1979; Van Elsacker et al., 1995; White, 1988; Wrangham, 1986). In this chapter I refer to a unit group or community as a group and to a temporary association of group members as a party.

Early studies on the ecology of bonobos suggested that their party sizes were larger than those of chimpanzees (Kano, 1992; Kuroda, 1979). However, in a recent comparative study using data available from various study sites and groups, I showed that there is considerable within-species variation in party size and that there is no significant difference between the two species (Furuichi, 2009). On the other hand, this study also confirmed that there is a significant difference between these species in the average number of individuals in a party relative to the total number of individuals belonging to the group. This proportional measure is called the relative party size (Boesch, 1996) or the attendance ratio (Furuichi et al., 2008). Boesch showed that the relative party size was larger for bonobos than for chimpanzees (Boesch, 1996). I showed the same pattern highlighting a significant difference in the two species' attendance ratios (27 per cent to 51 per cent for bonobos versus 9 per cent to 30 per cent for chimpanzees) (Furuichi, 2009).

Figure 2.1 illustrates these differences by comparing group and party sizes between the E1 group of bonobos at Wamba, in the Luo Scientific Reserve, Democratic Republic of the Congo (DRC), and the M group of chimpanzees in Kalinzu, Uganda. In both studies we followed the largest parties that we could observe at the time, using the same one-hour-party method developed for the comparison of party sizes and compositions across different species and sites (Hashimoto et al., 2001). As noted, the total numbers of individuals in one-hour parties were not significantly different between the two species. However, while almost half of all the group members were found in the one-hour parties of bonobos, a smaller proportion of members were found in those of chimpanzees. In particular, there is a marked between-species difference in female attendance ratio, with less than one-tenth of the female chimpanzees but almost two-thirds of the female bonobos found in parties.

Although the term 'fission–fusion' is usually used to describe the grouping patterns of both chimpanzees and bonobos, the species differ substantially in their cohesiveness. As seen in an example from Mahale, Tanzania (Nishida, 1968), fission and fusion of parties occurs frequently among chimpanzees. Because different parties may range in distant areas, only rarely can a researcher observe all members of the group within the same day (Ito and Nishida, 2007). In Kalinzu, where we have studied chimpanzees since 1992, several years elapsed between observations of certain females that usually ranged in the periphery of the home range. These females were probably not observed during pregnancy or lactation periods because they were primarily ranging with their dependent infants and juveniles, apart from other members of the group (Hashimoto and Furuichi, unpublished data).

In contrast, most members of the bonobo group at Wamba can be observed daily. Prior to 1996, when our study was interrupted by political instability, we occasionally provisioned the bonobos, either at sleeping sites or at a permanent provisioning site. During this period, we observed an average of 88.9 per cent of adult or adolescent females and 87.6 per cent of adult or adolescent males (Furuichi, 1987).

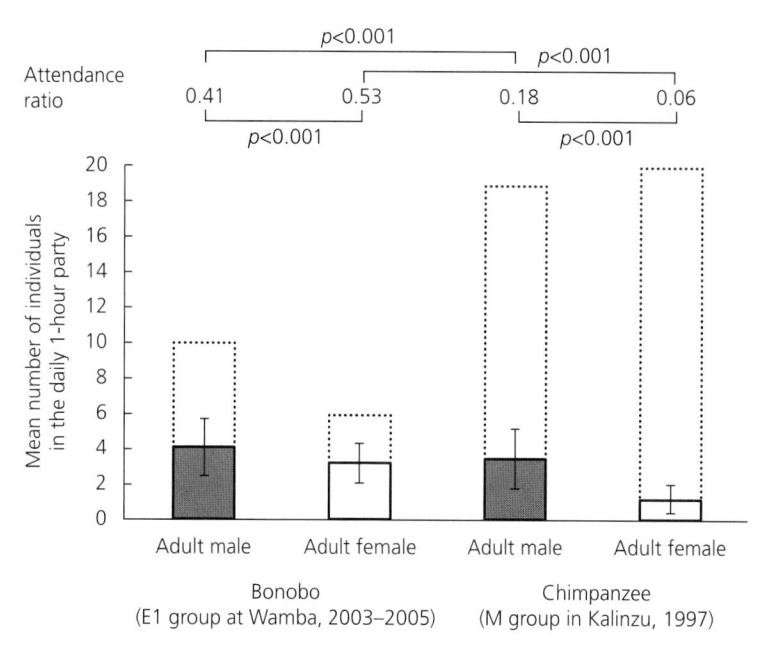

Figure 2.1 Comparisons of party size and composition in the M group of chimpanzees in Kalinzu and the E1 group of bonobos at Wamba. We recorded individuals observed in each one-hour period while following a party, which approximates the number of individuals ranging and feeding together. The length of bars with solid lines represents the average number of individuals found in the one-hour party; the length of bars with dotted lines represents the number of all individuals belonging to the unit-group. See Mulavwa and others (2008) for more details.
(Comparaisons de tailles de groupe et composition entre le groupe M de chimpanzés à Kalinzu et le groupe E1 de bonobos à Wamba. Nous avons enregistré les individus observés dans chaque période d'une heure pendant la suite d'un groupe, qui se rapproche le nombre d'individus allant et nourrissant ensemble. La longueur des bandes avec les lignes solides représente la taille moyenne d'un groupe. Voire Mulavwa et autres (2008) pour plus de détails).

Since 2003, in studies of the same group under completely natural conditions, all group members were present on many observation days, especially during the high-fruiting season. For example, during the 12 months of 2008, all the adult group members were observed on 35 of the 124 days on which parties of the study group were observed from sleeping site to sleeping site. Eighty-three percent (SD 27 per cent, median 100 per cent) of adult females and 79 per cent (SD 26 per cent, median 89 per cent) of adult males were observed during each of the 124 days (Sakamaki and co-workers, unpublished data). Although bonobos split into several parties during the day, group members ranged in adjacent areas and travelled in a similar direction, exchanging vocalizations, so that many of them appeared at least once a day in the party we were following.

Various hypotheses have been offered to explain the tendency of female chimpanzees to range alone or in small parties (Chapman et al., 1995; Janson and Goldsmith, 1995; Pusey and Packer, 1987; Wrangham, 1979; 2000). For example, in a food patch, females may be subject to larger costs from contest competition than are males, due to their lower dominance status. Foraging in a larger party may increase the frequency of shifts between food patches and thus impose larger costs from scramble competition on females because their lower velocity leads to longer travel times between, and late arrival to, food patches and feeding sites. If these hypotheses are true, then females may avoid large parties in favour of small ones, as was shown for chimpanzees in Kibale, Uganda (Wrangham, 2000). This prediction, however, was not verified in bonobos, among which the attendance ratio of females was always higher than that of males, irrespective of party size (Figure 2.2). This suggests that the relationship between social grouping and food supply may differ for bonobos and chimpanzees. For example, bonobos often forage on the abundant terrestrial herbs, fruits and young

Table 2.1 Estrus sex ratio of chimpanzees and bonobos. (Sex-ratio au cours de l'œstrus chez les chimpanzés et les bonobos.)

	Chimpanzee (Mahale)	Chimpanzee (Gombe)	Bonobo (Wamba)
Proportion of a cycling period during an interbirth interval[a]	0.16	0.11	0.77
Proportion of days in estrus during a mestrual cycle[b]	0.40	0.38	0.35
Proportion of days in estrus during an interbirth interval (a*b)[c]	0.064	0.042	0.270
Adult sex ratio (# adult males/# adult females)[d]	0.27	0.51	0.75
Estrus sex ratio (# adult males/# adult females showing estrus) (d/c)	4.2	12.3	2.8

becomes 4.2 for Mahale chimpanzees and 12.3 for Gombe chimpanzees. On the other hand, Wamba bonobos have the highest ratio of adult males to adult females (0.75), but the estrus sex ratio is still estimated to be as low as 2.8. Therefore, even for the Mahale chimpanzees, where the number of adult males is less, the estrus sex ratio is still higher than that among Wamba bonobos. In Gombe, where there are closer numbers of males and females, the estrus sex ratio is much higher than that of Wamba bonobos.

Although it ignores substantial between-group and temporal variation, Figure 2.4 illustrates the general differences in the sexual relations of chimpanzees and bonobos, using the number of adult chimpanzees in Mahale M group in 1984 (Nishida et al., 1990) and the number of adult bonobos and the number of estrous adult females in the Wamba E1 group in 1987–88 (Furuichi and Hashimoto, 2002). If we use the estimated proportion of days in estrus, only 2.2 of 35 female chimpanzees, on average, are expected to show estrus at a given time, while 10 adult males compete for access to these females (Fig. 4, top). Although the proportions vary for different groups and study sites, chimpanzee mating behaviors include possessive mating by high-ranking males, opportunistic and promiscuous mating, and mating during consortships (Boesch and Boesch-Achermann, 2000; Boesch et al., 2006; de Waal, 1982; Hashimoto and Furuichi, 2006; Hasegawa and Hiraiwa-Hasegawa, 1983; Muller et al., 2007; Nishida, 1979; Reynolds, 2005; Stumpf and Boesch, 2005; Tutin, 1979; Wrangham, 1993; Watts, 1998). In many groups, the alpha males and/or males allied with them may have priority in mating access, so that females cannot usually refuse their copulation attempts. During opportunistic mating, females are sometimes severely attacked by males that attempt copulations (Muller et al., 2007). Thus, copulations are, to a large extent, influenced

Chimpanzee
Higher estrus sex ratio (10♂/2.2♀=4.5)
Severe intermale sexual competition
Very limited female choice

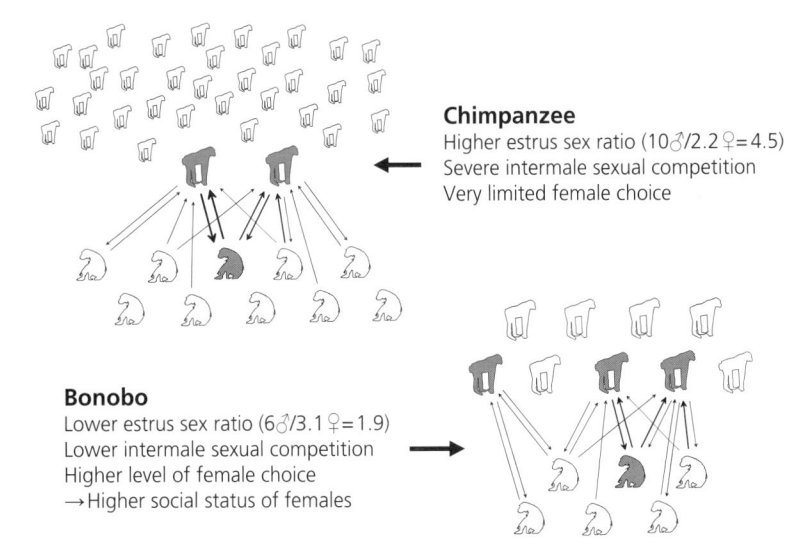

Bonobo
Lower estrus sex ratio (6♂/3.1♀=1.9)
Lower intermale sexual competition
Higher level of female choice
→Higher social status of females

Figure 2.4 Sexual relations among chimpanzees and bonobos. Females are drawn in upper parts, males in lower parts. Dark colours show estrous females and alpha males. Arrows show the solicitation or acceptance of copulations. (Les relations sexuelles entre les chimpanzés et les bonobos. Les femelles figurent dans les parties au-dessus, les mâles en-dessous. Les couleurs sombres montrent les femelles en œstrus et les mâles alpha. Les flèches montre la sollicitation ou le consentement aux copulations.)

by power games among males, with female choice of mating partners being limited.

Among bonobos, in contrast, although there were only 9 females in the group a greater number of females (3.1 on average) showed estrus at any given time (Figure 2.4, bottom) (Furuichi and Hashimoto 2002). In such situations, it is difficult for an alpha male to monopolize all estrous females. Therefore other males may be able to approach estrous females and solicit them for copulation more freely than can male chimpanzees (Furuichi, 1992; Furuichi and Hashimoto, 2004; Kano, 1992). Copulations are not frequently disturbed by other males at Wamba, although inter-male aggression is known to be more frequently observed in the mating context at Lomako (Hohmann and Fruth, 2003) and Lui Kotale (Surbeck et al., 2011). Under such circumstances, the most important thing for males is not to dominate other males but rather to be preferred by females as copulation partners. This may be why males rarely attack or attempt to sexually coerce females. Females can easily ignore solicitations by alpha males, and the occurrence of copulation largely depends on whether females accept a male's solicitation (Furuichi, 1992, 1997; Furuichi and Hashimoto, 2004; Hohmann and Fruth, 2003; Kano, 1992). Thus, with prolonged pseudo-estrus, females can reduce both excessive sexual competition among and harassment by males.

High social status of females

If females aggregate in the central part of a mixed-sex party, form close social associations and are active in mate choice, one naturally assumes that they have high social status. This assumption is true but the question of whether their social status is equal to that of males, or whether they are actually dominant over males, remains. This has been a controversial topic for quite some time.

In the wild, dominance between males and females is equal or equivocal but females seem to be dominant over males where feeding is concerned. Unfortunately, we have only a few reports on the frequency of agonistic interactions between males and females in wild bonobo populations. In a study I conducted on the E1 group over a 7–month period (528 hr and 56 min over 97 days), males were dominant over females in 27 cases, while females were dominant over males in 25 agonistic interactions, showing that males and females had relatively equal status (Furuichi, 1997). Most of the male-dominant cases involved display behaviors in which males ran around emitting excited vocalizations, dragging branches and dashing towards females. Females fled from these males, but such behavior rarely involved physical attacks. On the other hand, most of the female-dominant cases represented retreats by males following approaches by females. Some of the cases occurred in feeding situations. For example, when females approached males who were feeding in a preferred position at a feeding site, males yielded their positions to late-arriving females. Furthermore, males usually waited at the periphery of the feeding site until females finished eating. When overt conflict occurred at feeding sites, allied females sometimes chased males but males never formed aggressive alliances against females. It is interesting to note that even the alpha male might retreat when approached by middle- or low-ranking females. At Lomako, males were dominant over females in all 11 cases of dyadic agonistic interactions between adults and sub-adults for which dominance between the participants was decided. However, females had priority of access to food in terms of the order in which individuals entered food patches (White and Wood, 2007). Another study at Lomako showed that the frequency of female aggression against males was more than double that of male aggression against females, findings opposite those of the study at the same site mentioned earlier (Hohmann and Fruth, 2003).

On the other hand, many reports show that female bonobos are dominant over males in captivity (de Waal, 1995; Parish, 1994; 1996; Vervaecke et al., 2000). In a comparative study of various captive populations, Stevens and colleagues (Stevens et al., 2007) showed that the linearity and steepness of dominance vary among captive populations, and that female dominance is not exclusive, which means that not all females are dominant over all males. However, they also showed that the highest-ranking individuals were females and the lowest-ranking individuals were males in all of the populations studied. Thus, there really does appear to be a difference in this respect between the social tendencies of wild and captive bonobos.

As noted earlier, females can express dominance in situations involving competition for food. Even when feeding on hunted animals, a rare and valuable food resource, females in many cases maintained possession of the kill (Hohmann and Fruth, 1993; 2007; Surbeck and Hohmann, 2008), although among chimpanzees males typically monopolize such resources (Boesch and Boesch, 1989; Stanford et al., 1994; Takahata et al., 1984). Therefore, the higher status of females in captivity might be explained by the fact that competition over food occurs more frequently in captivity. However, this will need to be tested through further study of wild populations.

Influence of mothers on the social status of adult males

Interestingly, female bonobos congregate, maintain close social associations and control social relationships in the group, but they usually do not behave politically or forge tactical alliances in the manner of male chimpanzees. However, there is one important exception to this generalization: given the chance to raise the status of their adult or adolescent sons, mothers can behave quite aggressively, as we observed in the changes of the alpha female and the alpha male in 1983–84 (Figure 2.5) (Furuichi, 1997; Furuichi and Ihobe, 1994).

By 1983, when I started my study, the oldest female, Kame, was the alpha female; her oldest son, Ibo, was the alpha male. Kame had two other younger sons, and Kame and her three sons always kept close associations with each other. Ten, an adolescent son of the second-ranking female, Sen, approached the age of adulthood and began displaying at Kame's three sons. Ibo did not show submissive behaviors and, at the end of many such interactions, instead mounted Ten. However, he sometimes left the area to avoid the persistent provocations of the younger Ten.

When males are involved in agonistic interactions, mothers sometimes join in to support their sons. However, Kame, who was both pregnant and aged, rarely intervened. In contrast, Sen sometimes attacked Kame's sons in order to support Ten. On

	Alpha female	Alpha male
– 1983	Kame	Ibo
1983 – 1992	Sen	Ten
1994 – 1996	Halu*	Ten
2002 – 2007	Nao*	Tawashi
2008 –	Kiku	Nobita

Mother-son relations

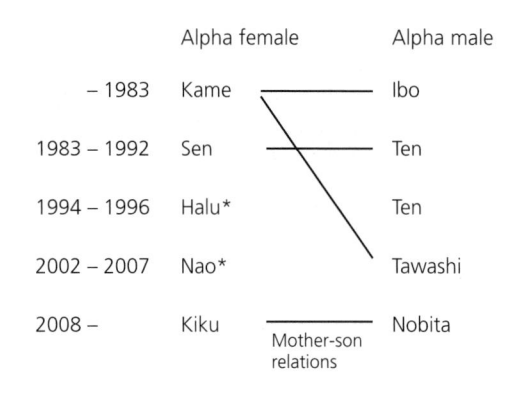

Figure 2.5 Alpha females and alpha males in each study period. Lines show the mother–son relationships. Asterisks show that those females did not have adult or adolescent sons during the periods of their alpha-female status.
(Les femelles et mâles alpha dans chaque période d'étude. Les lignes montrent les relations mère-fils. Les astérisques montrent que ces femelles n'ont pas eu de fils adolescents ou adultes durant leur périodes de statut alpha-femelle.)

one occasion, Ibo fled from Sen after an intense physical fight. A fight between the two mothers occurred five days later, with Sen emerging victorious. After this, fights between these females occurred several times but Kame never defeated Sen. Five days later, Ten approached Ibo, emitting display vocalizations, as was usual at the time. At first Ibo stood bipedally to fight Ten, but then turned his back to present his rump instead, at which point the two males performed rump to rump contact (Kuroda, 1980). From that time on, Sen and Ten behaved as the alpha female and the alpha male. Agonistic interactions between these two families became infrequent (Figure 2.6, Plate 2).

It seemed that this entire series of incidents was triggered by Ten's challenge but he could take the alpha position only when his mother overtook the mother of the previous alpha male and became herself the alpha female. Indeed, Sen was very supportive of her sons. After Ten took the alpha-male position, she persistently supported her seven-year-old second son. He sometimes behaved dominantly over adults in the group, and Sen threatened them when they resisted.

When we resumed our study in 1994, after a two-year break caused by political unrest, Sen had died and the second-ranking female, Halu, had become the alpha female. Because Halu's eldest son had

Figure 2.6 (Plate 2) An ex-alpha female, Kame (left), is groomed by the ex-alpha male, Ibo (centre), and other offspring.
Une femelle ex-alpha, Kame (gauche), est toilettée par un mâle ex-alpha, Ibo (milieu), et autres progénitures.)

died a few years earlier, she had no adult or adolescent son. Thus, Ten continued to be the alpha male. However, Halu was apparently dominant over Ten, so that the highest ranking individual in the E1 group during this period was a female.

Our research was again interrupted by a civil war in 1996. When we resumed our study in 2002, Halu was not in the group and was presumed to be dead. Although Nao had become the alpha female, she also had no adult or adolescent son to become alpha male. Instead, Tawashi, the third son of the deceased ex-alpha female, Kame, was in the alpha-male position.

In the 2008 study period, we found that Kiku, who had been the second-ranking female, had become the alpha female, though we do not know when the rank reversal occurred. During this period, a young adult male, Nobita, became the alpha male. When we resumed our study in 2002, we had tentatively given all adolescent males new names because it was difficult to confirm the identity of immature males after a six-year break. However, DNA analyses confirmed that Nobita was identical to Kikuo, the first son born to Kiku (Hashimoto et al., 2008). Therefore, it appears that another mother–son pair became the alpha female and male.

Thus, throughout our study of E1 group, which began in 1976, five prime or old adult females took the alpha-female position. Three of these had adult or adolescent sons at the time and those males took the alpha-male position. When the other two females without adult or adolescent sons took the alpha position, other males maintained the alpha position but the alpha female was apparently dominant over the alpha male in one of these cases. Stated differently, three out of four males that occupied the alpha position had a mother in the alpha position at the time of their acquisition of this rank.

It is clear that females do not reach the alpha-female position owing to the high status of their sons because males never support their mothers in agonistic interactions among females. Rather, females support their sons in agonistic interactions among males. Among chimpanzees, males at a prime adult age with strong male allies tend to take the alpha position (de Waal, 1982; Goodall, 1986; Nishida, 1983; Nishida and Hosaka, 1996; Reynolds, 2005). In contrast, male bonobos in E1 group tend to obtain the alpha position during late adolescence or early adulthood (estimated ages in years: Ibo = < 20, Ten = 13, Tawashi = 22–28, Nobita = 20). This may have occurred because, except in the case of Tawashi whose mother had already died, their mothers were at a prime age at the time of their ascent to alpha-male status. In contrast, many prime adult males occupied lower ranks among adult males in both the E1 and E2 groups of bonobos at Wamba, although there were some exceptions (Tawashi was the alpha male in the E1 group and Kuma was the alpha male in the E2 group during particular periods). This tendency may be partly explained by the fact that the mothers of those prime adult males had already died (Furuichi, 1997; Furuichi and Ihobe, 1994).

We may ask, then, why some female bonobos intentionally support their maturing sons' bid for alpha status, despite the fact that they do not often fight with one another or behave politically. I hypothesized that females may compete to increase their number of grand-offspring through the support of their sons (Furuichi, 1997). Although females cannot increase the number of offspring they themselves produce by fighting with other females, they may be able, by raising the social status of their sons, to increase the number of their sons' offspring. Gerloff and colleagues (Gerloff et al., 1999) reported that at Lomako two males that attained the highest paternity success were sons of high-ranking females. Surbeck and others (2011) reported that females at Lui Kotale frequently intervened in the mating attempts of unrelated males or provided their sons with agonistic support when unrelated males tried to interfere with their sons' mating activities. As a result, middle- or low-ranking males had increased mating success when their mothers were present in the party. This kind of support by mothers seems to be a common feature among wild bonobos.

Female role in peaceful encounters between groups

It is well known that intergroup encounters among chimpanzees are aggressive (Wrangham and Peterson, 1996). At Mahale, during conflicts between groups males of the K group disappeared one by one, supposedly killed by males of the M group which finally took over both the home range and the females of the extinct K group (Nishida et al., 1985). A similar incident occurred between Kasakela and Kamaha groups at Gombe (Goodall, 1983; Williams et al., 2004). After splitting into the two groups, the larger Kasakela group killed the males of the Kahama group, which finally became extinct. During this process, several fatal attacks were observed. Males of the Kahama group sometimes patrolled boundary areas, attacking and killing Kahama males that were ranging alone (Goodall, 1983). Male patrolling behavior and fatal agonistic interactions have also been observed at Kibale (Mitani et al., 2010; Watts et al., 2006; Wilson and Wrangham, 2003; Wrangham, 1999) and Kalinzu (Hashimoto and Furuichi, 2005) in Uganda, and at Taï in Côte d'Ivoire (Boesch et al., 2008).

Intergroup encounters among bonobos are also stressful. When they hear vocalizations of other groups bonobos usually climb up trees and carefully look in the direction of the vocalizations. They sometimes quietly change their direction to avoid other groups. In most cases, however, they respond to such vocalizations with a chorus of loud calls. The two groups gradually approach each other, exchanging vocalizations. When they finally meet, males often display but do not usually fight. Although males do not usually merge with each other when groups meet, after a time females do move beyond the boundary and begin to interact with females of the other group, engaging in genito–genital rubbing or grooming. It appears as if they have found old associates or relatives in the other group. After the initial excitement has passed the groups often start feeding in the same tree. Although males tend to stay behind the boundary, it becomes difficult to identify the boundary between the two groups (Idani, 1990; Kano, 1992).

In the 1986–87 study period, Idani observed 25 encounters between P group and the main study group, E1, over 3 months (Idani, 1990). In all encounters, both groups remained together for long periods. During such encounters, copulations frequently occurred between males and females of the different groups. Figure 2.7 shows the frequency of copulation by adult females of the E1 group. E1 females copulated with males of P group with considerable frequency, though less frequently than with males of the own E1 group. Considering only those encounters, E1 females copulated with P males more frequently than they did with E1 males, suggesting that females are eager to copulate with males of different groups. As noted, females move beyond the boundary without hesitation, but males tend to stay within the range of their own group. Therefore, males cannot do anything to restrict females even if

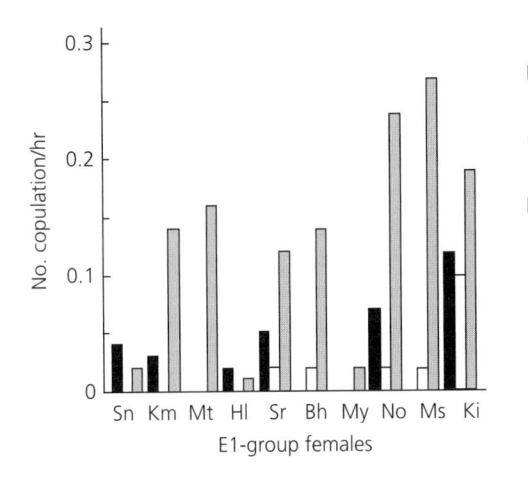

Figure 2.7 Frequency of inter- and intra-group copulations by E1-group females. Modified from Fig. 12 in Idani (1990). (Fréquence de copulations inter- et intra-groupe par les femelles du groupe E1. Modifié de la Fig. 12 dans Idani (1990).)

they copulate with males of the other group. On the other hand, males can copulate with females from the other group that come into their range without any disturbances from the males of that other group.

Encounters between two groups of bonobos can be repeated intermittently for several days (Idani, 1990). It seems that such encounters are stressful for males; they occasionally move away, vocalizing in an apparent attempt to encourage females to terminate the encounter. However, if females do not separate from the other group then males return to the encounter. Thus, these peaceful intergroup encounters apparently are led by females. Such encounters appear to hold no risk for females and they gain an opportunity to copulate with extra-group males or socially interact with extra-group females that they may have known before transferring to their current groups.

Bonobos seem to establish peaceful relationships with the groups they encounter readily, though such relationships are not formed for all combinations of the groups. After the study group E split into E1 and E2 in 1983, the two new groups continued to have peaceful encounters with each other. E1 bonobos also had peaceful encounters with another neighbouring group, K, but tended to vacate the area when they heard vocalizations of another neighbouring group, B. In the 1980s, the E1 group more frequently had peaceful encounters with P group as they extended their range into the south after splitting off from the original group. After we resumed our study in 2003, E1 group extended its range into the east and experienced peaceful encounters with a newly identified group in 2010. Peaceful encounters were also reported from Lomako (Hohmann and Fruth, 2002). Groups at that site displayed to one another and the frequency of agonistic interactions increased during the encounters. However, members of the different groups also copulated with and groomed one another.

Conclusion

This comparison of the social structures of chimpanzees and bonobos illustrates how the nature of societies may change depending on which sex controls behavioral initiatives. Female bonobos are highly gregarious despite living in male-philopatric social groups. They form close social associations and cooperatively defend their high social status against males, competing ferociously only for the high dominance rank of their sons. The high social status of females and their initiative in social, sexual and ranging behaviors seem to contribute to the peaceful nature of the bonobo society.

The characteristic features of female bonobos appear to be interrelated and together contribute to the high attendance of females in mixed-sex parties (Fig. 8). First, bonobo habitats, compared with those of chimpanzees, have a higher density of food patches, including large fruit trees and smaller food items on which they forage while travelling between larger trees. Such ecological conditions may decrease the travel distance between food patches (Furuichi, 2009; Malenky and

Stiles, 1991; White and Wrangham, 1988; Wrangham, 2000), thus reducing the cost for the slower-moving females. Second, if females can exercise the initiative in ranging, they can avoid incurring larger costs associated with travel than males do, which may promote the aggregation of females in mixed-sex parties. Third, with a prolonged pseudo-estrus, females mitigate any potential for excessive sexual competition among males and thereby avoid male harassment. In addition, the long periods of pseudo-estrus may prevent infanticide through paternity confusion. The high social status of females may contribute to their aggregation in at least two ways. One way is that females can have priority of access to food and, because of their high rank, avoid both infanticide and harassment by males. Another way is that, through their close association with and dominance support for their adult sons, they can take the initiative in ranging. On the other hand, the high degree of aggregation displayed by females can also contribute to their high social status, through formation of coalitions. The social status of females may also be enhanced by the extended female choice in mating partners resulting from prolonged pseudo-estrus.

Recent genetic studies have shown that chimpanzees and bonobos diverged within the last million years or so (Becquet and Przeworski, 2007; Hey, 2010), more recently than was previously thought. It is not surprising, therefore, that chimpanzees and bonobos share many common traits, including physical attributes and their male-philopatric residence patterns. However, something changed for bonobos, probably during a bottleneck period, which altered the sexuality and social status of females. Since chimpanzees, gorillas and orangutans all share common traits, such as a limited estrus period among females and male dominance, this change seems to be specific to bonobos. Considering the recent divergence of bonobos and chimpanzees, we may infer that small genetic changes occurred in one or a few key features and thus invoked development of the whole social system represented in Figure 2.8, rather than that the various features evolved independently. For example, if genetic changes occurred in the physiology of females, causing them to show pseudo-estrus during non-conceptive periods, this whole social system may have developed in an environment with abundant and dense food resources, without requiring many other genetic changes, at least in the early stages. I expect that genetic studies will clarify the small but important genetic differences that can explain the large differences in sexuality and society of the two species in the genus *Pan*.

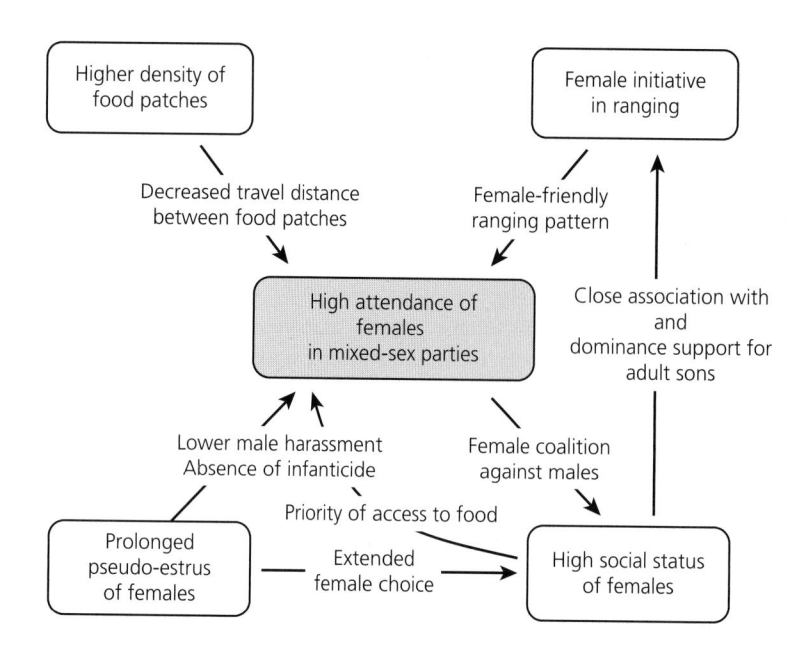

Figure 2.8 Characteristic features of female bonobos and relationships among them.
(Traits caractéristiques des femelles bonobos et leur relations entre elles-mêmes.)

Acknowledgments

All the research reported here on bonobos at Wamba and chimpanzees in the Kalinzu Forest was done with Takayoshi Kano, Suehisa Kuroda, Kohji Kitamura, Akio Mori, Evelyn Ono-Vineberg, Gen'ichi Idani, Tomoo Enomoto, Hiroshi Ihobe, Naobi Okayasu, Chie Hashimoto, Ellen J. Ingmanson, Yasuko Tashiro, Hiroyuki Takemoto, Tetsuya Sakamaki and other researchers who visited Wamba for short periods. Others who conducted research with us were Ekam Wina, Mwanza N. Ndunda, Mbangi N. Mulavwa, Kumugo Yangozene, Mikwaya Yamba-Yamba and Balemba Motema-Salo of the Research Center for Ecology and Forestry, Ministry of Scientific Research and Technology, D. R. Congo. I give special thanks for their great contribution to this work. I also thank Toshisada Nishida, Juichi Yamagiwa and Tetsuro Matsuzawa for their continued support of our studies, and members of the Laboratory of Human Evolution Studies and Primate Research Institute of Kyoto University for providing valuable suggestions. This study was mainly supported by the Japan Society for the Promotion of Science (JSPS) Grants-in-aid for Scientific Research (02041049, 06041064, 09041160, 10640613 to Kano; 58041025, 60041020, 62041021 to Nishida; 10CE2005 to Takenaka; 12575017, 17255005, 22255007 to Furuichi; 17570193, 19405015 to Hashimoto; 19107007 to Yamagiwa; 21255006 to Ihobe); the National Geographic Fund for Research and Exploration (7511–03 to Furuichi); JSPS Core-to-Core program (15,001 to Primate Research Institute); Toyota Foundation (D04-B-285 to Furuichi); JSPS International Training Program (2009–8 to Primate Research Institute); JSPS Asia-Africa Science Platform Program (2009–8 to Furuichi); JSPS Institutional Program for Young Researcher Overseas Visits (2009–37 to Primate Research Institute); Japan Ministry of the Environment Global Environment Research Fund (F-061 to Nishida); Charitable Trust Africa Support Fund (2010 to Support for Conservation of Bonobos); and Japan Ministry of the Environment Research and Technology Development Fund (D-1007 to Furuichi). Original version of this chapter was published in Evolutionary Anthropology, 20, 131–142 (2011). Minor corrections were made to the original version.

References

Beach, F.A. (1976). Sexual attractivity, proceptivity, and receptivity in female mammals. *Hormones and Behavior*, 7, 105–38.

Becquet, C. and Przeworski, M. (2007). A new approach to estimate parameters of speciation models with application to apes. *Genome Research*, 17, 1505–519.

Boesch, C. (1996). Social grouping in Taï chimpanzees. In: McGrew, W.C., Marchant, L.F., and Nishida, T. (eds). *Great Ape Societies*. Cambridge: Cambridge University Press, pp. 101–13.

Boesch, C. and Boesch, H. (1989). Hunting behavior of wild chimpanzees in the Taï National Park. *American Journal of Physical Anthropology*, 78, 547–73.

Boesch, C. and Boesch-Achermann, H. (2000). *The Chimpanzees of the Taï Forest: Behavioral Ecology and Evolution*. New York, NY: Oxford University Press.

Boesch, C., Crockford, C., Herbinger, I., Wittig, R., Moebius, Y., and Normand, E. (2008). Intergroup conflicts among chimpanzees in Taï National Park: lethal violence and the female perspective. *American Journal of Primatology*, 70, 519–32.

Boesch, C., Kohou, G., Nene, H., and Vigilant, L. (2006). Male competition and paternity in wild chimpanzees of the Taï Forest. *American Journal of Physical Anthropology*, 130, 103–15.

Chapman, C.A., Wrangham, R.W., and Chapman, L.J. (1995). Ecological constraints on group size: an analysis of spider monkey and chimpanzee subgroups. *Behavioral Ecology and Sociobiology*, 36, 59–70.

Dixson, A.F. (1998). *Primate Sexuality: Comparative Studies of the Prosimians, Monkeys, Apes, and Human Beings*. Oxford: Oxford University Press.

de Waal, F.B.M. (1982). *Chimpanzee Politics: Power and Sex among Apes*. London: Jonathan Cape.

de Waal, F.B.M. (1995). Bonobo sex and society. *Scientific American Library Series*. 272, 82–8.

Furuichi, T. (1987). Sexual swelling, receptivity, and grouping of wild pygmy chimpanzee females at Wamba, Zaïre. *Primates*, 28, 309–18.

Furuichi, T. (1989). Social interactions and the life history of female *Pan paniscus* in Wamba, Zaire. *International Journal of Primatology*, 10, 173–97.

Furuichi, T. (1992). The prolonged estrus of females and factors influencing mating in a wild unit-group of bonobos (*Pan paniscus*) in Wamba, Zaire. In: Itoigawa, N., Sugiyama, Y., Sackett, G., and Thompson, R. (eds). *Topics in Primatology: Behavior, Ecology, and Conservation*, Vol. 2. Tokyo: University of Tokyo Press, pp. 179–90.

Surbeck, M. and Hohmann, G. (2008). Primate hunting by bonobos at LuiKotale, Salonga National Park. *Current Biology*, 18, R906–907.

Surbeck, M., Mundry, R., and Hohmann, G. (2011). Mothers matter! Maternal support, dominance status and mating success in male bonobos (*Pan paniscus*). *Proceedings of the Royal Society B*, 278, 590–8.

Takahata, Y., Hasegawa, T., and Nishida, T. (1984). Chimpanzee predation in the Mahale Mountains from August 1979 to May 1982. *International Journal of Primatology*, 5, 213–33.

Takahata, Y., Ihobe, H., and Idani, G. (1996). Comparing copulations of chimpanzees and bonobos: do females exhibit proceptivity or receptivity? In: McGrew, W.C., Marchant, L.F., and Nishida, T. (eds). *Great Ape Societies*. Cambridge: Cambridge University Press, pp. 146–55.

Thompson-Handler, N., Malenky, R.K., and Badrian, N. (1984). Sexual behavior of *Pan paniscus* under natural conditions in the Lomako Forest, Equateur, Zaire. In: Susman, R.L. (ed.). *The Pygmy Chimpanzee: Evolutionary Biology and Behavior*. New York, NY: Prenum, pp. 347–68.

Tutin, C.E.G. (1979). Mating patterns and reproductive strategies in a community of wild chimpanzees (*Pan troglodytes schweinfurthii*). *Behavioral Ecology and. Sociobiology*, 6, 29–38.

Tutin, C.E.G. and McGinnis, P.R. (1981). Chimpanzee reproduction in the wild. In: Graham, C.E. (ed.). *Reproductive Biology of the Great Apes: Comparative and Biomedical Perspectives*. New York, NY: Academic Press, pp. 239–64.

Van Elsacker, L., Vervaecke, H., and Verheyen, R.F. (1995). A review of terminology on aggregation patterns in bonobos (*Pan paniscus*). *International Journal of Primatology*, 16, 37–52.

Vervaecke, H., de Vries, H., and Van Elsacker, L. (2000). Dominance and its behavioral measures in a captive group of bonobos (*Pan paniscus*). *International Journal of Primatology*, 21, 47–68.

Wallis, J. (1997). A survey of reproductive parameters in the free-ranging chimpanzees of Gombe National Park. *Journal of Reproduction and Fertility*, 109, 297–307.

Watts, D.P. (1998). Coalitionary mate guarding by male chimpanzees at Ngogo, Kibale National Park, Uganda. *Behavioral Ecology and Sociobiology*, 44, 43–55.

Watts, D.P., Muller, M., Amsler, S.J., Mbabazi, G., and Mitani, J.C. (2006). Lethal intergroup aggression by chimpanzees in Kibale National Park, Uganda. *American Journal of Primatology*, 68, 161–80.

White, F.J. (1988). Party composition and dynamics in *Pan paniscus*. *International Journal of Primatology*, 9, 179–93.

White, F.J. and Wood, K.D. (2007). Female feeding priority in bonobos, *Pan paniscus*, and the question of female dominance. *American Journal of Primatology*, 69, 837–50.

White, F.J. and Wrangham, R.W. (1988). Feeding competition and patch size in the chimpanzee species *Pan paniscus* and *Pan troglodytes*. *Behaviour*, 105, 148–64.

Williams, J.M., Oehlert, G.W., Carlis, J.V., and Pusey. A.E. (2004). Why do male chimpanzees defend a group range? *Animal Behavior*, 68, 523–32.

Wilson, M. and Wrangham. R. (2003). Intergroup relations in chimpanzees. *Annual Reviews in Anthropology*, 32, 363–92.

Wrangham, R.W. (1979). Sex differences in chimpanzee dispersion. In: Hamburg, D.A. and McCown, E.R. (eds). *The Great Apes*. Menlo Park, CA: Benjamin/Cummings, pp. 481–89.

Wrangham, R.W. (1986). Ecology and social relationships in two species of chimpanzee. In: Rubenstein, D.I. and Wrangham, R.W. (eds). *Ecological Aspects of Social Evolution on Birds and Mammals*. Princeton, NJ: Princeton University Press, pp. 352–78.

Wrangham, R.W. (1993). The evolution of sexuality in chimpanzees and bonobos. *Human Nature*, 4, 47–79.

Wrangham, R. (1999). Evolution of coalitionary killing. *Yearbook of Physical Anthropology*, 42, 1–30.

Wrangham, R.W. (2000). Why are male chimpanzees more gregarious than mothers? A scramble competition hypothesis. In: Kappeler, P.M. (ed.). *Primate Males: Causes and Consequences of Variation in Group Composition*. Cambridge: Cambridge University Press, pp. 248–58.

Wrangham, R. and Peterson, D. (1996). *Demonic Males: Apes and the Origins of Human Violence*. Boston, MA: Houghton Mifflin.

CHAPTER 3

Affiliations, aggressions and an adoption: Male–male relationships in wild bonobos

Martin Surbeck and Gottfried Hohmann

Abstract The nature of the relationships between males is a characteristic trait of many multi-male group living species with implications for the individuals. In our study population of bonobos, certain male dyads exhibit clear preferences for ranging in the same party and sitting in proximity. These preferences are not reflected in the frequency of aggression towards each other and only to some extent in their affiliative and socio-sexual behaviors. While bonobo males at LuiKotale clearly do not benefit from close relationships in the way chimpanzee males do (cooperative hunting, territorial patrol, mate competition), some relationships might result from close associations between their mothers. In some particular situations, these male relationships can be very important as in the case of an orphan adopted by his older maternal brother.

La nature des relations entre mâles est un trait caractéristique de plusieurs groupes qui ont plusieurs mâles, avec des implications au niveau d'individus. Dans notre étude des populations de bonobos, certains dyades mâles montrent une préférence à aller dans le même groupe et s'asseoir proche l'un de l'autre. Cette préférence n'est pas reflétée dans la fréquence d'agression entre eux et est seulement lié, à degrés, à leur comportements socio-sexuels et d'appartenance. Tandis que les mâles bonobos à LuiKotale ne profitent pas de leur fortes relations comme les chimpanzés mâles (chasse coopérative, patrouille territoriale, compétition pour compagnon), ils peuvent aider leur partenaires à supporter le stress de la vie en groupe et peuvent en conséquence contribuer au bien-être des individus. Quelques proches associations entre les mâles peuvent provenir d'associations entre leurs mères. Dans quelques situations particulières, ces relations mâles prouvent leur importance comme dans le cas d'un orphelin adopté par son grand frère maternel.

Introduction

Sociality and close social bonds have been shown to affect individual fitness in group living species (Schülke et al., 2010; Silk et al., 2003, 2009, 2010). Bonding between individuals promotes the formation of agonistic alliances and cooperation, increasing an individual's access to limiting resources (Alberts, 2010; Schülke et al., 2010). Furthermore, close bonds serve to mediate stress through social contact and affiliation and therefore play an important role in the well-being of individuals (Aureli et al., 1999; Carter, 1998; Crockford et al., 2008; Engh et al., 2006). While these mutual benefits of bonding for both partners could lead to its occurrence between non-relatives (Clutton-Brock, 2009), close bonds should form preferentially between relatives because according to kin selection theory, both partners can gain additionally indirect fitness benefits (Hamilton, 1964). In most mammals, males disperse and females remain with their relatives in the natal group (Handley and Perrin, 2007). Thus, strong bonds are expected to form preferentially between female relatives (Silk, 2009). In species in which the males remain in their natal community, close bonds are expected to occur between related males. While some researchers have pointed out the inherent difficulty of bonding and cooperation between males in theory (van Hooff and van Schaik, 1994), this does not seem to hold true for male-philopatric chimpanzees.

Surbeck, M. and Hohmann, G., *Affiliations, aggressions and an adoption: male-male relationships in wild bonobos.* In: *Bonobos: Unique in Mind, Brain, and Behavior.* Edited by Brian Hare and Shinya Yamamoto: Oxford University Press (2017). © Oxford University Press. DOI: 10.1093/oso/9780198728511.003.0003

In chimpanzees, both sexes are known to form close social relationships with certain community members that are characterized by affiliative behaviors and, in the case of males, by extensive cooperation during territorial border patrol, mate guarding and hunting (Langergraber et al., 2007; Mitani et al., 2000; Muller et al., 2005; Watts, 1998). Close male associates are often relatives, whereas close female associates appear generally unrelated (Langergraber et al., 2007; Langergraber et al., 2009; Mitani, 2009).

Bonobos, like chimpanzees, exhibit differentiated relationships among group members (Furuichi et al., 1998; Hohmann and Fruth, 2002). Close female dyads are characterized by high levels of association and grooming, and exhibit low frequencies of genito–genital contacts (Hohmann and Fruth, 2000). Furthermore, close female relationships might facilitate agonistic coalition formation during within-group conflicts but it is not clear whether closely associated females are more likely to support each other (Parish, 1996; but Jaeggi et al., 2010; Surbeck and Hohmann, 2013). On the other hand, little is known about the nature and strength of relationships among bonobo males. Field and captive data suggest an apparent lack of agonistic coalitions among males (Ihobe, 1992; Surbeck et al., submitted; Vervaecke et al., 2000). The close associations among males that are characteristic of chimpanzees seem to be substituted in bonobos by long-lasting bonds between females and their adult sons, which manifest during competition for high ranks and access to females and in affiliative behaviors (Furuichi, 1997; Ihobe, 1992; Surbeck et al., 2011). Indirect evidence for the possible substitution of male relationships by mother–son bonds comes from the observation that males engage in close relationships when strong bonds with their mothers are absent (Ihobe, 1992). Close relationships between sons and mothers have the advantage that they do not lead to conflicting reproductive interests (van Hooff and van Schaik, 1994).

While close male associations in bonobos do not serve as a basis for extended cooperation as observed in chimpanzees, they might still offer benefits for the involved males. Grooming is a behavior that reduces social stress by decreasing cortisol levels (Gust et al., 1993; Shutt et al., 2007) and grooming with close associates possibly leads to a stronger decrease in cortisol as compared to other individuals (Crockford et al., 2013; Wittig et al., 2008). Another behavior associated with stress reduction in bonobos is socio-sexual behavior (Hohmann and Fruth, 2000), but little is known about its relationship with association patterns among males. Thus, even though there may be benefits to forming close male associations in bonobos, it is also possible that association patterns among bonobo males are a phylogenetic remnant of their last common ancestor with chimpanzees and that close associations are no longer functionally important in modern bonobo society.

In this chapter we aim to explore the range of variation in association patterns within male dyads and how differentiated associations translate into affiliative behaviors such as grooming and socio-sexual behaviors. While high amounts of grooming are associated with close associations in several species (e.g. chimpanzees: Boesch and Boesch-Ackermann (2000), baboons: Silk et al., (2003)), the role of socio-sexual behavior in bonobo society is controversial. Most studies concentrate on the occurrence of these behaviors among females and while some authors hypothesize socio-sexual behavior to be prominent in the formation of close associations (Furuichi et al., 1998), other studies found that avoidance relationships are characterized by the use of socio-sexual behavior (Hohmann and Fruth, 2000). So far nothing is known about how socio-sexual behavior among males is linked to the associations between them.

This chapter investigates the strength and nature of male relationships in bonobos. This is done by first comparing the range of male–male party affinity to that of male–female and female–female dyads. Second, we identify close and avoidant associates among males based on party and proximity affinity and test whether they differ according to measurements of affiliative and aggressive behavior. Third, we investigate how maternal kinship influences the formation of such dyads and discuss possible explanations for the occurrence of differentiated relationships among male bonobos in the light of additional behavioral observations.

Methods

Study site and subjects

Behavioral data were collected between May and August 2007 and between December 2007 and July 2009 on the Bompusa bonobo community at the LuiKotale field site, Democratic Republic of Congo (DRC) (Hohmann and Fruth, 2003). The study community consisted during this period of nine males (older than ten years; for specific age estimates see Table 1). All nine males were included in this study because all engaged in competition over access to maximally tumescent females and actively participated in dominance interactions with other males (Surbeck et al., 2011). In addition, there were 11 parous females, and up to 5 nulliparous immigrant females. The ages were estimated based on physical features such as body size, dentition and (in females) genital swellings (Furuichi et al., 1998).

Behavioral data collection

Parties preferentially containing males were followed from the time they left their sleeping site in the morning until they constructed their nest in the evening. Every full hour we recorded the party composition ($N = 2112$). We recorded all occurring aggression, socio-sexual behavior, food sharing and grooming events. In addition we performed 10-minute focal follows of individual males (470 hours of focal follows, Table 1). At the start and at the end of each focal follow we recorded all individuals within proximity (5 m) of the focal individual.

Aggressive behavior between males consisted of contact ($N = 83$; combining all occurrences and focal follows) and non-contact aggression including directed displays ($N = 577$; combining all occurrences and focal follows). Agonistic alliances were recorded if two males showed simultaneously aggressive behavior towards a third individual.

Non-aggressive behavior included socio-sexual behavior and grooming between males. Socio-sexual behavior involved mounting ($N = 28$; combining all occurrences and focal follows), rump–rump contacts ($N = 4$; combining all occurrences and focal follows) and ventro–ventro rubbing ($N = 2$; combining all occurrences and focal follows). A grooming bout involved continuous grooming between two individuals that did not stop for longer than five minutes and was not interrupted by a third individual ($N = 91$; combining all occurrences and focal follows). Food-sharing events between males consisted of occasions in which one male was in possession of a monopolizable food (*Treculia africana*, *Annonidium manni*) and another male obtained pieces of this food item ($N = 5$).

Data analysis

For the analysis, we calculated different measurements of spatial affinity between individuals: (1) a standardized pairwise affinity value (PAV) based on party association in order to compare affinity of different sex dyads. (2) A combined pairwise affinity value (CPAV) based on party and proximity association in order to compare affinity within male dyads.

Pairwise affinity value (PAV)

In order to compare association patterns of male–male dyads with male–female and female–female

Table 3.1 Ages and dominance ranks of the males, as well as individual observation time. * indicate males with a living mother in the community.
(Âges et rangs de dominance des mâles, et le temp de leur observation individuelles. * indique les mâles avec une mère en vie dans la communauté.)

Individual	Age class [years]	Rank	Observation hours (absence/presence of maximally tumescent females)	Number of 10-min focals (absence/presence of maximally tumescent females)
CA*	26–35	1	466/668	168/254
TI*	26–35	2	434/600	157/206
JA	16–25	3	431/541	175/193
DA	> 35	4	446/524	165/184
BE*	16–25	5	536/640	202/239
ap*	10–15	6	342/473	122/149
pn*	10–15	7	256/413	76/93
em*	10–15	8	323/462	100/131
Mx	10–15	9	279/353	91/97

dyads, we calculated a pairwise affinity value (PAV) based on the hourly party compositions. This index was derived by first calculating the observed simple ratio index of a given dyad (SRI_{obs}) and then subtracting the expected simple ratio value of this dyad (SRI_{exp}), derived by randomization considering individual differences in observation time and characteristics of data structure such as party sizes. Finally, the outcome was standardized to a value ranging from −1 (together as little as possible) to +1 (together as much as possible). Positive values indicate association preference and negative values indicate avoidance.

We calculated $SRI_{obs} = Pa(AB)/(Pa(A) + Pa(B) - Pa(AB))$, with $Pa(AB)$ = number of parties containing both A and B, $Pa(A)$ = number of parties containing A, and $Pa(B)$ = number of parties containing B.

The value of PAV was then derived as

$$PAV = \frac{(SRI_{obs} - SRI_{exp})}{(1 - SRI_{exp})}, \text{ if } (SRI_{obs} - SRI_{exp}) > 0$$

and as

$$PAV = \frac{(SRI_{obs} - SRI_{exp})}{SRI_{exp}}, \text{ if } (SRI_{obs} - SRI_{exp}) < 0$$

The randomization procedure consisted of randomly drawing as many individual per party as were present in the original party, with a probability corresponding to their overall frequency of appearance in the whole dataset. Such a randomization keeps constant the observation frequencies of individuals and the observed party sizes. The results of this randomization procedure are highly correlated with the results from another randomization which keeps constant both the total duration and temporal autocorrelation in an individual's party attendance, and the frequency distribution of the duration of its party attendances. The dyadic SRI_{exp} was the mean SRI value of 1000 of such randomizations (for further details, see Surbeck et al., 2011).

Combined pairwise affinity value (CPAV)

In order to refine the measurement of pairwise affinity for the analysis of only male–male dyads we considered proximity measurements within a party in addition to party associations to calculate a combined pairwise affinity value (CPAV). The CPAV is based on (1) the tendency of a given dyad to associate at a party level (party affinity), and (2) the tendency of a certain dyad to be in proximity (5 m), given the males within the same party (proximity affinity). Note that the second measure is independent of the first one.

(1) Party affinity

To measure the tendency of a given male dyad to be in the same party, we first calculated a standardized pairwise affinity value (see earlier) using only male party attendances, but instead of standardizing the values from −1 to +1, we z-transformed the values to a mean of 0 and a standard deviation of 1.

(2) Proximity affinity

In order to quantify whether a male dyad, given the males in the same party, has a preference to associate in proximity, we calculated a proximity affinity. Therefore we established first the frequency with which two individuals were observed within 5 m distance of each other (PI_{obs}).

We calculated $PI_{obs} = Pr(AB) / (Pr(A) + Pr(B) - Pr(AB))$, with $Pr(AB)$ = number of scans of focal A with B in proximity, $Pr(A)$ = number of scans of focal A and $Pr(B)$ = number of scans of focal A with B in party. We then subtracted from the PI_{obs} of a given dyad an expected proximity index given its observation frequency within the same party (PI_{exp}). Positive values at this point indicate that a dyad was observed more in proximity than expected (preferred proximity partners), negative values indicate that a dyad was observed less in proximity than expected (avoidant proximity partners). We finally z-transformed the outcome values to get the proximity affinity of a given dyad.

The algorithm to calculate the PI_{exp} for each dyad considered party attendance during the focal follows to exclude preferred proximity to occur as a by-product of party composition. To get PI_{exp} we first randomly draw a number of males from the actual party of the proximity scan whereby the number of males drawn was equal to the actual number of males in close proximity to the focal subject. To account for differences between males in how often they were in proximity to other males we draw males present in the party with likelihood proportionate to their total appearance in proximity scans. Since, using this approach, individual A can have a slight different proximity affinity to B than vice versa, we took the mean value for each dyad.

We compared party affinity and proximity affinity using a Mantel test as described earlier. The combined pairwise affinity value (CPAV) is the sum of proximity affinity and the party affinity.

In order to exclude the possibility that the strength of male associations is driven by the presence of a maximally tumescent female in a given party, we furthermore analysed the periods during which maximally tumescent females were absent separately and compared the results with a Mantel test with Spearman's rho as test statistics while enumerating all possible permutations (Sokal and Rohlf, 1995).

Close associates

Close associates were defined as dyads that (1) are observed significantly more frequently than expected by chance within the same party in the absence of maximally tumescent females in the community (to exclude close associates that are a by-product of common interest) AND (2) have mutual positive values from observed-expected proximity indices ($PI_{obs} - PI_{exp}$): in other words, individuals who are not only more often within the same party but also prefer to sit close to each other. Alternatively, dyads were ranked according to their CPAV rank and the upper third ($n = 12$) was arbitrarily categorized as close associates.

Avoidant relationships

Avoidant relationships were defined as dyads that (1) are observed significantly less frequently than expected by chance within the same party in the absence of maximally tumescent females in the community, AND (2) have mutual negative values from observed-expected proximity indices ($PI_{obs} - PI_{exp}$). In other words, dyads which avoid travel in the same party and, in case they end up in the same party, avoid staying within proximity.

Alternatively, dyads were ranked according to their CPAV rank and the lower third ($n = 12$) was arbitrarily categorized as avoidant relationships.

Linking associations patterns and behavior

There is controversy as to whether it is accurate to control for association frequency when comparing the occurrence of certain social behaviors such as grooming between dyads in fission–fusion societies. To avoid controversy we analysed both: the frequencies of behaviors within a dyad based on the time both individuals have been observed within the same party and the total amount of observed behavior within a dyad.

All data processing and analysis was done in R (Baayen, 2008) using the package gtools for the matrix correlation (Gregory et al., 2013).

Genetic analysis

Males and females were genotyped at 19 autosomal loci which allowed identifying maternal half-brothers in the presence of the mother. Maternity assignments for all sub-adult and adult males ($N = 9$) were conducted using both a likelihood approach implemented in CERVUS 3.0 (Kalinowski et al., 2007) and confirmation that the parent–offspring pair shared an allele at every locus. Genotyping followed guidelines used by Arandjelovic and colleagues (2009). Six males were assigned a mother at the 99 per cent confidence level and shared an allele with the most likely candidate mother at every locus compared while mismatching with the second-best candidate mother at two or more loci. No compatible mothers were found for the remaining three males.

Consequently, for 33 of the 36 male dyads, the existing data allowed us to infer whether the dyads are maternal half-brothers or not. For the three remaining dyads we cannot exclude maternal relatedness based on the small number of loci typed (Schubert et al., 2011).

Results

Diversity of male–male associations

The results show a great diversity in the strength of male–male associations (Figure 3.1). The range of standardized pairwise affinities value (PAV) in male–male dyads (min = –0.52, max = 0.56) was larger than of female–female (min = –0.25, max = 0.20) or

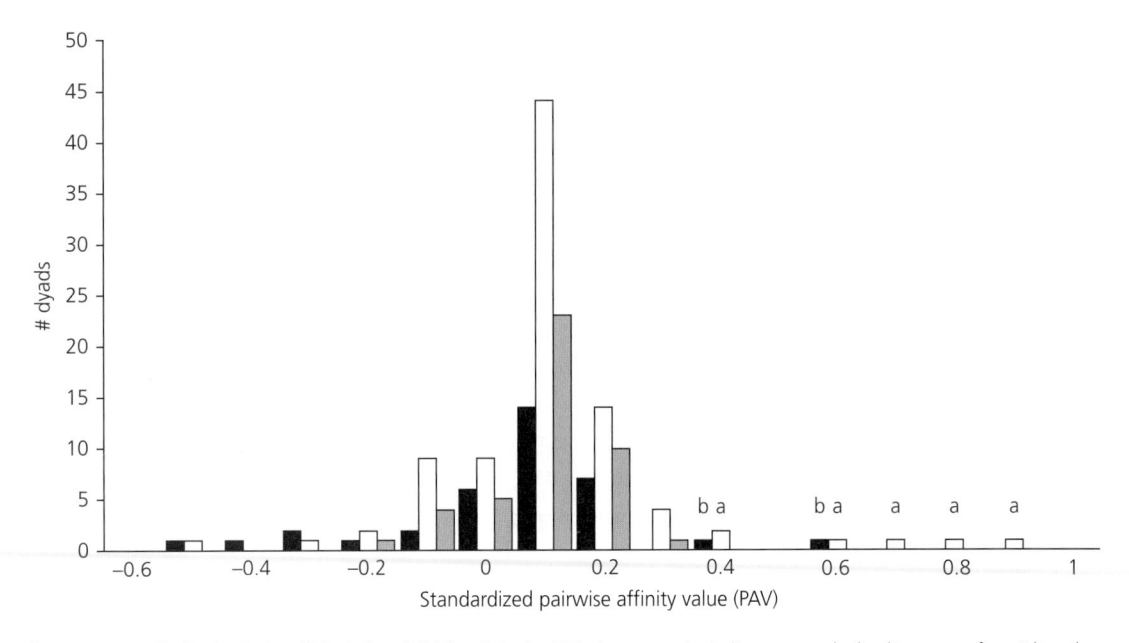

Figure 3.1 Standardized pairwise affinity indices (PAV) for all dyads within the community. Indices are standardized to a range from 0 (together as little as possible) to 1 (together as much as possible) per dyad. Black bars represent male–male dyads, grey bars represent female–female dyads and white bars represent mixed sex dyads. [a] indicates mother–son pairs, [b] indicates maternal brothers, and [c] indicates a male dyad which we cannot exclude from being maternal brothers.
(Indices d'affinité appariées standardisés (PAV) de touts les dyades dans la communauté. Les indices sont normalisés entre 0 (ensemble le moins possible) et 1 (ensemble le plus possible) pour chaque dyade. Les bandes noires représentent les dyades entre mâles, les bandes grises représentent les dyades entre femelles et les bandes blanches représentent les dyades de sexe mixte. a indique les paires mère-fils, b indique les frères maternelles et c indique un dyade mâle qu'on ne peut pas exclure d'être frères maternelles.)

male–female dyads when mother–son pairs are excluded (all male–female dyad: min = −0.51 max = 0.80; mother–son dyads excluded: min = −0.51, max = 0.29), but the standard deviation of male PAV (0.21) was not significantly different from female PAV (0.10) (matrix permutation test: $p = 0.12$), nor from male–female PAV (matrix permutation test: $p = 0.67$). PAV of males calculated over the whole study period correlate strongly with the values obtained when omitting time periods with maximally tumescent females present in the party (Mantel-like Matrix correlation: Rho = 0.94, $N = 9, p < 0.001$). PAV between mothers correlate significantly with the PAV between their sons (same result if either of the two sons of the only female with two sons is taken: Mantel-like matrix correlation: Rho = 0.87, $N = 5, p = 0.02$). Male PAV correlate significantly with proximity affinities (Mantel-like matrix correlation: Rho = 0.46, $N = 9, p = 0.006$).

Seven dyads consisting of eight males fulfil the criteria of close associates in that they are seen significantly more often in the same party than expected by chance, and in that they mutually prefer to be in proximity to one another (Figure 3.2). The number of dyads in which a given male is included ranges from 0 to 3. Nine dyads consisting of eight males are avoidant relationships in that they spend significantly less time in the same party than expected by chance and they mutually avoid sitting in each other's proximity when within the party (Figure 3.2). The number of dyads in which a given male is included ranges again from 0 to 3. Neither male rank nor age correlated with the amount of preferred or avoided dyads of a given male (rank: preferred: Spearman's rank correlation: $R_s = -0.40, N = 9, p = 0.29$; avoided: Spearman's rank correlation: $R_s = 0.23, N = 9, p = 0.55$; age: preferred: Spearman's rank correlation: $R_s = 0.47, N = 9, p = 0.20$; avoided: Spearman's rank correlation: $R_s = -0.31, N = 9, p = 0.40$). Avoidant and preferred associates do not differ in rank differences within the dyads (Mann–Whitney U test:

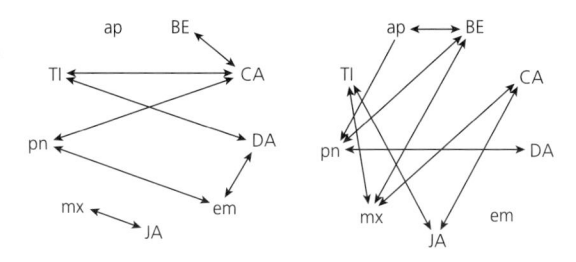

Figure 3.2 (left) Dyads categorized as close associates; (right) dyads categorized as avoidant relationships. Adult males are abbreviated in capital and subadult males in minor letters. The male ranks are CA(1), TI(2), JA(3), DA(4), BE(5), ap(6), pn(7), em(8), mx(9).
((gauche) Dyades catégorisés comme associés proches; (droite) dyades catégorisés comme évitant relations. Les mâles adultes sont abrégés en capital et les mâles subadultes en minuscule. Les rangs des mâles sont CA(1), TI(2), JA(3), DA(4), BE(5), ap(6), pn(7), em(8), mx(9).)

$U = 28$, $N = 16$, $p = 1$). Close associates always have a combined pairwise affinity index (CPAV) within the top third of the CPAV values and all but one avoidant relationships are among the lowest third of CPAV values.

Association, grooming, socio-sexual behavior and aggression

Close associates neither groom each other more often nor do they groom each other more frequently or longer than avoidant relationships when in the same party (Total amount: Mann–Whitney U test: $U = 18$, $N = 16$, $p = 0.15$; frequency: Mann–Whitney U test: $U = 16.5$, $N = 16$, $p = 0.11$; duration: Mann–Whitney U test: $U = 10$, $N = 8$, $p = 0.57$). This holds for frequency and duration of grooming, but the total amount of grooming differs significant between male dyads with the highest CPAV (upper third) and those with the lowest CPAV (lower third) (Total amount: Mann–Whitney U test: $U = 36$, $N = 24$, $p = 0.03$; frequency: Mann–Whitney U test: $U = 49$, $N = 24$, $p = 0.19$; duration: Mann–Whitney U test: $U = 24.5$, $N = 15$, $p = 0.80$) (Figure 3.3). Furthermore close associations do not influence the occurrence frequency of socio-sexual behavior (Mann–Whitney U test: $U = 40$, $N = 16$, $p = 0.33$) and the third of dyads with the highest CPAV did not have a higher number or frequency of socio-sexual behavior than the third of dyads with the lowest CPAV (Total amount: Mann–Whitney U test: $U = 80.5$, $N = 24$, $p = 0.62$; frequency: Mann–Whitney U test: $U = 86$, $N = 24$, $p = 0.39$) (Figure 3.3).

When comparing the ratio of events of grooming to socio-sexual behavior between close associates and avoidant relationships, close associates use significantly more grooming over socio-sexual behavior as compared to avoidant associations (Mann–Whitney U test: $U = 3$, $N = 10$ (dyads which interacted with at least one of those behaviors), $p = 0.047$) (Figure 3.4). The same results are reflected in the trend when comparing the third of dyads with the highest CPAV to the third of dyads with the lowest CPAV (Mann–Whitney U test: $U = 22$, $N = 18$, $p = 0.10$).

Mounting was in 23 out of 28 cases indicative of dominance relationships within the dyad in that the high-ranking individual was mounting the low-ranking one. Aggression frequency does not differ between close associates and avoidant relationships (Mann–Whitney U test: $U = 24$, $N = 16$, $p = 0.47$), nor between dyads of the upper and the lower third of CPAV values (Mann–Whitney U test: $U = 55$, $N = 24$, $p = 0.35$).

Association, food sharing and agonistic alliances

Of the six agonistic alliances that were observed exclusively between males, all were formed between two different dyads made up of three males with positive CPAV, but only one of these two dyads were close associates (half-brother pair). Food sharing occurred five times among four different dyads which all had positive CPAV, but only two dyads were close associates.

Association and maternal kinship

Of the nine males, only two could be assigned to the same mother. This male–male dyad is among the close associates and has the highest CPAV value. Of the six other close associates, five can be excluded from being maternal brothers while the dyad with the second largest CPAV cannot be excluded from being maternal brothers. All avoidant relationships are between individuals which are not maternal brothers.

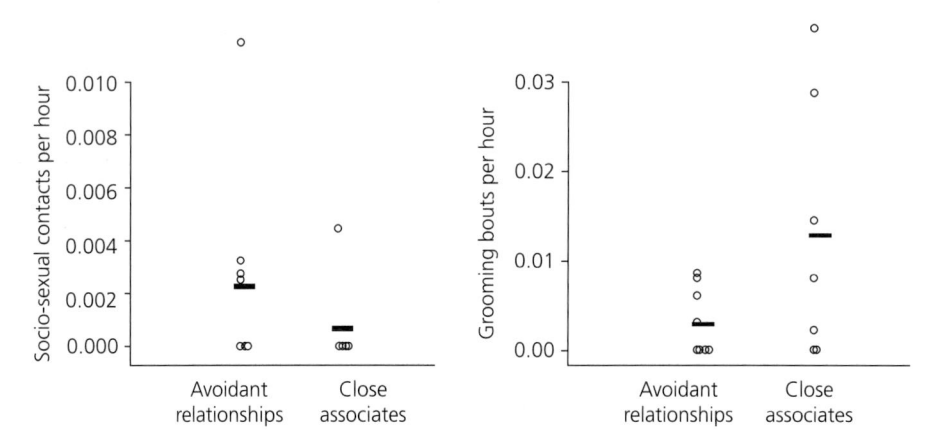

Figure 3.3 Hourly grooming-bout frequency and socio-sexual contacts of male–male dyads from close associates ($N = 7$) and avoidant relationships ($N = 9$). Horizontal lines indicate means.
(La fréquence de périodes de pansage horaire et de contacts socio-sexuels des dyads mâles par leur proches associés ($N =7$) et leur relations évitantes ($N = 9$). Les lignes horizontales montrent les moyennes.)

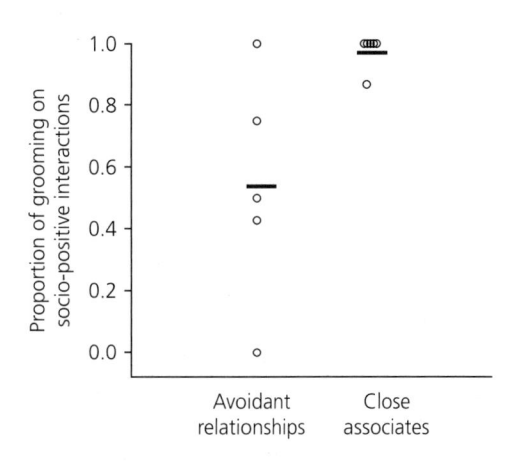

Figure 3.4 Proportion of grooming during non-aggressive interactions (grooming and socio-sexual behaviors) between dyads of close associates ($N = 4$) and avoidant relationships ($N = 5$). Horizontal lines indicate means.
(Proportion de pansage pendant les interactions non-agressives (pansage et comportement socio-sexuel) entre les dyads de proches associés ($N = 4$) et relations évitantes ($N = 5$). Les lignes horizontales montrent les moyennes.)

Discussion

This study found that the variation in male–male standardized pairwise affinity indices based on party membership (PAV) is comparable to the values of other sex dyads, and that male association preference is reflected in their preference for whom to sit in proximity to within a party. While close associates (ranked in the upper third of the composite pairwise affinity indices groom each other overall more often than avoidant relationships (ranked in the lower third of the CPAV), they do not groom each other more frequently than avoidant relationships within a given party. Dyads of the upper and lower third of CPAVs do not differ significantly in their frequency and overall amount of socio-sexual behavior, nor in their frequencies of aggressive behavior. However, when comparing non-aggressive interactions, we find a trend for a stronger preference to use grooming over socio-sexual behavior among the close associates as compared to avoidant relationships. These differences in preferences is, however, significant when comparing close associates and avoidant relationships based on a more rigid definition (significant preference/ avoidance in party associations AND mutual preference/avoidance in proximity). Agonistic support and food sharing were rare among males, and the few observed cases consist of dyads which rank in the upper third of CPAV values but did not all classify as close associates based on the more rigid criteria. The male dyad with the highest CPAV value consists of maternal brothers, and for the dyad with second-highest CPAV value, we cannot exclude them from being maternal brothers. Consequently, close associations among males seems to be partially, but not exclusively, based on kinship, or maternally related bonobos seem to associate when available but such partners might not always be around, which is in line with findings from chimpanzees (Langergraber et al., 2007; Mitani, 2009). Taken together, the results paint a picture of differentiated relationships among male bonobos based on association which is only weakly reflected in the differentiated behaviors within dyads characteristic of bonding in other species (Langergraber et al., 2007; Silk et al., 1999). Furthermore, male bonobos rarely affiliate via grooming and cooperate via food sharing unlike male chimpanzees where these behaviors likely are crucial for maintaining bonds (Mitani, 2009).

The finding that socio-sexual behaviors are more frequent in non-aggressive interactions within avoidant male relationships corresponds to a similar finding of the use of socio-sexual behavior within female dyads (Hohmann and Fruth, 2000). This indicates that socio-sexual behavior in bonobos does not seem to be an expression of friendships and close social ties. Furthermore, given its low frequency of occurrence, it might be less important for the management of social relationships as compared to socio-sexual behavior among females.

Our result that dyads with higher CPAV groom more often than dyads with low CPAV might only be a by-product of more time spent together, especially as these differences disappear if we compare the frequencies of grooming when observed within the same party. Nevertheless, it can be argued that it is the absolute amount of observed behavior within a dyad that ultimately matters, especially as individuals in a fission–fusion society can influence with whom they associate.

It is important to note, however, that the fact that males preferentially associate with their mothers may offer another explanation of how differentiated association patterns could emerge in the absence of direct male–male preference. Accordingly, the close associations might only reflect social preferences of their mothers. This idea is supported by our finding that sons' preferences for each other are similar to the spatial preferences of their mothers for each other. Furthermore, the highest association was found among maternal brothers with a living mother in the community, but as there are dyads that include males without mothers among the ones that classify as close associates and avoidant relationships, male associations do not purely reflect maternal associations. Studies in sanctuaries such as at Lola ya Bonobo on adult orphan males could help further answer the question regarding the importance of maternal influence.

While we observed few cases of male agonistic alliances and food sharing, bonobo males seem to differ very clearly in the frequency of occurrence of such cooperative behaviors from chimpanzees. The absence of such cooperative male behavior in bonobos has been noted before (Ihobe, 1992), and has been linked to reductions of aggressive competition between males within and between communities (Aureli et al., 2006; Manson et al., 1991; van Hooff and van Schaik, 1994).

So, while association patterns seem to suggest that there are social preferences among males, we find neither clear biases in grooming behavior typically indicative of social bonds between males in other species nor frequent occurrence of cooperative behavior among male dyads. This can have several potential causes. First, there might be a decrease in the potential benefits of cooperative behaviors in contexts such as during mate competition or intergroup encounters which lower the general interest to invest in the maintenance of close male relationships. In order to improve our understanding of the possible benefits of male coalitions in bonobos, future studies will have to address their role in within- and between-group competition. For example, males in other male-philopatric societies such as chimpanzees and spider monkeys collectively defend their territory (Watts and Mitani, 2001; Wallace, 2008). While encounters between bonobo communities are less violent than in chimpanzees (Idani, 1990; Kano, 1992; Wilson et al., 2014) they still involve aggressive behavior (Hohmann and Fruth, 2002).

Second, because females occupy high dominance ranks in bonobos, they may represent an alternative to males as coalition partners (especially mothers, given their overlap in interests due to kin selection), and consequentially males invest more in relationships with females. While further studies will have to address these questions, there are examples of adaptive significance of bonds between males when their common mother dies. We now present an observed case of adoption of an infant male by his older maternal brother.

Coinciding with the presence of poachers in the home range of the Bompusa community in April 2010, the adult female Hannah disappeared together with her sub-adult son Apollo (est. age 10–15 years) and her 4-year-old male infant Hugo. Infants at that age are often still carried by their mothers and sleep in their mother's nest (B. Fruth, G. Hohmann and M. Surbeck, unpublished data). On 22 September 2011, approximately 1.5 years after their disappearance, Apollo and Hugo once again appeared in the home range of the Bompusa community. During the following days, both individuals shared the same night nest, when travelling Apollo carried his maternal half-brother Hugo on his back (Plate 3) and their inter-individual distance never exceeded 5 metres. When resting on the ground they maintained physical contact and Apollo was seen to share food with Hugo (allowing Hugo to take pieces of *treculia africana* out of his mouth). While Hugo appeared to be in good physical condition, the fur of the older male Apollo was sparse which is thought to indicate poor physical condition. One month later, on 26 November, the two males were seen for the first time travelling together with other members of the Bompusa community (observation L. Goldberg) and from then on have remained within the community.

As paternal care has not been described for wild bonobos, it seems extraordinary that a male was able to perform the necessary behaviors in order to raise his younger brother and it confirms the behavioral flexibility exhibited by great apes (Malone et al., 2012). This observation is also the first record of adoption in wild bonobos (for chimpanzees see Boesch et al., 2010) and indicates that close bonds between maternal brothers may exceed the lifespan of their mothers. From this perspective, it seems reasonable to infer that male–male relationships may derive indirect fitness benefits and therefore, may be determined by kinship. Whether this behavior is restricted to male–male dyads cannot be inferred from this single observation and more observations are needed to draw meaningful conclusions.

Conclusion

A previous study on the Bompusa bonobo community found that male relationships can be characterized by the outcome of agonistic interactions such that a linear dominance hierarchy could be constructed, thus indicating a despotic social organization for the group (Surbeck et al., 2011). The high predictability of the outcome of male–male conflicts contrasts the much less predictable outcome of intersexual conflicts (Surbeck and Hohmann, 2013). This study now shows that agonistic interactions are not linked to spatial association patterns among the males. While certain male dyads at LuiKotale exhibit a clear preference for ranging in the same party and sitting in proximity, these preferences are not reflected in aggression frequencies and only to some extent in affiliative behavior and the use of socio-sexual behaviors. The bonobo males at LuiKotale do not seem to benefit from close relationships in the same way as chimpanzee males and they rarely engage in cooperative behaviors such as agonistic coalitions or collective hunting and these relationships may partially reflect their mothers' ranging patterns or non-adaptive phylogenetic inertia. Nevertheless, in some particular cases these male relationships can be very important, as demonstrated by the case of an infant being adopted by his older maternal brother. Further studies in the field in concert with captive experiments/observations will help us to improve our understanding of the factors underlying bonobo male sociality.

Acknowledgements

Thanks to ICCN for access, SNF, Basler Stiftung fuer Biologische Forschung for granting financial support, Lucas Goldman, Isaak Schamberg for assistance in the field, Barbara Fruth for support during various stages of the project, Kevin Langergraber, Mimi Arandjelovic, Brian Hare and an anonymous reviewer for helpful comments on an earlier draft of the chapter.

References

Alberts, S.C. (2010). Primatology: 'A Faithful Friend Is the Medicine of Life'. *Current Biology*, 20, R632–R634.

Arandjelovic, M., Guschanski, K., Schubert, G., Harris, T. R., Thalmann, O., Siedel, H., and Vigilant, L. (2009). Two-step multiplex polymerase chain reaction improves the speed and accuracy of genotyping using DNA from noninvasive and museum samples. *Molecular Ecology Resources*, 9, 28–36.

Aureli, F., Preston, S.D., and De Waal, F.B.M. (1999). Heart rate responses to social interactions in free-moving rhesus macaques (*Macaca mulatta*): A pilot study. *Journal of Comparative Psychology*, 113, 59–65.

Aurelli, F., Schaffner, C.M., Verpooten, J., Slater, K., and Ramos-Fernandez, G. (2006). Raiding parties of male spider monkeys: Insights into human warfare? *American Journal of Physical Anthropology*, 131, 486–97.

Baayen, R. (2008). *Analyzing Linguistic Data*, Cambridge, Cambridge University Press.

Boesch, C. and Boesch-Ackermann, H. (2000). *The Chimpanzees of the Tai Forest. Behavioural Ecology and Evolution*. Oxford, Oxford University Press.

Boesch, C., Bolé, C., Eckhardt, N., and Boesch, H. (2010). Altruism in Forest Chimpanzees: The Case of Adoption. *PLoS One*, 5, e8901.

Carter, S.C. (1998). Neuroendocrine perspectives on social attachment and love. *Psychoneuroendocrinology*, 23, 779–818.

Clutton-Brock, T. (2009). Cooperation between non-kin in animal societies. *Nature*, 462, 51–7.

Crockford, C., Wittig, R.M., Langergraber, K., Ziegler, T.E., Zuberbühler, K., and Deschner, T. (2013). Urinary oxytocin and social bonding in related and unrelated wild chimpanzees. *Proceedings of the Royal Society B*, 280.

Crockford, C., Wittig, R.M., Whitten, P.L., Seyfarth, R. M., and Cheney, D.L. (2008). Social stressors and coping mechanisms in wild female baboons (*Papio hamadryas ursinus*). *Hormones and Behavior*, 53, 254–65.

Engh, A.L., Beehner, J.C., Bergman, T.J., Whitten, P.L., Hoffmeier, R.R., Seyfarth, R.M., and Cheney, D.L. (2006). Behavioural and hormonal responses to predation in female chacma baboons (*Papio hamadryas ursinus*). *Proceedings of the Royal Society B*, 273, 707–12.

Furuichi, T. (1997). Agonistic interactions and matrifocal dominance rank of wild bonobos (*Pan paniscus*) at Wamba, Zaire. *International Journal of Primatology*, 18, 855–75.

Furuichi, T., Idani, G., Ihobe, H., Kuroda, S., Kitamura, K., Mori, A., Enomoto, T., Okayasu, N., Hashimoto, C., and Kano, T. (1998). Population dynamics of wild bonobos (*Pan paniscus*) at Wamba. *International Journal of Primatology*, 19, 1029–44.

Gregory, R.W., Bolker, B., and Lumley, T. (2013). gtools: Various R programming tools. R package version 2.7.1. http://CRAN.R-project.org/package=gtools.

Gust, D.A., Gordon, T.P., Hambright, M.K., and Wilson, M.E. (1993). Relationship between social factors and pituitary–adrenocortical activity in female rhesus monkeys (*Macaca mulatta*). *Hormones and Behavior*, 27, 318–31.

Hamilton, W.D. (1964). The genetical evolution of social behavior. *Journal of Theoretical Biology*, 7, 1–52.

Handley, L.J.L. and Perrin, N. (2007). Advances in our understanding of mammalian sex-biased dispersal. *Molecular Ecology*, 16, 1559–78.

Hohmann, G. and Fruth, B. (2000). Use and function of genital contacts among female bonobos. *Animal Behaviour*, 60, 107–20.

Hohmann, G. and Fruth, B. (2002). Dynamics in social organisation of bonobos (*Pan paniscus*). In: Boesch, C., Hohmann, G., and Marchant, L.F. (eds) *Behavioural Diversity in Chimpanzees and Bonobos*. Cambridge: Cambridge University Press, pp. 138–150.

Hohmann, G. and Fruth, B. (2003). Lui Kotal—A new site for field research on bonobos in the Salonga National Park. *Pan African News*, 10, 25–7.

Idani, G. (1990). Relations between unit-groups of bonobos at Wamba, Zaire: Encounters and temporary fusions. *African Study Monographs*, 11.

Ihobe, H. (1992). Male–male relationships among wild bonobos (*Pan paniscus*) at Wamba, Republic of Zaire. *Primates*, 33, 163–79.

Jaeggi, A.V., Stevens, J.M.G., and Van Schaik, C.P. (2010). Tolerant food sharing and reciprocity is precluded by despotism among bonobos but not chimpanzees. *American Journal of Physical Anthropology*, 143, 41–51.

Kalinowski, S.T., Taper, M.L., and Marshall, T.C. (2007). Revising how the computer program CERVUS accommodates genotyping error increases success in paternity assignment. *Molecular Ecology* 16, 1099–106.

Kano, T. (1992). *The Last Ape: Pygmy Chimpanzee Behavior and Ecology*, Stanford, CA: Stanford University Press.

Langergraber, K., Mitani, J., and Vigilant, L. (2009). Kinship and social bonds in female chimpanzees (*Pan troglodytes*). *American Journal of Primatology*, 71, 840–51.

Langergraber, K.E., Mitani, J.C., and Vigilant, L. (2007). The limited impact of kinship on cooperation in wild chimpanzees. *Proceedings of the National Academy of Sciences*, 104, 7786–90.

Malone, N., Fuentes, A., and White, F. J. (2012). Variation in the social systems of extant hominoids: comparative insight into the social behavior of early hominins. *International Journal of Primatology*, 33, 1251–77.

Manson, J.H., Wrangham, R.W., Boone, J.L., Chapais, B., Dunbar, R.I.M., Ember, C.R., Irons, W., Marchant,r L.F., McGrew, W.C., Nishida, T., Paterson, J.D., Smith, E.A., Stanford, C.B., and Worthman, C.M. (1991). Intergroup aggression in chimpanzees and humans [and Comments and Replies]. *Current Anthropology*, 32, 369–90.

Mitani, J.C. (2009). Male chimpanzees form enduring and equitable social bonds. *Animal Behaviour*, 77, 633–40.

Mitani, J.C., Merriwether, D.A., and Zhang, C. (2000). Male affiliation, cooperation and kinship in wild chimpanzees. *Animal Behaviour*, 59, 885–93.

Muller, M.N., Mitani, J.C., Peter, J.B. Slater, C.T.S.T.J.R. H.J.B., and Marc, N. (2005). Conflict and cooperation in wild chimpanzees. *Advances in the Study of Behavior*. New York, NY: Academic Press.

Parish, A.R. (1996). Female relationships in bonobos (*Pan paniscus*)—Evidence for bonding, cooperation, and female dominance in a male-philopatric species. *Human Nature*, 7, 61–96.

Schubert, G., Stoneking, C.J., Arandjelovic, M., Boesch, C., Eckhardt, N., Hohmann, G., Langergraber, K., Lukas, D., and Vigilant, L. (2011). Male-Mediated Gene Flow in Patrilocal Primates. *PLoS One*, 6, e21514.

Schülke, O., Bhagavatula, J., Vigilant, L., and Ostner, J. (2010). Social bonds enhance reproductive success in male macaques. *Current Biology*, 20, 2207–210.

Shutt, K., Maclarnon, A., Heistermann, M., and Semple, S. (2007). Grooming in Barbary macaques: better to give than to receive? *Biology Letters*, 3, 231–33.

Silk, J.B. (2009). Nepotistic cooperation in non-human primate groups. *Philosophical Transactions of the Royal Society B*, 364, 3243–54.

Silk, J.B., Alberts, S.C., and Altmann, J. (2003). Social bonds of female baboons enhance infant survival. *Science*, 302, 1231–4.

Silk, J.B., Beehner, J.C., Bergman, T.J., Crockford, C., Engh, A.L., Moscovice, L.R., Wittig, R.M., Seyfarth, R.M., and Cheney, D.L. (2009). The benefits of social capital: close social bonds among female baboons enhance offspring survival. *Proceedings of the Royal Society B*, 276, 3099–104.

Silk, J.B., Beehner, J.C., Bergman, T.J., Crockford, C., Engh, A.L., Moscovice, L.R., Wittig, R.M., Seyfarth, R.M., and Cheney, D.L. (2010). Strong and consistent social bonds enhance the longevity of female baboons. *Current Biology*, 20, 1359–61.

Silk, J.B., Seyfarth, R.M., and Cheney, D.L. (1999). The structure of special relationships among female savanna baboons in Moremi Reserve, Botswana. *Behaviour*, 136, 679–703.

Surbeck, M. and Hohmann, G. (2013). Intersexual dominance relationships and the influence of leverage on the outcome of conflicts in wild bonobos (*Pan paniscus*). *Behavioral Ecology and Sociobiology*, 67, 1767–80.

Surbeck, M., Mundry, R., and Hohmann, G. (2011). Mothers matter! Maternal support, dominance status and mating success in male bonobos (*Pan paniscus*). *Proceedings of the Royal Society B*, 278, 590–98.

Van Hooff, J. and Van Schaik, C. (1994). Male bonds: affiliative relationships among nonhuman primate males. *Behaviour*, 130, 309–37.

Vervaecke, H., De Vries, H., and Van Elsacker, L. (2000). Function and distribution of coalitions in captive bonobos (*Pan paniscus*). *Primates*, 41, 249–65.

Wallace, R.B. (2008). Towing the party line: territoriality, risky boundaries and male group size in spider monkey fission–fusion societies. *American Journal of Primatology*, 70, 271–81.

Watts, D P. (1998). Coalitionary mate guarding by male chimpanzees at Ngogo, Kibale National Park, Uganda. *Behavioral Ecology and Sociobiology*, 44, 43–55.

Watts, D.P. and Mitani, J.C. (2001). Boundary patrols and intergroup encounters in wild chimpanzees. *Behaviour*, 138, 299–327.

Wilson, M.L., Boesch, C., Fruth, B., Furuichi, T., Gilby, I.C., Hashimoto, C., Hobaiter, C.L., Hohmann, G., Itoh, N., Koops, K., Lloyd, J.N., Matsuzawa, T., Mitani, J.C., Mjungu, D.C., Morgan, D., Muller, M.N., Mundry, R., Nakamura, M., Pruetz, J., Pusey, A.E., Riedel, J., Sanz, C., Schel, A.M., Simmons, N., Waller, M., Watts, D.P., White, F., Wittig, R.M., Zuberbuhler, K., and Wrangham, R.W. (2014). Lethal aggression in *Pan* is better explained by adaptive strategies than human impacts. *Nature*, 513, 414–17.

Wittig, R.M., Crockford, C., Lehmann, J., Whitten, P.L., Seyfarth, R.M., and Cheney, D L. (2008). Focused grooming networks and stress alleviation in wild female baboons. *Hormones and Behavior*, 54, 170–77.

PART II

Social Development

CHAPTER 4

Bonobo baby dominance: Did female defense of offspring lead to reduced male aggression?

Kara Walker and Brian Hare

Abstract The dominance style of bonobos presents an evolutionary puzzle. Bonobos are not male dominant but female bonobos do not show traits typical of female-dominant species. This chapter proposes the offspring dominance hypothesis (ODH) as a potential solution. ODH suggests the social system of bonobos evolved as a defence against infanticide and is not due to pressure to monopolize resources. Females that prevented aggression towards offspring and preferred mating with less aggressive males were most successful. Supporting ODH, during observations at Lola ya Bonobo Sanctuary it was found that: 1) adult male bonobos are rarely aggressive towards offspring with mothers, 2) some mother-reared juvenile bonobos attain rank higher than adult males and 3) mother-reared offspring often socially interact with adult males without their mothers nearby. These preliminary findings provide initial support that the bonobo social system evolved due to fitness advantages of effectively protecting offspring against consequences of male aggression.

Le style de dominance des bonobos présente un puzzle évolutionnaire. Les bonobos ne sont pas dominés par les mâles mais les bonobos femelles ne montrent pas les traits caractéristiques d'une espèce dominée par femelles. On propose l'hypothèse de dominance de progéniture (ODH) comme une solution potentielle. La ODH suggère que le système social des bonobos a évolué en défense contre l'infanticide et pas sous pression pour la monopolisation des ressources. Les femelles qui préviennent l'agression vers leur progéniture et leur préférence d'accouplement avec des mâles moins agressives étaient très efficaces. À l'appui de la ODH on a trouvé pendant nos observations à Lola ya Bonobo Sanctuary que: 1) les mâles adultes bonobos agressent rarement vers les bébés avec mères, 2) quelques adolescents bonobos qui furent élevés par leurs mères atteignent un rang plus haut que les mâles adultes et 3) la progéniture élevée par la mère interagissent avec avec d'adultes mâles sans la présence de leur mère. Ces trouvailles préliminaires donnent appuie à l'hypothèse que le système social des bonobos a évolué par les avantages corporelles de la protection de la progéniture contre les conséquences de l'agression mâle.

Introduction

Bonobos represent a socio-ecological puzzle. Throughout nature, larger individuals are often dominant over those that are smaller (Davies et al., 2012). Within the primate order, when reproductive access or resources are contested, weight is often related to the outcome and dominance between sexes is related to sexual dimorphism (Mitani et al., 1996). Male primates are frequently larger, initiate more aggression and are usually the dominant sex within the social group (Bernstein, 1981). In species where this relationship is reversed, females tend to be of similar size or larger than males and frequently initiate aggression (Kappeler, 1990; Petty and Drea, 2015a); this includes species of lemurs and marmosets that show female dominance (Jolly, 1985). Females of these species tend to be dominant over all or most males in their group. Across mammals, this type of female dominance can also be associated with varying degrees of masculinization (e.g. hyenas and lemurs: Drea, 2009; Petty and Drea, 2015b). Meanwhile, in primate species characterized as co-dominant, where both sexes are equally dominant and aggressive,

Walker, K. and Hare, B., *Bonobo baby dominance: did female defense of offspring lead to reduced male aggression?* In: *Bonobos: Unique in Mind, Brain, and Behavior*. Edited by Brian Hare and Shinya Yamamoto: Oxford University Press (2017). © Oxford University Press. DOI: 10.1093/oso/9780198728511.003.0004 males and females tend to be of similar size (e.g. Callitrichids and Hylobates, Leigh and Shea, 1995).

Bonobos are a potential outlier to this strong phylogenetic signal across a range of species. Bonobos and their close relatives, chimpanzees, share many physical features, including a similar degree of sexual dimorphism (Leigh and Shea, 1995; Smith and Jungers, 1997), but they differ markedly in social characteristics. Importantly, in contrast to male-dominated chimpanzees, no mixed age and sex group of bonobos in the wild or captivity has ever been observed to have an alpha male (Hare et al., 2012). This has led to a debate in the literature regarding exactly how the dominance style of bonobos is best characterized (Stevens et al., 2007). *The female dominance through coalitions* (FDCH) has been proposed to suggest that smaller bonobo females are able to defend each other through coalitionary behavior directed at coercive males (Parish, 1996; Parish et al., 2000; Tokuyama and Furuichi, 2016; White and Wood, 2007). According to this hypothesis, females are the dominant sex through cooperative defense against male aggression. This explanation is unconventional since bonobos, like chimpanzees, are patrilocal and would require unrelated, immigrant females to become coalition partners without the kinship bonds typically seen in other species (Sakamaki et al., 2015). This contrasts with male chimpanzee coalitions that result from life-long bonds between unrelated group mates and/or support of male kin in this patrilocal, male-dominated species (Langergraber et al., 2007; Mitani et al., 2000). Two other main alternatives have been suggested. The *observer bias hypothesis* (OBH) suggests that initial data supporting female dominance in bonobos are skewed by reliance on a small sample of captive observations and the cultural bias of the researchers (Stanford, 1998). The *co-dominance hypothesis* (CDH), proposed in a variety of forms, essentially suggests that male and female bonobos share social power and presumably predicts a relatively random distribution of males and females throughout the hierarchy, with males and females both potentially taking on the role of alpha (Paoli et al., 2006; Stevens et al., 2007; Surbeck and Hohmann, 2013).

In this chapter, we review the most recent literature relevant to re-examining these hypotheses given the surge in bonobo publications over the last decade. We also propose and test the *offspring dominance hypothesis* (ODH), which we present as a potential solution to the tension in the literature.

Male bonobos are not dominant

In the past decade, a host of publications has confirmed that males are not the dominant sex in bonobos. Unlike chimpanzees, male bonobos do not form coalitions against females with other males in their group (Kanō, 1992). Male bonobos instead rely on their female relatives, especially their mothers, for social support (Furuichi, 2011; Kanō, 1992). Male bonobos range together with their mothers and grandmothers more than chimpanzees, further supporting the importance of having female relatives present in the group (Surbeck et al., 2011). Their closest affiliative bond is with their mother and her kin and both mothers and grandmothers support their sons in conflicts between other males (Kanō, 1992; Surbeck et al., 2011). Moreover, female bonobos are not coerced into mating and most likely prefer non-coercive males related to their female friends (Surbeck et al., 2012a). Aggressive males tend to receive the most aggression from females and may not obtain more copulations (Hohmann and Fruth, 2003; Tokuyama and Furuichi, 2016), though higher-ranking males are generally more aggressive than lower-ranking males. It is inferred that male reproductive success is highly dependent on the support of female relatives in conflicts with other males and in aiding males in gaining access to their female kin's social network (Surbeck et al., 2012b; Surbeck et al., 2011). Finally, no male bonobo has been observed committing infanticide (Fowler and Hohmann, 2010; Hare et al., 2012; Wilson et al., 2014). In light of this evidence, it is hard to dispute that male bonobos are not dominant over females, especially since no multimale, multi-female bonobo group has ever been observed with an alpha male (Hare et al., 2012). These patterns have been observed in multiple wild and captive populations and in our view effectively refute the observer bias hypothesis. This, however, does not necessarily lead to the conclusion that bonobo females are the dominant sex.

Female bonobos are not like other female-dominant primates

Bonobos do not share traits associated with female dominance in other primate species. Female dominance is often accompanied by masculinization in morphology, physiology and behavior (Drea, 2009). Females in lemur species that are characterized by female dominance are generally larger than males (Hemelrijk et al., 2008) and have elevated testosterone and aggression levels when compared with females from closely related species that are not female dominant (Petty and Drea, 2015). Females in these species easily displace males in dyadic interactions and high levels of testosterone are thought to mediate the escalated aggression initiated by females of these species. Species with dominant females tend to be highly territorial and engage in infanticide (Digby, 1995; Morelli et al., 2009). In contrast, bonobos show similar sexual dimorphism as chimpanzees and are not larger than males. Differences in sexual dimorphism could be explained by phylogenetic inertia, though other features of bonobo female dominance are not as easily explained. Testosterone levels in female bonobos are similar to those seen in female chimpanzees. While male and female bonobos have more similar testosterone levels than in chimpanzees, this difference is thought to be due to reduced levels and testosterone reactivity in male bonobos compared to chimpanzees (Behringer et al., 2014; Sannen et al., 2004; Surbeck et al., 2012a; Wobber et al., 2013). It is also likely that both sexes in bonobos are exposed to fewer pre-natal androgens than chimpanzees, as reflected in their 2D:4D finger ratios. No appreciable sex differences among bonobos were observed in these digit ratios—again an observation inconsistent with increases in female aggression (McIntyre et al., 2009). Bonobo females also appear to initiate aggression at similar frequencies observed in chimpanzee females (Hohmann and Fruth, 2003; Pusey et al., 2008; Stevens et al., 2006). Crucially, unlike chimpanzees and macaques, bonobos lack any formal dominance signals that are typically associated with the dominant sex (Clay et al., 2015; de Waal, 1988). Furthermore, bonobo females are not known to commit infanticide, something that occurs in chimpanzees as well as in many female-dominated species (Fowler and Hohmann, 2010; Wilson et al., 2014). Morphologically, female bonobos are not masculinized in any way in comparison to female chimpanzees. In contrast, bonobo cranial and dental morphology has been described as peadomorphic or 'juvenilized' relative to that of chimpanzees (Hare et al., 2012). Overall, bonobos do not fit any traditional definition of female dominance that has previously been proposed.

The offspring dominance hypothesis

The proposal of competing hypotheses is understandable since bonobos do not fit traditional definitions of male or female dominance. As a solution, a number of different descriptions of bonobos as being co-dominant have been articulated (Stevens et al., 2007). Co-dominance remains an interesting possibility, but none of the varied proposals of co-dominance explain the observed sexual dimorphism in bonobos, the juvenile features of males, the complete absence of alpha males in any mixed-sex bonobo group, or the functional significance of this particular social system. Meanwhile, all of these categories of dominance style largely ignore the contribution of an important faction within bonobo social groups: juveniles and infants. Growing animals are generally conceptualized as entering the dominance hierarchy when they approach adult size and reproductive age (Davies et al., 2012). However, despite their diminutive size, these younger individuals may have an outsized effect on interactions among adult bonobos. Nothing can motivate aggression in a female vertebrate more than a threatened offspring and the key to understanding bonobo sociality may be the response of females to those that threaten their offspring.

We propose the *offspring dominance hypothesis* to explain the unusual dominance style observed in bonobos. The hypothesis proposes that infant and juvenile bonobos evolved to play a central role in motivating conflicts of interests that lead to female aggression in this species. For example, female coalitions most frequently occur in retaliation to male aggression against juveniles (Surbeck and Hohmann, 2013; Tokuyama and Furuichi, 2016). Aggression such as this by females towards males may function to reinforce an evolved psychological predisposition

Table 4.2 Rank in 2012–13.
(Rang en 2012–13).

Group 1			Group 2		
Rank	Name	Age/Sex Class	Rank	Name	Age/Sex Class
High	Bandundu	Adult Female	High	Maya	Adult Female
High	Lomami	Adult Male	High	Kisantu	Adult Female
High	Manono	Adult Male	High	Lisala	Adult Female
High	**Elykia** (Semendwa)	Juvenile Female	High	Bandaka	Adult Male
Middle	Kalena	Adult Female	Middle	Kalena	Adult Female
Middle	Lisala	Adult Female	Middle	Isiro	Adult Female
Middle	Opala	Adult Female	Middle	Likasi	Adult Female
Middle	Salonga	Adult Female	Middle	Keza	Adult Male
Middle	Semendwa	Adult Female	Middle	Max	Adult Male
Middle	Api	Adult Male	Middle	Bili	Adult Male
Middle	Fizi	Adult Male	Middle	**Nyota** (Lisala)	Juvenile Female
Middle	Matadi	Adult Male	Middle	**Malaika** (Kalena)	Juvenile Female
Middle	Kikwit	Adult Male	Middle	**Liyaka** (Kisantu)	Juvenile Female
Middle	Kasongo	Adult Male	Middle	**Mayele** (Maya)	Juvenile Male
Middle	**Nyota** (Lisala)	Juvenile Female	Middle	Waka	Orph Juv Female
Middle	**Pole** (Opala)	Juvenile Male	Middle	Sake	Orph Juv Female
Middle	**Wongolo** (Bandundu)	Juvenile Male	Low	Yolo	Adult Male*
Middle	Katako	Orph Juv Female	Low	Eleke	Adult Male*
Low	Dilolo	Adult Male	Low	**Bissengo** (Maya)	Juvenile Male
Low	Ilebo	Adult Male	Low	Masisi	Orph Juv Female
Low	Mabali	Adult Male	NR	Muanda	Adult Female*
Low	**Kimya** (Salonga)	Juvenile Female	NR	**Bolingo** (Kalena)	Juvenile Male
Low	**Malaika** (Kalena)	Juvenile Female	NR	Elonga	Juvenile Female
NR	**Makasi** (Semendwa)	Juvenile Male			
NR	Moseka	Juvenile Female			

Focal individuals are listed in bold with their mother listed in parentheses for reference. NR = Not Ranked.
*Individuals entered adulthood in 2013.

Two other individuals, Wongolo and Sanza, moved from Group 2 to Group 3. Observations on Wongolo and Sanza ceased when transferred to Group 3. The remaining subjects were in the same groups for the duration of the study.

Procedure

We conducted 30-minute focal follows on each subject and accumulated 206 total observation hours (412 total observation sessions ranging between 6–53 observations per subject for a mean of 34 per subject). When in their large enclosures, bonobos are only visible when close to the fence line so observations were restricted to the morning (08:00–10:30), noontime (12:00–14:00) and evening hours (16:00–18:30), when bonobos routinely relaxed and foraged near the fence line. Proximity information was recorded every minute; the identity of and the distance to the nearest bonobo who was not the subject's mother was noted. It was also noted whether the infant was within or beyond arm's reach of his/her mother. Mating between opposite sex adults that included penetration and aggressive interactions involving any target individual were recorded *ad libitum*. Aggression consisted of threats, chases and physical altercations. Threats occurred when an individual lunged and/or waved an arm in the direction of a second individual, usually accompanied by a threat grunt/bark. Chases occurred when the recipient fled and/or screamed in reaction to aggression, and physical altercations included one of the following elements: hit, trample, shove or bite. Subjects were scored as giving aggression or displacements when they initiated an interaction. Subjects were coded as receiving aggression or displacements when they were the target of the giver's aggression. For use in calculating dominance, subjects were scored as winning aggressive interactions or successfully displacing a group member when they initiated a behavior that resulted in the recipient showing fear or moving away from them. Focal subjects were rotated through opportunistically to ensure equal coverage and no subject was observed for multiple consecutive focals. In each year, subjects needed to be followed for a minimum of 180 minutes to be included in the dataset. Reliability assessments were conducted before the observations began. The first author conducted the observations together with an assistant she trained. For reliability assessments both observers independently conducted 6 focal follows on the same subject after which time they had 90 per cent agreement and actual data collection began.

Analysis

Nearest neighbour: In each year, we first assessed the proportion of time an individual's mother was his/her nearest neighbour and the proportion of time an individual was within arm's reach of his/her mother. To account for the mixed-sex, wide age range of our sample and the presence of two social groups, we assessed if both measures were affected by age, sex and group using a linear mixed model where we controlled for a random effect of subject identity (ID). We also calculated the proportion of time an offspring's nearest neighbour was an adult male and compared this value to the expected proportion given the number of males in the group. We used a one-sample t-test to compare the difference between expected minus observed with zero. We also assessed if age and sex affected proportion of time with an adult male as nearest neighbour using a linear mixed model controlling for a random effect of subject ID.

Aggression: Aggression, including displacements, between adult males and focal mother-reared offspring was rare and thus only descriptive statistics are reported here. We compare clear cases of aggression and displacements with winners by age in years and sex class.

Rank: Group composition stabilized in 2011 after individuals moved between groups and a large number of males reached adulthood. For this reason, we only calculate rank for the period including 2012 and 2013. One mother–offspring pair switched groups after this time

and are ranked in both groups based on their interactions therein. At the time of our observations there was still no recognized formal dominance signal or submissive vocalization in bonobos for use in determining rank. Instead we used the outcome of both aggression and displacements with clear winners to achieve the maximum possible number of interactions. Aggressive interactions are more prevalent in feeding contexts, especially when resources can be monopolized. Our observations centred on times during provisioning or the hour or so prior or following a provisioning event. Therefore, despite limited observation hours we observed 211 displacements and aggressive interactions between group members, a number that exceeds observed interactions in some long-term studies of dominance (e.g. in female chimpanzees; Murray et al., 2006). We then calculated rank using the normalized modified David's Score (de Vries et al., 2006). Due to the paucity of our target interactions and the presence of dyads who never interacted we assigned individuals to rank categories rather than ranking them in a linear hierarchy. Individuals with a score more than one standard deviation above the mean where classified as high ranking and those one standard deviation below the mean as low ranking and the remainder as middle ranking. Five individuals could not be categorized because they did not participate in any aggressive events or displacements. Four were infant bonobos and the fifth was a young adult female (see Table 4.2).

Copulation and association: In each year we calculated the proportion of time each adult male was the subject's nearest neighbour by dividing the number of times the two were nearest neighbours with the total time observed in that period. We then tallied the number of copulations that occurred between that individual's mother and that particular male. Observations were conducted during times when most group members were together around provisioning times such that male–female association rates were near equal for each dyad. We used a linear mixed model to assess if the proportion of time an offspring spends with an adult male was predictive of the copulation frequency between that male and the infant's mother. Many mother–male pairs never copulated resulting in a large number of zeros in this analysis. To simplify the analysis we then coded if a pair copulated or not and compared infant-male association rates between the two groups using a non-parametric Kruskal–Wallis test.

Results

Prediction 1: Juveniles frequently associate with other group members, including adult males, regardless of the location of their mother.

Mother-reared juveniles spent the majority of their time (55 per cent) beyond arm's reach of their mother and with a nearest neighbour who was not their mother (54 per cent). Modelling found that the amount of time bonobos spent within reach of their mother was related to their age in years (T = −7.29, $p < 0.0001$) but not by sex (T = −0.21, p = 0.84) or group (T = −0.97, p = 0.33). Younger bonobos spent more time within reach of their mother and the proportion dramatically decreased with age (Figure 4.1). The amount of time juveniles spent with their mother as their nearest neighbour also varied with age (T = −4.03, $p < 0.001$; Figure 4.1) but not with sex (T = −1.29, p = 0.24) or group (T = 0.59, p = 0.56). Younger bonobos spent more time with their mother as their nearest neighbour. Prior to age three, infants spent most of their time within reach and next to their mother. After age three this pattern reversed and declined with age.

Mother-reared juveniles spent, on average, 20 per cent of their time with an adult male as their nearest neighbour. Taking into account changing group composition, males accounted for 27 per cent, on average, of group members with mother-reared offspring spending significantly less time than expected by chance with adult males (One sample t-test: T = −5.03, $p < 0.0001$). Modelling found that neither age (T = 0.94, p = 0.35) nor sex (T = −0.46, p = 0.65) affected time spent with an adult male as nearest neighbour.

Prediction 2: Mother-reared offspring can displace adult males. Aggression against juveniles by adult males is rare and if it occurs, the infant's mother is quick to respond.

Mother-reared offspring displaced other group members on 31 occasions. They displaced adult

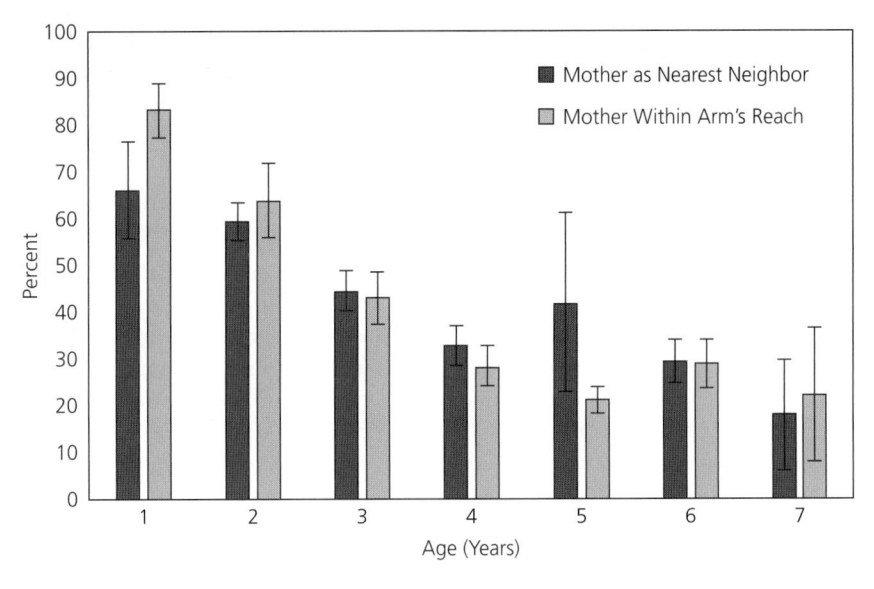

Figure 4.1 Mean amount of time juveniles spent with their mother as their nearest neighbour and with their mother within arm's reach. Error bars represent standard error.
(Durée moyenne que les adolescents ont passé avec leurs mères comme leur plus proche voisin et avec leur mère à portée de main. Les bandes d'erreur représentent l'erreur standard.)

males 11 times, the majority of which (*N* = 8) occurred when the juvenile's mother was over 5 metres away. Mother-reared offspring were displaced 63 times overall but adult males were responsible for a minority of cases (*N* = 9). We recorded 92 aggressive events between offspring and other group members. These offspring received aggression on 54 occasions and were aggressive towards others on 38 occasions (Figure 4.2). Mother-reared juveniles were most frequently attacked by adult females (*N* = 32), and offspring only initiated aggression against adult females on one occasion. In contrast, mother-reared offspring most frequently attacked other infants and juveniles (*N* = 30). Mother-reared offspring initiated aggression against adult males on seven occasions and received aggression from adult males in five instances. The remaining cases involved aggression among juveniles. In four of the seven cases where mother-reared offspring behaved aggressively towards adult males, they were five or more metres away from their mothers. In five instances we observed mothers coming to the aid of offspring when harassed or attacked by orphan juveniles, but we did not observe any instances of mothers coming to the aid of offspring when harassed or attacked by adult males. In one instance of a mother aiding her offspring against an orphan, she was joined by a coalition partner.

Prediction 3: Mother-reared juveniles will outrank older individuals, including some adult males.

We observed 227 displacements and aggressive encounters among all dyads in Group 1 and Group 2 and were able to rank all but five individuals (Table 4.2). In Group 1, one female offspring was high ranking, another middle ranking while two more were low ranking. Two male offspring were middle ranking while another did not participate in any dominance interactions and could not be ranked. Three adult males were low ranking and thus outranked by several mother-reared offspring. In Group 2, all three juvenile females were middle ranking. One juvenile male was middle ranking, one was low ranking and one could not be ranked because he did not participate in any dominance interactions. Two adult males were low ranking and outranked by most of the mother-reared offspring.

Prediction 4: Mating is more frequent between males and the mother of juveniles with high association rates.

Males who associated frequently with offspring copulated more frequently with their mothers (Figure 4.3). Males who copulated with adult females had significantly higher association rates with the mother's offspring than males who never copulated (Kruskal–Wallis, Z = 3.55, *p* < 0.001). Modelling again found that copulation rates varied significantly with

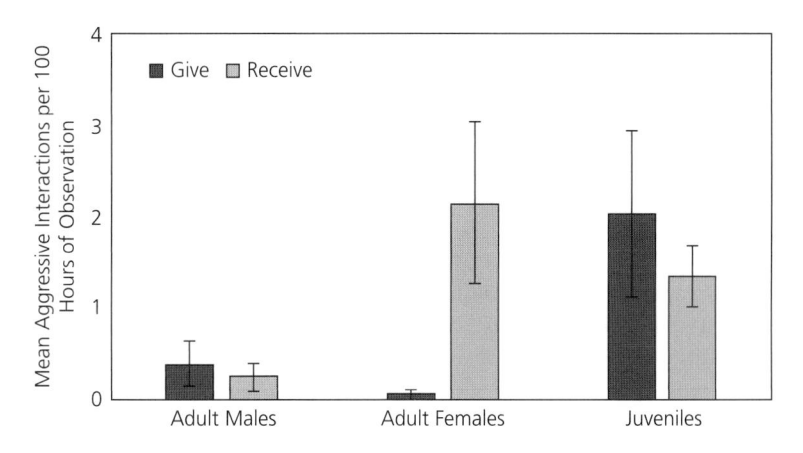

Figure 4.2 Mean rate of aggression given and received per 100 observation hours across the entire study period from adult males, adult females and juveniles. Error bars represent standard error.
(Moyenne de fréquence d'agression donnée et reçue par 100 heures d'observations à travers la période complète d'étude des mâles adultes, femelles adultes et adolescents. Les bandes d'erreur représentent l'erreur standard.)

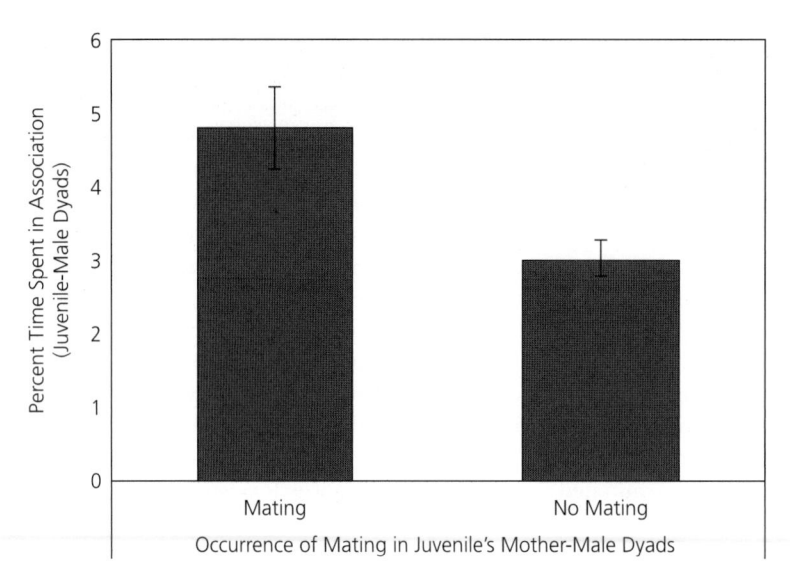

Figure 4.3 Mean time spent in association between juvenile-male dyads in instances where mating occurred between the juvenile's mother and adult male and in instances where mating did not occur. Error bars represent standard error.
(Moyenne de temps passé en association entre dyades adolescent-mâle pendant des instances d'accouplement entre le mâle et la mère de l'adolescent et pendant d'instances oú il n'y avait pas d'accouplement. Les bandes d'erreur représentent l'erreur standard.)

association rates. Males who associated at high rates with offspring copulated more frequently with their mothers (T = 3.34; $p = 0.001$).

Discussion

Our initial test of the offspring dominance hypothesis supports the idea that the social system of bonobos evolved, at least in part, as a response to the pressure of male infanticide on female reproductive success. The ODH predicts that bonobo offspring can freely interact with and even outrank adult males as a result of female intolerance for male aggression towards offspring. Consistent with its predictions, we observed mother reared bonobos readily interacting with all members of their group and by the age of three spending the majority of time in proximity to others including adult males. Unlike juvenile chimpanzees that rarely interact with adult males, mother-reared bonobos spent 20 per cent of their time in close proximity to the adult males in their group (Figure 4.4, Plate 4). Our dominance hierarchy, which includes infants and juveniles, reveals a pattern different from that observed in chimpanzees. Young mother-reared bonobos were present in each rank category and in some cases outranked adult individuals, including some adult males. For the most part, mother-reared juveniles were middle ranking while orphan juveniles were middle or low ranking (i.e. replicating the pattern seen for orphans in Clay and Zuberbuehler, 2010). Adult females clustered in the upper half of the hierarchy while adult males appeared in each rank class. Older juvenile female orphans also outranked some adult males, suggesting males are generally fearful of maturing females. This is very different from hierarchies seen in chimpanzees where females and their offspring are uniformly lower ranking than adult males and female rank is correlated with age (Kahlenberg et al., 2008; Murray et al., 2007; Wittig and Boesch, 2003). We also found evidence that adult males may benefit from amicable relations with a mother's offspring since males that interacted with offspring were more likely than other males to copulate with that offspring's mother. These preliminary results suggest that the ODH has the potential to help resolve the tension in the literature regarding bonobo and chimpanzee social systems since the differences between the two species come into starkest contrast when the interactions of offspring are considered.

Figure 4.4 Adult males interacting with mother-reared infants. Top: Adult male Makali plays/grooms with a mother-reared juvenile. Bottom (Plate 4): Adult male Keza fear grimacing to mother-reared juvenile.
(Mâles adultes interagissant avec des nourrissons élevés par leur mère. En haut : le mâle adulte Makali jouant avec/toilettant un jeune élevé par sa mère. En bas (plaque 4) : le mâle adulte Keza faisant une grimace destinée à effrayer un jeune élevé par sa mère.)Photos by J. Tan and used with permission.

As predicted based on previously observed high levels of play and reduced self-control in bonobos, mother-reared juveniles showed little inhibition in interactions with adult group mates. They spent significant time away from their mothers and near adult males. They interact with adult males less than expected by chance (it may even be males avoiding infants given their relative ranks in some cases), but their level of interaction far exceeds what would be expected in chimpanzees (with the exception of male adoptions that occasionally occur, most notably in Tai National Park (Boesch et al., 2010)). Also consistent with previously observed high rates of prosocial interactions, aggressive encounters between adult males and mother-reared offspring were relatively rare. Mother-reared offspring and adult males occasionally displaced each other, but at similar rates. Interestingly, most of the displacements of adult males occurred regardless of the location of an infant's mother and aggression was even more rare than displacements.

Despite being the smallest members of their group (i.e. typically even smaller than the bottle-fed same-aged orphans), some mother-reared offspring attain a high rank. The hierarchical structure where individuals of different ages and sexes are scattered throughout the rank classes and males rarely, if ever, act aggressively towards offspring is in stark contrast to patterns seen in chimpanzees, where all males outrank all females and aggression towards juveniles by males accounts for 12 per cent of their total aggression in Kanyawara (Muller, 2002).

Derived rank most likely accounts for much of the observed interactions between offspring and adult males. When in proximity to their mothers, young bonobos may be emboldened if adult males defer to their more dominant mother. Adult males may also fear reprisal, not only from the offspring's mother but also her potential coalition partners. However, our data also suggest that mothers may not always need to be present to help their offspring. Some offspring were able to win contests in the absence of their mothers. Vocal support could not be ruled out and should be considered in future study but several observed interactions occurred when mothers were over 15 metres away and sometimes out of sight. Moreover, when infant/juveniles and adult males competed, mothers never intervened on behalf of their offspring and instead seemed to ignore most interactions. This contrasts with their more frequent support during altercations between their offspring and other juveniles that were orphans. Unlike previous studies of wild bonobos (i.e. Surbeck and Hohmann, 2013), we also did not observe any instances where mothers attacked adult males that were harassing their offspring. In these cases females did not demonstrate intolerance towards male aggression. One interpretation might be that males typically do not view young bonobos as adversaries and mothers do not see males as particularly threatening given the rare cases of aggression are so restrained. This may be why even orphaned juveniles occasionally won contests against adult males. Correspondingly, mother-reared juveniles outranked some adult males and in one case, a juvenile female outranked several adult females, including her own mother (e.g. Elykia over Semendwa). Typically, young offspring dependent on their mother's rank do not also out rank their mother. Together these observations do not tightly fit traditional definitions of dependent rank in which an offspring's rank completely relies on the physical presence of her mother (Kawai, 1958).

One potential benefit of adult males avoiding aggression towards offspring is increased reproductive access. Bonobo males, unlike chimpanzees, are known to be less aggressive when females are maximally swollen and are hypothesized to gain reproductive benefits through affiliative behaviors with females (Hohmann and Fruth, 2003; Ryu et al., 2015). Here we have observed that this may also extend to behavior towards a female's offspring. Males that copulated at least once with a particular female had higher association rates with her infant than those who did not. Males may have benefited evolutionarily by trading off status and the risk of injury associated with mate guarding and infanticide for reproductive access via affiliative relationships with females and their kin (Hare et al., 2012; Surbeck et al., 2012a; Tokuyama and Furuichi, 2016).

To our knowledge we have observed the largest sample of mother-reared infant and juvenile bonobos ever studied. Our subjects also live in large forested enclosures and are semi-free ranging within large multi-male, multi-female groups. This context presents a unique opportunity to study the aggressive interactions of offspring and adults. However, we had no adult kin since all our adults were orphans and, relative to typical observational studies, we conducted limited hours of observations and instead focused on times when social interactions were frequent. Moreover, we were unable to compare interactions between orphan juveniles and adult males with those between mother-reared juveniles and adult males, a comparison that would clarify the importance of mothers in juvenile–adult male relationships. This means some of our null findings could be attributed to our lack of adult kin or underpowered analyses. Despite these limitations, we were able to detect differences in the way mother-reared offspring behave towards adults—and males in particular. This effect may have been detected precisely because of the captive conditions under which the bonobos were observed, where conflicts during provisioning occurred frequently. The varied diet and fission–fusion social system of wild bonobos make it difficult to observe dyadic competitive interactions between all group members routinely. On the other hand, sanctuary bonobos are provisioned multiple times a day, providing the perfect backdrop to observe a large number of dominance interactions. This allowed us to rank categorically nearly every member of the group in a short time period and was sufficient for a preliminary understanding of juvenile–adult male dominance interactions. Rank differences can be further clarified with a larger sample that allows for individuals to be ranked rather than categorized. Unfortunately, we were also unable to collect similar data with a population of mother-reared chimpanzees since we are not aware of another African sanctuary with so many mother-reared infants. This means all of our species comparisons rely on qualitative comparisons based on previous research from zoos and the wild. Hopefully, future research in sanctuaries, zoos or the wild will allow for direct quantitative species comparisons as tests of the ODH (De Lathouwers and Van Elsacker, 2006). The ODH predicts very different results in chimpanzees when observed under similar circumstances as the bonobos studied here. For example, ODH predicts that all juvenile chimpanzees, regardless of the presence or absence of a mother, would rank below all adult females who would in turn rank below all adult males.

Another challenge to address in future research is the causal direction of affiliative relationships between adult males, mothers and their offspring. There remain multiple explanations for the increased copulations between the males who also associate with a female's offspring. Infants may affiliate with particular males because the male is already friendly with their mother, making the mother's relationship with the male the best predictor of copulation rates. However, given that juveniles spent a large proportion of their time away from their mothers and often interacted with males, we expect that this relationship is at least partially driven by the adult males and juveniles themselves. Examining who initiates proximity, play behavior and grooming interactions between these dyads will clarify which infants or mothers drive these associations. In particular, attention should be paid to both the agonistic and affiliative relationships between juveniles and adult males when their mother is and is not present. If males do curry favour with females through their offspring, relationships should be maintained regardless of the location of an infant's mother. On the other hand, if males merely fear reprisal from a juvenile's mother then relationships may break down when a juvenile's mother is not in the immediate vicinity. Here, examining adult male behavior towards orphaned juveniles in comparison to mother-reared juveniles may also clarify the importance of the mother in adult male-juvenile interactions.

We have provided a preliminary test of the *offspring defence hypothesis* and found initial support of its predictions. We have presented evidence that juvenile bonobos interact with adult individuals of the group and can displace and win aggressive interactions with adult males, even sometimes ranking above them. Juveniles were sometimes even able to win these dyadic interactions without physical interference from their mothers or other group members. All of these findings contrast sharply with what has been observed in captive and wild chimpanzees. While still preliminary, this work provides a testable hypothesis that can be investigated in other populations of captive and wild bonobos and highlights the need for further quantitative species comparisons that includes all age and sex categories. Based on our initial results, the offspring defence hypothesis may help solve the paradox of bonobo sociality that is difficult to categorize using traditional dominance definitions. The solution to whether bonobos are female or co-dominant may simply be that they are neither and both. Males have been selected to be less aggressive and juvenilized while female bonobos have evolved to defend offspring against infanticide and maximize female mate choice (Hare et al., 2012; Marvan et al., 2006). As a result, male bonobos may compete physically against each other for access to females but have evolved to rely on affiliative relationships with females and their kin, and not overall status to maximize their reproductive success. The ultimate puzzle that should also inform the evolution of our own species social systems is what ecological pressures allow for such a dramatic shift in sociality between such close phylogenetic relatives (Hare, 2017). We will never know without more concentrated research on bonobos as well as chimpanzees.

Acknowledgements

We thank Suzy Kwetuenda and Delphin Bilua for assistance in collecting data. We thank Claudine Andre, Dominique Morel, Fanny Mehl, Pierrot Mbonzo and the animal caretakers at Lola ya Bonobo and the Ministry of Research and the Ministry of Environment in the Democratic Republic of Congo (permit MIN.RS/SG/004/2009) for hosting our research. We thank J. Tan for providing photographs. This research followed the laws of the country in which it was carried out, was approved by Duke IACUC (A035-140-02), and supported in part by National Science Foundation grant NSF-BCS-10-25172 to OBH.

References

André, C., Kamate, C., Mbonzo, P., Morel, D., and Hare, B. (2008). *The Conservation Value of Lola ya Bonobo Sanctuary. The Bonobos.* New York, Springer.

Behringer, V., Deschner, T., Deimel, C., Stevens, J.M., and Hohmann, G. (2014). Age-related changes in urinary testosterone levels suggest differences in puberty onset and divergent life history strategies in bonobos and chimpanzees. *Hormones and Behavior*, 66, 525–33.

Bernstein, I.S. (1981). Dominance: the baby and the bathwater. *Behavioral and Brain Sciences*, 4, 419–29.

Boesch, C., Bole, C., Eckhardt, N., and Boesch, H. (2010). Altruism in forest chimpanzees: the case of adoption. *PLoS One*, 5, e8901.

Clay, Z., Archbold, J., and Zuberbühler, K. (2015). Functional flexibility in wild bonobo vocal behaviour. *PeerJ*, 3, e1124.

Clay, Z. and Zuberbühler, K. (2012). Communication during sex among female bonobos: effects of dominance, solicitation and audience. *Scientific Reports*, 2.

Davies, N.B., Krebs, J.R., and West, S.A. (2012). *An Introduction to Behavioural Ecology.* New York, John Wiley & Sons.

De Lathouwers, M. and Van Elsacker, L. (2006). Comparing infant and juvenile behavior in bonobos (*Pan paniscus*) and chimpanzees (*Pan troglodytes*): a preliminary study. *Primates*, 47, 287–93.

De Vries, H., Stevens, J.M., and Vervaecke, H. (2006). Measuring and testing the steepness of dominance hierarchies. *Animal Behaviour*, 71, 585–92.

de Waal, F.B. (1988). The communicative repertoire of captive bonobos (*Pan paniscus*), compared to that of chimpanzees. *Behaviour*, 106, 183–251.

Digby, L. (1995). Infant care, infanticide, and female reproductive strategies in polygynous groups of common marmosets (Callithrix jacchus). *Behavioral Ecology and Sociobiology*, 37, 51–61.

Drea, C.M. (2009). Endocrine mediators of masculinization in female mammals. *Current Directions in Psychological Science*, 18, 221–26.

Fowler, A. and Hohmann, G. (2010). Cannibalism in wild bonobos (*Pan paniscus*) at Lui Kotale. *American Journal of Primatology*, 72, 509–14.

Furuichi, T. (2011). Female contributions to the peaceful nature of bonobo society. *Evolutionary Anthropology: Issues, News, and Reviews*, 20, 131–42.

Hare, B. (2017). Survival of the friendliest: *Homo Sapiens* evolved via selection for prosociality. *Annual Review of Psychology*, 68.

Hare, B., Wobber, V., and Wrangham, R. (2012). The self-domestication hypothesis: evolution of bonobo psychology is due to selection against aggression. *Animal Behaviour*, 83, 573–85.

Hemelrijk, C.K., Wantia, J., and Isler, K. (2008). Female dominance over males in primates: Self-organisation and sexual dimorphism. *PLoS One*, 3, e2678.

Hohmann, G. and Fruth, B. (2003). Intra-and inter-sexual aggression by bonobos in the context of mating. *Behaviour*, 140, 1389–413.

Jolly, A. (1985). The Evolution of Primate Behavior: A survey of the primate order traces the progressive development of intelligence as a way of life. *American Scientist*, 73, 230–9.

Kahlenberg, S.M., Thompson, M.E., and Wrangham, R.W. (2008). Female competition over core areas in *Pan troglodytes schweinfurthii*, Kibale National Park, Uganda. *International Journal of Primatology*, 29, 931–47.

Kanō, T. (1992). *The Last Ape: Pygmy Chimpanzee Behavior and Ecology*. Stanford, CA: Stanford University Press.

Kappeler, P.M. (1990). The evolution of sexual size dimorphism in prosimian primates. *American Journal of Primatology*, 21, 201–14.

Kawai, M. (1958). On the rank system in a natural group of Japanese monkey (I). *Primates*, 1, 111–30.

Langergraber, K.E., Mitani, J. C., and Vigilant, L. (2007). The limited impact of kinship on cooperation in wild chimpanzees. *Proceedings of the National Academy of Sciences*, 104, 7786–90.

Leigh, S.R. and Shea, B.T. (1995). Ontogeny and the evolution of adult body size dimorphism in apes. *American Journal of Primatology*, 36, 37–60.

Marvan, R., Stevens, J., Roeder, A. D., Mazura, I., Bruford, M.W., and De Ruiter, J. (2006). Male dominance rank, mating and reproductive success in captive bonobos (*Pan paniscus*). *Folia Primatologica*, 77, 364–76.

Mcintyre, M.H., Herrmann, E., Wobber, V., Halbwax, M., Mohamba, C., De Sousa, N., Atencia, R., Cox, D., and Hare, B. (2009). Bonobos have a more human-like second-to-fourth finger length ratio (2D: 4D) than chimpanzees: a hypothesized indication of lower prenatal androgens. *Journal of Human Evolution*, 56, 361–5.

Mitani, J.C., Gros-Louis, J., and Richards, A.F. (1996). Sexual dimorphism, the operational sex ratio, and the intensity of male competition in polygynous primates. *American Naturalist*, 966–80.

Mitani, J.C., Merriwether, D.A., and Zhang, C. (2000). Male affiliation, cooperation and kinship in wild chimpanzees. *Animal Behaviour*, 59, 885–93.

Morelli, T.L., King, S.J., Pochron, S.T., and Wright, P.C. (2009). The rules of disengagement: takeovers, infanticide, and dispersal in a rainforest lemur, Propithecus edwardsi. *Behaviour*, 146, 499–523.

Muller, M.N. (2002). Agonistic relations among Kanyawara chimpanzees. *Behavioural Diversity in Chimpanzees and Bonobos*, 112–24.

Murray, C.M., Mane, S.V., and Pusey, A.E. (2007). Dominance rank influences female space use in wild chimpanzees, Pan troglodytes: towards an ideal despotic distribution. *Animal Behaviour*, 74, 1795–804.

Murray, C.M., Eberly, L.E., and Pusey, A.E. (2006). Foraging strategies as a function of season and rank among wild female chimpanzees (*Pan troglodytes*). *Behavioral Ecology*, 17, 1020–8.

Paoli, T., Palagi, E., and Tarli, S. (2006). Reevaluation of dominance hierarchy in bonobos (*Pan paniscus*). *American Journal of Physical Anthropology*, 130, 116–22.

Parish, A.R. (1996). Female relationships in bonobos (*Pan paniscus*). *Human Nature*, 7, 61–96.

Parish, A.R., de Waal, F., and Haig, D. (2000). The other 'closest living relative': How bonobos (*Pan paniscus*) challenge traditional assumptions about females, dominance, intra-and intersexual interactions, and hominid evolution. *Annals of the New York Academy of Sciences*, 907, 97–113.

Petty, J.M. and Drea, C.M. (2015). Female rule in lemurs is ancestral and hormonally mediated. *Scientific Reports*, 5, 9631.

Pusey, A., Murray, C., Wallauer, W., Wilson, M., Wroblewski, E., and Goodall, J. (2008). Severe aggression among female Pan troglodytes schweinfurthii at Gombe National Park, Tanzania. *International Journal of Primatology*, 29, 949–73.

Ryu, H., Hill, D.A., and Furuichi, T. (2015). Prolonged maximal sexual swelling in wild bonobos facilitates affiliative interactions between females. *Behaviour*, 152, 285–311.

Sakamaki, T., Behncke, I., Laporte, M., Mulavwa, M., Ryu, H., Takemoto, H., Tokuyama, N., Yamamoto, S., and Furuichi, T. (2015). Intergroup transfer of females and social relationships between immigrants and residents in bonobo (*Pan paniscus*) societies. *Dispersing Primate Females. Primatology Monographs*, Springer.

Sannen, A., Van Elsacker, L., Heistermann, M., and Eens, M. (2004). Urinary testosterone-metabolite levels and dominance rank in male and female bonobos (*Pan paniscus*). *Primates*, 45, 89–96.

Smith, R.J. and Jungers, W.L. (1997). Body mass in comparative primatology. *Journal of Human Evolution*, 32, 523–59.

Stanford, C.B. (1998). The social behavior of chimpanzees and bonobos: empirical evidence and shifting assumptions 1. *Current Anthropology*, 39, 399–420.

Stevens, J.M., Vervaecke, H., De Vries, H., and Van Elsacker, L. (2006). Social structures in *Pan paniscus*: testing the female bonding hypothesis. *Primates*, 47, 210–17.

Stevens, J.M., Vervaecke, H., De Vries, H., and Van Elsacker, L. (2007). Sex differences in the steepness of dominance hierarchies in captive bonobo groups. *International Journal of Primatology*, 28, 1417–30.

Surbeck, M., Deschner, T., Schubert, G., Weltring, A., and Hohmann, G. (2012a). Mate competition, testosterone and intersexual relationships in bonobos, *Pan paniscus*. *Animal Behaviour*, 83, 659–69.

Surbeck, M., Deschner, T., Weltring, A., and Hohmann, G. (2012b). Social correlates of variation in urinary cortisol in wild male bonobos (*Pan paniscus*). *Hormones and Behavior*, 62, 27–35.

Surbeck, M. and Hohmann, G. (2013). Intersexual dominance relationships and the influence of leverage on the outcome of conflicts in wild bonobos (*Pan paniscus*). *Behavioral Ecology and Sociobiology*, 67, 1767–80.

Surbeck, M., Mundry, R., and Hohmann, G. (2011). Mothers matter! Maternal support, dominance status and mating success in male bonobos (*Pan paniscus*). *Proceedings of the Royal Society of London B*, 278, 590–8.

Tokuyama, N. and Furuichi, T. (2016). Do friends help each other? Patterns of female coalition formation in wild bonobos at Wamba. *Animal Behaviour*, 119, 27–35.

White, F.J. and Wood, K.D. (2007). Female feeding priority in bonobos, *Pan paniscus*, and the question of female dominance. *American Journal of Primatology*, 69, 837–50.

Wilson, M.L., Boesch, C., Fruth, B., Furuichi, T., Gilby, I.C., Hashimoto, C., Hobaiter, C.L., Hohmann, G., Itoh, N., and Koops, K. (2014). Lethal aggression in Pan is better gestures and facial expressions. After that, we will investigate similarities and differences in play behavior between chimpanzees and bonobos in order to explore the connection linking adult play behavior and social tolerance. Finally, we will provide evidence that the results on play behavior and tolerance obtained in bonobos can be theoretically expanded to other species sharing tolerant social systems.

What is play and what are its roles?

There is no one single way to play; there are unlimited ways to play. It can be either solitary or social and, in both cases, it may also rely on objects. However, even if categorization helps scholars to identify, measure, describe and quantify the different playful activities of animals in a more or less standardized way, it must be noted that these are just ideal groupings that are often mixed together in extremely complex and variable sequences forming a single natural category (Pellis and Pellis, 2010).

Both solitary and social play are widespread among mammals (Fagen, 1981) but they can assume different functions according to species, sex, age, ranking position, habitat and affinitive or agonistic relationships shared by the players (Cordoni, 2009; Pellegrini et al., 2007). The pervasiveness of play in mammalian species makes it a wonderful behavior to make cross-species comparisons that can be extremely fruitful for a wide array of disciplines (Palagi, 2007).

Everybody recognizes when two or more animals are playing, but very few scholars attempted to define this kind of social interaction scientifically (Miller, 1973) so that there remains no agreed definition of play. Social play is often described via *litotes*, a figure of speech consisting of an understatement in which an affirmative is expressed by negating its opposite. Therefore, play is defined as *what it is not* more than *what it is*: play is *not* a predatory act, play does *not* have a reproductive function, play *cannot* be considered as an agonistic interaction, and so on. Play borrows many motor patterns from 'serious' contexts, but such acts are *not* exhibited as they are when performed under serious circumstances. The two main criteria used to define social play are linked to its functional and/or operational aspects (Bekoff, 2001; Pellis and Pellis, 1996). Among various attempts to define play, the most accepted theoretical definition was provided by Burghardt (2005) who listed five criteria involved in recognizing a playful activity. According to him, a playful behavior is: 1) incompletely functional as it is not linked to immediate survival; 2) self-rewarding and voluntary; 3) structurally modified (exaggeration in intensity and/or duration), or temporally modified (anticipation in development compared to functional behaviors); 4) performed in a repeated manner; 5) initiated in a relaxed context and when the animal is generally adequately fed, healthy and free from stress or intense competition.

Looking at these criteria it becomes clear why a universally operational and theoretical definition of play was (and still is) so hard to find: in fact, what differentiates play from 'serious' behaviors relies only in the *way* in which behavioral patterns are performed, not in their mere presence.

Moreover, the essence of play is its versatile and ephemeral nature. The main features characterizing play are its multi-functionality, the complexity in the execution of its motor patterns and the massive presence of communicative signals.

The correct management of a playful interaction requires complex social abilities. For this reason, playing socially, particularly in a *play-fighting* session, has been considered a powerful means to gain social competence. Fine-tuning the intensity of playful patterns according to the different playmates, correctly deciphering each communicative signal and adopting self-handicapping strategies to balance the session are only some of the abilities required to prolong the session and avoid the risk of an escalation into real aggression. Indeed, two opposing components coexist in play: cooperation and competition. This cooperative/competitive interaction serves to test a partner's willingness to invest in a relationship and, simultaneously, to demonstrate one's own willingness to accept vulnerability. A good player must balance these two forces acting and communicating coherently. The most important role in maintaining a balanced interaction is expressed by communicative signals. Facial expressions, body postures, gestures and vocalization function as a sort of *control knob* that regulates the relative intensity of these two forces.

Masters of playful communication: facial expressions and gestures in bonobos

The behavioral freedom essential for play implies that animals must face unpredictable and rapid sequences of actions. Therefore, the correct management of a playful interaction passes through three main steps: pay attention, interpret correctly and respond appropriately. For this reason, play requires such highly complex communicative skills that it has been suggested that one of its functions is to learn decoding facial and body signals (Fagen, 1981; Pellis and Pellis, 1996). Playful signals are multifunctional in triggering the session, in maintaining a playful mood and in avoiding the risk of an escalation into overt aggression (Burghardt, 2005; Palagi et al., 2016; Pellis and Pellis, 2010).

Solitary play represents per se a signal, informing the audience of the motivation and intention of the player. It is common that a solitary play session gives rise to a dyadic playful interaction that, in turn, can beckon other players, thus becoming a polyadic interaction (Palagi, 2008). It appears clear that play can produce more play by functioning as a means of aggregation among group members. All the playful patterns can be easily read thanks to their exaggeration, emphasis and repetition, so it is the modality of the performance of motor patterns that conveys the playful intent.

Aside from the intrinsically infective nature of play behavior itself, playful communication also relies upon context-specific signals. In primates, the relaxed open-mouth display, or play face, is the typical playful facial expression (van Hooff and Preuschoft, 2003). In some species, such as bonobos, this playful facial expression can be displayed in two different configurations: Play Face (PF) and Full Play Face (FPF). In the PF (Figure 5.1), the mouth is opened and only the lower teeth are exposed, whereas in the FPF (Figure 5.2) the upper teeth are also visible (Palagi, 2006). The FPF, which is a more intense facial expression, can be associated with a pant-like vocalization (Provine, 2000) that has been considered to be homologous to human laughter (Davila-Ross et al., 2009). The two variants of the play face are present in the two *Pan* species, although they are used to different degrees, especially by adults. Immature subjects of the two species display PF and FPF at comparable frequencies. As adults, however, bonobos almost exclusively perform the FPF, while adult chimpanzees almost exclusively perform the PF, and even this variant occurs at a very low frequency. The difference in the use of playful signals by the adults of the two species could be due to the higher propensity and roughness of play that is typical of bonobos (Palagi, 2006). Nowadays, there is passionate debate about the proximate factors leading to the production of facial expressions, with scholars arguing if facial expressions are intentionally or emotionally driven signals. Darwin (1872) stated that facial expressions are the inevitable counterpart of felt emotions, because they are linked to specific internal states. More recently, some studies confirmed the presence not only of a neuro-anatomical pathway controlling the emission of 'emotional' facial expressions but also of another pathway controlling the emission of 'intentional' facial expressions (Sherwood et al.,

Figure 5.1 During a face-to-face interaction, an immature male (on the right) performs a play face in a locomotor play session with an immature female. (Photo by E. Demuru)
(Pendant une interaction face à face, un mâle immature (droite) effectue un visage de jeu dans une session de jeu locomoteur avec une femelle immature.)

Figure 5.2 An adult female plays with an unrelated infant by pulling its arms. The full play face of the infant is clearly visible. (Photo by E. Demuru)
(Une femelle adulte joue avec un enfant, qui n'a aucun rapport avec elle, et lui tire le bras. Le visage de jeu de l'enfant est visible.)

2004, 2005). Cattaneo and Pavesi (2014) demonstrated that the two systems are not so clear-cut but are, rather, connected, even though the degree to which they intermingle for the emission of a given signal is unknown.

An anecdote describing the emotional nature of PFs reported in gorillas may clarify the novel levels of control over play signals seen in apes (Tanner and Byrne, 1993). To avoid the possibility that group members could perceive its playful facial expressions, a female gorilla repeatedly concealed her face with the hand. That is, she could not refrain from producing the emotionally charged play face, but she could use a cognitive strategy to attenuate the signal. The spontaneous nature of playful facial expressions and the difficulty in inhibiting them suggest that they can be driven by emotional states.

Yet, the expression of play faces can be affected by both the audience present and the context in which they are performed, suggesting also an intentional dimension, a dimension that may not be exclusive to the great apes. Geladas (*Theropithecus gelada*) and chimpanzees, for example, can modify the performance of playful expressions according to the audience attending the session (Flack et al., 2004; Palagi and Mancini, 2011). Moreover, geladas increase the frequency of the emission of PF during those sessions occurring immediately after an agonistic event (Palagi and Mancini, 2011). These two findings support the view that playful facial expressions are also driven by intentionality since under a context of social tension it is necessary to use redundant signals to avoid the risk to be misinterpreted and to convey a positive message in the most effective way.

Play, both in humans and great apes, is characterized by another kind of intentional signal: gestures (Genty et al., 2009; Liebal et al., 2006; Pollick and de Waal 2007). Different from facial expressions that are specific to play, gestures are used flexibly and are disengaged from specific behavioral contexts. Therefore, the 'meaning' of a gesture has to be interpreted by the receiver by evaluating the environmental and social conditions in which it is produced (Pollick and de Waal, 2007). In the African great apes, a vast majority of the gestural repertoire occurs during play (Genty et al., 2009; Genty and Byrne, 2010; Pika et al., 2005; Tomasello et al., 1994) and, therefore, the playful activity may provide a 'training ground' to test the effectiveness of old and new gestures as there is evidence to suggest that gestures during play are found in various contexts, and thus, gestures during play may serve as a practice for non-play behavior (Pollick, Jeneson and de Waal, 2008).

Data on bonobos show that they have an impressive use of communicative signals during playful interactions. Particularly, Demuru and co-workers (2014) found that bonobos increase the rate of both facial expressions and gestures during contact play and especially during *play fighting*. The reason for such an increase in signalling may be that when players are in physical contact there is an increased risk of harming or scaring the playmate.

Moreover, these authors found that the number of players significantly affected the rates of gestures but not those of facial expressions; in particular, dyadic play was characterized by a larger number of gestures compared to polyadic play. The intentional component of gestures can explain this finding, since the gesture reaches its maximum efficacy when a single receiver is attending. One of the criteria of intentional communication is that the signal must be directed towards a specific receiver and this condition is more rarely met during polyadic play sessions.

In bonobos, gestures are not only adjusted according to the presence of a receiver but also according to its attentional state. Whereas the emission of acoustic or tactile gestures was not affected by the attentional state of the receiver, Demuru and her team (2014) demonstrated that both visual gestures and playful facial expressions were more often performed when the receiver was visually attentive. Therefore, the authors suggested, facial expressions may be emitted intentionally by the playmates to manipulate and manage the play session as it progresses.

Facial expressions and gestures are finely tuned as a function of the features of the session, probably influencing the degree of emotionality and/or intentionality characterizing the signals. Thus, play may provide critical information on the ontogenetic and evolutionary pathways characterizing non-human and human communication. In particular, given the rewarding nature of play, it represents a good opportunity to search for the possible dichotomy between intentions and emotions within communication systems.

Play and the importance of tolerance in the *Pan* genus

Rough-and-tumble play

'Rough-and-Tumble' play (hereafter, R&T) is defined as a physically vigorous set of social behavioral patterns, such as chase, jump, bite, slap, stamp on, accompanied by a positive affect between players (Burghardt, 2005; Pellis and Pellis, 2010). R&T was first academically named by Karl Groos in his book *Play of Animals* (1898). R&T, also named *play fighting* (Aldis, 1975), can reliably be distinguished from aggression and other competitive activities. Although it may be the most fundamental form of play, it has received little experimental attention in humans and non-human species. The research carried out on R&T play in children and young non-human primates, however, indicates that this playful activity creates valuable scenarios for complex social interactions that animals need to undertake in order to become competent, socially mature adults (Fagen, 1981). Indeed, play behavior is a feature of ontogeny in many species and is widely believed to have an important role in the assembly of adult behavior (Palagi, 2011). Pellis and Pellis (2007) suggested that R&T experience affects the ability to regulate emotional responses and this, in turn, affects the ability to perform appropriate actions in the appropriate context, thus increasing social competence. In many primate species, behavioral plasticity is an essential component to track an environment that frequently changes both socially and ecologically. Play is one of the best means to obtain and maintain such plasticity. Both immature and adult animals have to cope with new situations and social challenges (Špinka et al., 2001), therefore play may have an important role in adulthood as well (Palagi et al., 2006; Tacconi and Palagi, 2009). For this reason, a complete understanding of the potential roles of play, and especially of social play in its roughest version, can be reached by taking into account its occurrence during each phase of life.

Rough-and-tumble play in adults

Play involving only adults opens new scenarios regarding the potential role of this behavior. In this case, play may promote the establishment and maintenance of social relationships (Palagi, 2006; Palagi and Paoli, 2007) and, at the same time, animals can gather information on group members and/or competitors (Pellis and Iwaniuk, 2000). Adult play, in the form of R&T, is present in several species and in several contexts. Courtship play fighting, for instance, is prevalent in solitary species where adult males and females are unfamiliar with one another (the genera *Mirza, Daubentonia, Perodicticus* and *Pongo* are good examples) (Pellis and Iwaniuk, 1999, 2000). Outside reproductive contexts, R&T is generally more frequent in the species showing a complex and high degree of social aggregation (Pellis and Iwaniuk, 2000). However, the size of social groups per se does not seem to increase the frequency of play. Many species of cercopithecids living in very large troops do not show high levels of adult play (*Papio anubis*, Owens, 1975; *Papio hamadryas*, Kummer, 1995). Adult–adult play seems to be present in species living in social aggregations that are characterized by a fluid and loose composition (e.g. fission–fusion society, *Ateles, Cacajao* and *Pan*, Pellis and Iwaniuk, 2000).

A further key feature favouring the playful attitude in adults is the level of social tolerance of a certain species. In fact, when considering species characterized by egalitarian and tolerant societies, play among adults appears to embrace the function of regulating social activities (Palagi, 2006). Palagi (2006; 2007) demonstrated that, despite their phylogenetic closeness and similar social structure (fission–fusion society), chimpanzees and bonobos are characterized by striking differences in adult social play as well as in tolerance level.

Fourteen years of studies carried out on several chimpanzee and bonobo colonies by our research group demonstrate that the differences observed between the two *Pan* species are not linked to the idiosyncrasy of a single group. Therefore, we can reasonably conclude that adult bonobos play much more than adult chimpanzees (Demuru and Palagi, in preparation; Palagi, 2006; 2008).

Another peculiarity that differentiates bonobo play is that it includes a huge amount of sexual behaviors that are performed in all age, rank and sex combinations. The function of bonobos' socio-sexual contacts in lowering social tension and promoting affiliative bonding is well known in other contexts and it can be easily applied also to play. In this sense, play and socio-sexuality could be considered as the two sides of the same coin in bonobos' society. This view is supported not only by studying bonobo intra-group social dynamics but also by observing what happens during inter-group encounters. Recent studies reported that socio-sexuality and play are frequently observed when two neighbouring bonobo communities meet (Behncke, 2015; Furuichi, 2011). This pattern of interaction diverges strikingly from that observed during chimpanzee inter-group encounters with neighbouring communities engaging in real 'border wars' that can even end with the annihilation of the weaker group.

Development of play behavior

The divergence of the developmental timing of play between bonobos and chimpanzees occurs during the transitional phase from infancy to juvenility (Palagi and Cordoni, 2012), with the most striking divergence being evident in R&T. Chimpanzees engaged in fewer R&T sessions as their age increased, in contrast with bonobos, who maintained constant levels of play throughout infant, juvenile and adult periods. Compared to chimpanzees, R&T sessions in juvenile bonobos escalated less frequently into overt aggression, lasted longer and frequently involved more than two playmates concurrently (polyadic play, Figure 5.3). In juvenile bonobos, R&T contains many cooperative elements (Palagi, 2008; Palagi and Paoli, 2007), whereas in juvenile chimpanzees it seems to be based on more competitive interactions (Palagi and Cordoni, 2012). The level of competition and cooperation within a playful session can be measured by evaluating the symmetry/asymmetry between offensive (e.g. play bite, play push, play slap, etc.) and defensive patterns (e.g. shelter, play crouching, etc.) performed by the players. More symmetric play sessions highlight a high degree of cooperation, whereas asymmetric sessions reveal a more competitive intent (Pellis et al., 2010). Chimpanzees show a decrease in tolerance with age, whereas bonobos do not (Wobber et al., 2010). This idea is also supported by data derived from wild chimpanzees. Shimada and Sueur (2014) found that the frequency of participation in social play contributed positively to the development of affiliative social relationships within the chimpanzee group during the infant and early juvenile period, but did not have the same effect during adolescent

Figure 5.3 Polyadic R&T play involving an adult male (on the left), an adult female (on the right) and an immature subject (in the middle). (Photo by E. Demuru)
(Polyadic R&ampT jeu avec un mâle adulte (gauche), une femelle adulte (droite) et un sujet immature (milieu).)

and adult period. The social play network may allow chimpanzees to develop the social techniques necessary to acquire a central position in a society and enable them to develop affiliative relationships during the first phases of life. Bonobos, compared to chimpanzees, show a developmental delay in social inhibition that may be responsible for the retention of juvenile traits into adulthood (Wobber et al., 2010). The divergence in the development of social play of the two *Pan* species could be due to the low degree of social inhibitory control in bonobos (Wobber et al., 2010), which is essential for R&T to be maintained (Pellis and Pellis, 2010) (Figure 5.4).

Bonobo society is characterized by a rich set of social dynamics in which adults, especially females, negotiate and maintain their relationships both through alliances and pre- and post-conflict activities (Clay and de Waal, 2013; Palagi and Norscia, 2013). Also, chimpanzees form coalitions, alliances and show pre- and post-conflict interventions, but, differently from bonobos, these activities are up to males (Boesch, Hohmann and Marchant, 2002). Social assessment is a key factor that makes all these activities possible and it seems that bonobos can use play for social assessment due to their relaxed relationships. Moreover, differently from chimpanzees, in bonobos the high levels of undecided conflicts and the lack of formal submission displays suggest that hierarchical dynamics are not so codified and that social relationships can be frequently renegotiated (Demuru and Palagi, 2012; Furuichi, 2011). Alliances and some forms of support during both agonistic and affiliative interactions can include many different subjects whose behavior can change according to the presence of an audience and/or of additional participants.

Figure 5.4 Adult male playing with an infant. The infant pulls the ear of the adult male while performing a full play face. (Photo by E. Demuru)
(Mâle adulte joue avec un enfant. L'enfant tire l'oreille de l'adulte en faisant un visage de jeu.)

Play networks

The negotiation of social relationships implies not only the ability to manage dyadic interactions but also interactions involving more than two individuals. Polyadic play represents a good platform on which such dynamics can be tested and elaborated in a relaxed manner. During polyadic interactions there are more social variables and relationships to contend with than during dyadic encounters, and this is a strong factor in driving the acquisition of social competence.

Even though social play starts with two players, it is very frequent in bonobos to observe the immediate involvement of additional subjects who are motivated to join the session by simply perceiving the playful arousal of others. The contagious nature of social play can be revealed in this species by the high levels of polyadic playful interactions. These play sessions can include some or all adults among those interacting (Palagi, 2008) (Figure 5.5). The participation of adult subjects in polyadic play once again demonstrates the tolerant nature of bonobos and highlights the role of play in social assessment, especially given the high level of R&T play between adult females. Indeed, polyadic play potentially allows interacting individuals to evaluate their playmates strategically, enlarge their social networks and strengthen social bonds. The role of play as promoter and means to reinforce and test social bonds is demonstrated by the observation that, among numerous factors such as size, rank and age, the relationship quality measured through grooming is the only variable explaining a high occurrence of play (Palagi and Paoli, 2007).

The social dynamics observed in polyadic play sessions involving adult bonobos strongly differ from those in dyadic play (and it is not just a matter of the number of bonobos involved). Adult animals that rarely play together in dyadic interactions may find a contact point in polyadic play when it involves at least one immature subject. From this perspective, immature individuals may function as a sort of bridge between two socially untied adult individuals. This 'social bridge' resembles what has been found in chimpanzees (Palagi et al., 2004) where adults played with unrelated immatures frequently before food distribution, a period characterized by a high level of social tension. Palagi and colleagues (2004) also detected a positive correlation between the frequency of adult-unrelated immature play (pre-feeding period) and co-feeding rates (feeding period) between the adult involved in the play session and adults related to the immature playmate. The authors, therefore, hypothesized that adult–immature play could increase the tolerance between the adults around food and, in the long term, strengthen social bonds among them. In chimpanzees, the temporal separation between adult–immature play and affiliation between the adult playmate and the adult related to the immature is due to the virtual absence of adult social play in this species. Similarly, the playful side of bonobos permits the adults to affiliate through play.

Expanding the concept: play and tolerance in other species

Both in human and non-human animals, play at all ages can be viewed as a socially and culturally moulded behavior. To be fully expressed, play

Figure 5.5 An adult female jumps over an adult male while performing a full play face. (Photo by E. Demuru)
(Une femelle adulte saute par-dessus un mâle adulte en faisant un visage de jeu.)

requires plastic and versatile societies where the inter-individual relationships have a high degree of freedom and are not bridled into codified schemes or crystallized roles. Therefore, variability in play often reflects important parameters of social systems such as cooperation, parental behavior, tolerance and affiliation (Gosso et al., 2005; Norbeck, 1974; Palagi, 2011; Thierry et al., 2000).

Children differ in R&T with boys playing more frequently and roughly than girls (Pellegrini, 2009; Rose and Rudolph, 2006). This gender difference in play may reflect differences in gender norms in the society. In children living in industrial and many pre-industrial societies, gender differences in R&T become evident at about three years of age although the magnitude of sex difference in R&T varies across cultures (Eibl-Eibesfeldt, 2007; Gray, 2009; Whiting and Edwards, 1973). Fouts and co-workers (2013) found that gender segregation in play is strongly rooted in those systems based on high levels of rigidity in social relationships. For example, gender segregation becomes more prominent among the children of Bofi farmers (an African population subsisting on slash-and-burn horticulture and trading) than those of Bofi foragers (a semi-nomadic African population subsisting on net hunting and gathering). The two Bofi populations show profound cultural differences in managing social relationships probably due to their different patterns of subsistence. Being more egalitarian, the Bofi foragers do not typically base their status or power on age or gender. They support personal autonomy and value sharing by daily engaging in cooperative behaviors within and between households. From a cultural point of view, Bofi foragers expect individuals to share every item requested and support such sharing by rough joking and storytelling (Fouts et al., 2013). On the other hand, in Bofi farmers, physical and verbal play are discouraged or inhibited thus reflecting more crystallized social roles as a function of status and gender (Fouts et al., 2013). Therefore, it appears clear that sex differences in animal and human play are expected whenever males and females differ in their physical, behavioral or social features (Byers and Walker, 1995; Fagen, 1981; Špinka et al., 2001) since play can be used differently by adult males and females according to their role within the society (Palagi, 2006).

The higher the level of egalitarianism, the lower the play differences according to gender.

Similar sex differences in play are found in non-human primates. Generally, in catarrhine monkeys and apes, juvenile males engage more often in R&T play than females (Maestripieri and Ross, 2004; Mendoza-Granados and Sommer, 1995; Owens, 1975). This may reflect sex differences in adult behavior and morphology. No sex biases in rough-and-tumble play are exhibited in tamarins, *Saguinus oedipus* (Cleveland and Snowdon, 1984), and ring-tailed lemurs, *Lemur catta* (Gould, 1990; Norscia and Palagi, 2016; Palagi, 2009), which are female-centred species showing a virtual absence of sexual dimorphism.

The intra-sexual comparison between the two *Pan* species reveals a difference in play distribution. While bonobo and chimpanzee males show similar frequencies in their playful activity, bonobo females (mean hourly frequency 0.200 ±0.038 SE) play much more than chimpanzee females (mean hourly frequency 0.039 ±0.009 SE) (Demuru and Palagi, in preparation). This difference can be interpreted in the light of the social network characterizing the two *Pan* species which do not show high levels of sexual dimorphism. Since bonobos live in a female-bonded society, it is not surprising that female relationships with all other group members can also be mediated by play, an activity which gathers subjects in a highly relaxed context. That is the reason why we observed high frequencies of play especially between females (Palagi, 2006).

Similarly, in macaques play seems to reflect the nature of the differences in social networks of species. For example, *Macaca tonkeana* and *M. fuscata* are two species at the opposite extremes in the social tolerance gradient of the genus (Thierry et al., 2000), and they strongly differ in the distribution of social play according to the gender and age of the players. In the tolerant Tonkean macaque, play is frequently engaged, not only by immature subjects but also among adults, which do not show any preference for play partners according to their age. Play segregation as a function of sex occurs in the despotic *M. fuscata* but not in *M. tonkeana* (Ciani et al., 2012). In Tonkean macaques, immature social play is affected by the degree of the mothers' permissiveness which, in turn, is strictly linked to the adult tolerance levels

(Ciani et al., 2012). In Japanese macaques, mothers show high levels of protectiveness that limit their infants' social contacts with peers and adults, thus inhibiting R&T. This strict control on infants limits their relational sphere, thus creating the conditions for social canalization, a process that strongly limits the amplitude of an individual's social network (Berman, 1982). The limitation of social play interactions during the immature phase also leads to a reduced propensity to play later in life, both in juveniles and in adults. Ciani and colleagues (2012) demonstrated that Tonkean macaques, compared to Japanese macaques, retain some traits typical of the juvenile phase, such as play and tolerance, in adulthood. It would be extremely interesting to understand to what extent such behavioral differences are due to phylogenetic or ontogenetic processes. One way of pondering the relative weight of the genetic and environmental components of a specific behavioral trait is to perform cross-fostering experiments. This methodology consists in removing the newborns from their biological parents at birth and raising them with surrogates parents, normally belonging to a closely related species which shows differences in the behavior under investigation compared to the natural parents. Such an experiment has been performed with macaques by de Waal and Johanowicz (1993) in order to discover the role of social experience in determining the reconciliation levels. These authors chose to compare the stumptail macaque (*Macaca arctoides*), that shows high levels of reconciliatory contacts after conflicts, with the rhesus macaque (*M. mulatta*), in which this behavior is uncommon. Results showed a threefold increase in the proportion of reconciled fights in rhesus macaque infants raised by stumptail macaque parents. Therefore, de Waal and Johanowicz (1993) suggested that the social style of a despotic species can be modified through contact with a species characterized by a more relaxed dominance style. Would the same result be obtained for play behavior in Japanese and Tonkean macaques?

The natural selection for reduced aggressiveness within a species, or 'self-domestication' (Hare et al., 2012), could be responsible for the retention of juvenilized traits into adulthood. These authors suggest that variation in tolerance may account for the differences in behavior between bonobos (who exhibit the 'self domestication syndrome') and chimpanzees and suggested that this theoretical concept could also apply to the more egalitarian species of macaques. Indeed, Tonkean macaques share with bonobos many behavioral characteristics such as social cohesiveness (De Marco et al., 2014; Palagi et al., 2014; Thierry et al., 1994), low competitive relationships (Petit et al., 1992), a large use of facial expressions (Scopa and Palagi, 2016; Thierry et al., 1989) and a relaxed and playful mood of infant nurturing (Ciani et al., 2012; Petit et al., 2008; Reinhart et al., 2010).

Conclusions

The self-rewarding nature, the high degree of freedom in the performance of particular actions, the necessity to appropriately emit and decode emotional and intentional signals and the social competence required to correctly manage each session are all essential elements characterizing play. Data coming from different species suggest that the full expression of play behavior is allowed and enhanced in species living in tolerant societies to such an extent that tolerance and playful attitude cannot be disentangled. This assumption is reinforced by data coming from phylogenetically distant species that share common features in their lifestyle and the origin of such differences will most likely be revealed by the study of early development. Of particular value may be mother–infant pair bonding and the ontogeny of the first playful interactions between infants and mothers and between infants and other adult subjects, both related and unrelated. Early infancy is probably a critical time window in which the degree of trust in others grows and gives the infants the opportunity to freely develop their emotional competence. Moreover, within the same species, the variable maternal style depending on rank position of the mothers, their experience with infants (primiparous or multiparous females) and the degree of permissiveness towards the infant can be all factors strongly influencing the future playful attitude of the subjects as well as their social and cognitive skills. Therefore, the comparative approach applied to both closely and more distantly related species can provide valuable hints about particular behavioral phenomena, especially when those species share analogous social correlates mediated both by biology and culture.

Acknowledgments

We are grateful to Brian Hare and Shinya Yamamoto for their kind invitation to contribute to this book. We wish to thank Marc Bekoff, Gordon Burghardt, Giada Cordoni, Matthew Campbell, Sergio Pellis, Barbara Smuts, Hillary Fouts, Marek Špinka and Frans de Waal for sharing with us some basic ideas on the concept of tolerance, play, cooperation and fairness in animals. We want to thank the parks that permitted us to collect data on their bonobo colonies over years: Apenheul Primate Park (Apeldoorn, The Netherlands), La Vallée des Singes (Romagne, France), Wilhelma Zoo (Stuttgart, Germany) and Frankfurt Zoo (Frankfurt, Germany). Finally, Elisabetta Palagi is grateful to all the colleagues of the NIMBioS Working Group (www.nimbios.org/workinggroups/WG_play) for the stimulating input on one of the most controversial behaviors an ethologist can encounter.

References

Aldis, O. (1975). *Play Fighting*. New York, NY: Academic Press.
Behncke, I. (2015). Play in the Peter Pan ape. *Current Biology*, 25(1), R24–R27.
Bekoff, M. (2001). Social play behaviour. Cooperation, fairness, trust, and the evolution of morality. *Journal of Consciousness Studies*, 8(2), 81–90.
Berman, C.M. (1982). The ontogeny of social relationships with group companions among free-ranging infant rhesus monkeys: II. Differentiation and attractiveness. *Animal Behaviour*, 30, 163–70.
Boesch, C., Hohmann, G., and Marchant, L.F. (2002). *Behavioural Diversity in Chimpanzees and Bonobos*. Cambridge: Cambridge University Press.
Burghardt, G.M. (2005). *The Genesis of Animal Play: Testing the Limits*. Cambridge, MA: MIT Press.
Byers, J.A. and Walker, C. (1995). Refining the motor training hypothesis for the evolution of play. *American Naturalist*, 146(1), 25–40.
Cattaneo, L. and Pavesi, G. (2014). The facial motor system. *Neuroscience and Biobehavioral Reviews*, 38, 135–59.
Ciani, F., Dall'Olio, S., Stanyon, R., and Palagi, E. (2012). Social tolerance and adult play in macaque societies: a comparison with different human cultures. *Animal Behaviour*, 84(6), 1313–322.
Clay, Z. and de Waal, F.B.M. (2013). Bonobos respond to distress in others: consolation across the age spectrum. *PloS One*, 8(1), e55206.
Cleveland, J. and Snowdon, C.T. (1984). Social development during the first twenty weeks in the cotton-top tamarin (*Saguinus o. oedipus*). *Animal Behaviour*, 32(2), 432–44.
Cordoni, G. (2009). Social play in captive wolves (*Canis lupus*): not only an immature affair. *Behaviour*, 146(10), 1363–385.
Darwin, C. (1872). *The Expression of the Emotions in Man and Animals*. Oxford: Oxford University Press.
Davila-Ross, M., Owren, M.J., and Zimmermann, E. (2009). Reconstructing the evolution of laughter in great apes and humans. *Current Biology*, 19, 1106–111.
De Marco, A., Sanna, A., Cozzolino, R., and Thierry, B. (2014). The function of greetings in male Tonkean macaques. *American Journal of Primatology*, 76(10), 989–98.
Demuru, E., Ferrari, P.F., and Palagi, E. (2014). Emotionality and intentionality in bonobo playful communication. *Animal Cognition*, 18(1), 333–44.
Demuru, E. and Palagi, E. (2012). In bonobos yawn contagion is higher among kin and friends. *PLOS One*, 7(11), e49613.
Eibl-Eibesfeldt, I. (2007). *Human Ethology*. New York, NY: Transaction Publishers.
Fagen, R. (1981). *Animal Play Behavior*. New York, NY: Oxford University Press.
Flack, J.C., Jeannotte, L.A., and de Waal, F.B.M. (2004). Play signaling and the perception of social rules by juvenile chimpanzees (*Pan troglodytes*). *Journal of Comparative Psychology*, 118, 149–59.
Fouts, H.N., Hallam, R.A., and Purandare, S. (2013). Gender segregation in early-childhood social play among the Bofi foragers and Bofi farmers in Central Africa. *American Journal of Play*, 5(3), 333–56.
Furuichi, T. (2011). Female contributions to the peaceful nature of bonobo society. *Evolutionary Anthropology*, 20(4), 131–42.
Genty, E., Breuer, T., Hobaiter, C., and Byrne, R.W. (2009). Gestural communication of the gorilla (Gorilla gorilla): repertoire, intentionality and possible origins. *Animal Cognition*, 12, 527–46.
Genty, E. and Byrne, R.W. (2010). Why do gorillas make sequences of gestures? *Animal Cognition*, 13, 287–301.
Gosso, Y., Otta, E., Morais, M., and Ribeiro, F.B.V. (2005). Play in hunter–gatherer society. In: Pellegrini, A.D. and Smith, P.K. (eds). *The Nature of Play*. New York, NY: Guildford Press, pp. 213–54.

PART III
Mind and Communication

PART III

Mind and Communities of Work

CHAPTER 6

Does the bonobo have a (chimpanzee-like) theory of mind?

Christopher Krupenye, Evan L. MacLean and Brian Hare

Abstract Theory of mind—the ability to reason about the thoughts and emotions of others—is central to what makes us human. Chimpanzees too appear to understand some psychological states. While less is known about bonobos, several lines of evidence suggest that the social-cognitive abilities of the two sister taxa may differ in key respects. This chapter outlines a framework to guide future research on bonobo social cognition based on the predictions of two potentially complementary hypotheses. The self-domestication hypothesis suggests that selection against aggression and for prosociality in bonobos may have impacted the ontogeny of their social-cognitive skills relative to chimpanzees. The empathizing–systemizing hypothesis links degree of prenatal brain masculinization, a potential result of self-domestication, to adult cognition. Specifically, relative feminization may yield more flexible theory of mind skills in bonobos than chimpanzees. Finally, directions for future study, including development of new paradigms that maximize ecological validity for bonobos, are discussed.

La théorie de l'esprit—le pouvoir de raisonner les pensées et émotions des autres—est centrale à notre nature humaine. Il parait que les chimpanzés peuvent comprendre quelques états psychologiques. Tandis que nous savons moins des bonobos, plusieurs témoignages suggèrent que les capacités socio-cognitives des deux taxons soeur peuvent différer dans des aspects clefs. Nous traçons un cadre pour guider les prochaines recherches sur la cognition sociale des bonobos, basé sur les prédictions de deux hypothèses potentiellement complémentaires. L'hypothèse d'auto-domestication suggère que l'anti-agression et la prosocialité des bonobos a influé leur ontogenèse et leur capacités socio-cognitives relativement aux chimpanzés. L'hypothèse d'empathie systématique (Empathizing–Systemizing) forme un lien entre le degré de masculinisation prénatale du cerveau, le résultat potentiel d'auto-domestication, et la cognition adulte. Spécifiquement, la féminisation relative génère des théories de l'esprit plus flexibles chez les bonobos que chez les chimpanzés. Enfin, nous discutons le directions pour les prochaines études, inclut le développement de nouveaux paradigmes qui maximisent la validité écologique des bonobos.

Introduction

Humans are adept at making inferences about the psychological states of others, including their emotions, desires and beliefs. These inferences then inform our attempts to cooperate, resolve conflicts and communicate effectively. Such *theory of mind* (ToM) abilities, as they are known, were long believed to be unique to humans but there is now evidence that in some contexts chimpanzees are capable of attributing to others a range of psychological states (Hare, 2011; Tomasello and Call, 1997). Less work has been accomplished with bonobos but even the limited comparisons that are now possible between the two species are helping to test hypotheses about how such skills might evolve across species, including our own (Hare, 2011).

In 1978, Premack and Woodruff first posed the question, 'Does the chimpanzee have a theory of mind?' At the time, chimpanzees were a clear choice for the first comparative investigations of human social cognition. Humans have two closest relatives. However, bonobos were discovered much later than chimpanzees, in large part because they inhabit a much more restricted range, south of the Congo River, compared to chimpanzees'

Krupenye, C., MacLean, E. L., and Hare, B., *Does the bonobo have a (chimpanzee-like) theory of mind?* In: *Bonobos: Unique in Mind, Brain, and Behavior*. Edited by Brian Hare and Shinya Yamamoto: Oxford University Press (2017).
© Oxford University Press. DOI: 10.1093/oso/9780198728511.003.0006

Figure 6.1 Several lines of evidence suggest that bonobos may be more empathic than chimpanzees but further social cognitive comparisons are critical. (Plusieurs sources de données suggèrent que les bonobos montreraient plus d'empathie que les chimpanzés, mais d'autres comparaisons cognitives sociales relativisent cette observation.)

distribution throughout East, Central and West Africa. Bonobos were still barely known to science in 1978 while chimpanzees were relatively common in captivity and were already the subject of several captive and field research programs. However, even after field sites were established to study bonobos in the wild, they remained little studied in captivity.

We currently find ourselves at an exciting moment in bonobo research since there has been more work on bonobo cognition in the past ten years than the previous hundred. For the first time, quantitative and qualitative comparisons between bonobos and chimpanzees are possible. Here we summarize what has been learned about ToM in chimpanzees since Premack and Woodruff's seminal investigations. We then present two hypotheses relevant to the selective pressures that have shaped—and proximate mechanisms that underlie—ape social cognition. From these hypotheses, we generate predictions about possible differences in social cognition between chimpanzees and bonobos. Finally, we evaluate our hypotheses against existing data and discuss the most pressing directions for future research (Figure 6.1).

Theory of mind in chimpanzees

Historical perspective

Premack and Woodruff (1978) showed videos to a human-raised chimpanzee, Sarah. These videos depicted humans facing a variety of problems (e.g. a banana was out of reach, or the actor shivered because the heater was broken). Sarah was then allowed to choose among photographs of various objects, only one of which would solve the problem presented in the video (e.g. a stick to retrieve the banana, or a lit wick for the heater). Sarah consistently chose the correct photograph, leading the authors to conclude that she understood the goals of the actors in the videos. Later, Povinelli and colleagues (1990) reported a study in which chimpanzees discriminated between a knowledgeable and an ignorant informant and selectively followed pointing gestures from the knowledgeable informant in order to find hidden food. However, Heyes (1993) offered compelling alternative interpretations for existing data on animal mindreading, and in the years that followed a number of studies reported that chimpanzees did not follow the communicative cues of a knowledgeable human or conspecific when searching for food. These data led Tomasello and Call (1997) to conclude, in their landmark book *Primate Cognition*, that 'there is no solid evidence that nonhuman primates understand the intentionality or mental states of others' (p. 340). ToM was therefore an aspect of human cognition in which the differences between humans and other animals were thought to be qualitative, rather than quantitative, in nature.

This view began to change when Hare (2001) noted that primate social life is highly competitive and suggested that since primate cognitive abilities have most likely been adapted for such a setting, it may be in competitive situations that primates are most motivated to exhibit their cognitive skills. Additionally, Tomasello and colleagues (2003) suggested that chimpanzees' failure to follow the communicative cues of a knowledgeable informant may not owe to a lack of ToM abilities in general but rather to a specific inability to understand the cooperative intentions underlying such communicative cues. If chimpanzees do not produce cooperative communicative cues themselves, why should they understand such cues when others produce them? Through more ecologically valid paradigms—that used some combination of spontaneous behavioral measures, larger sample sizes and competitive social contexts—substantial evidence for chimpanzees' understanding of others' perception, knowledge and goals has accumulated.

Understanding of perception

Much work has explored what chimpanzees understand of others' perception. This research has largely focused on *seeing* and to some extent *hearing*. Like many species, chimpanzees are capable of following the gaze of conspecific or human social partners (Brauer et al., 2005; Povinelli and Eddy, 1996; Tomasello et al., 1998). Gaze following could reflect an appreciation of the other's visual perspective or it could be a simple reflexive response that evolved for its clear adaptive benefits (Tomasello et al., 1998). Co-orienting to match the target of another's attention may alert the gaze follower to important resources, such as food or mates, as well as threats, such as predators or aggressive conspecifics. In chimpanzees at least, gaze following appears to be more than just a reflexive response. This species follows gaze geometrically, around barriers and checks back with the actor when it cannot identify the target of her gaze (Brauer et al., 2005; Tomasello et al., 1999). Chimpanzees are also sensitive to an agent's attentional state. For example, when begging for food, they spontaneously use more gestures when their target's face or body is oriented towards them than when it is oriented away (Kaminski et al., 2004). When attempting to communicate with an agent who is facing away from them, chimpanzees will walk around to the front of the agent before attempting to communicate (Liebal et al., 2004). They also shift their use of gestural versus vocal communication depending on whether their target is facing towards or away from them (Hostetter et al., 2001, Leavens et al., 2004).

To test the idea that chimpanzees can assess what another individual can see, Hare and colleagues (2000) examined how chimpanzees compete against one another over food in an experimental setting. Two chimpanzees were situated in separate testing rooms on either side of a central room where food was hidden. Each could see her competitor on the other side of the central room. Dominant individuals are able to monopolize visible food. However, in each experiment, there was a critical condition in which both pieces of food were visible to the subordinate competitor but only one of the pieces was visible to the dominant in the pair. In these conditions, subordinates preferentially targeted the food that the dominant could not see. In these crucial tests subordinates were given a slight head-start, and could not see the dominant when they made their choice of direction to approach. This ensured that subjects were not simply responding to the behavior or gaze direction of their competitor. These studies provided the first compelling evidence that chimpanzees may be sensitive to others' visual perspectives.

Building on this initial success, competitive experiments were then designed to examine whether apes can intentionally manipulate what others perceive (Hare et al., 2006). Chimpanzees were given the opportunity to steal food from a human competitor who pulled the food out of reach if he detected the subject's approach. Chimpanzees were then given a choice between paths that either visually concealed or exposed their approach. The chimpanzees spontaneously preferred a hidden path avoiding the competitor's face, behind an opaque barrier rather than a transparent one, and behind a complete barrier rather than a partial one. They also used indirect approaches in which they initially moved away from the food and out of the view of their competitor before they approached the food behind an occluder. In a similar setup in which chimpanzees could remain concealed until the final step of their approach in which they needed to reach through an opaque or transparent tube to retrieve the food, they reliably chose the opaque tube that concealed their reach (Melis et al., 2006).

Further, Karg and colleagues (2015a) showed that chimpanzees strategically manipulate whether they themselves can be seen by another agent and what other features of the environment the agent can see. Food was located in several small trays. The chimpanzees could move the individual trays behind or in front of an occluder, but could not retrieve the food themselves. Subjects left food concealed more often when interacting with a competitor who would steal the food than when interacting with a cooperator who would retrieve it for them. They exposed more food to the cooperator than the competitor; however, they did not actively hide visible food from the competitor.

Research has also explored whether chimpanzees are sensitive to what others can hear. Melis and colleagues (2006) showed that, when visually concealed from their competitor, chimpanzees preferred to steal food using an approach that was silent over one that that was noisy. However, Brauer and co-workers (2008) found that when chimpanzees competed with conspecifics for hidden food in a manner similar to that reported by Hare and colleagues (2000), they did not appear to infer which piece of food their competitor was aware of based on what their competitor could hear. They were equally likely to approach a piece of food that their competitor could not see but could hear during baiting and one that their competitor could neither see nor hear.

To determine whether chimpanzees were responding to others' visual perspectives and not just using a behavior rule (e.g. that agents pursue anything in their environment as long as a direct line can be drawn between the agent and the target, Heyes, 1993; Heyes, 1998; Penn and Povinelli, 2007, but see Table 1 in Hare, 2011 for a summary of evidence against this and other alternative explanations), critics proposed giving subjects unique self-experience with unfamiliar equipment that alters one's perceptual access. Then these subjects could be tested to evaluate whether their self-experience aided their ability to correctly predict the effect of this same equipment on the visual experience of others (Heyes, 1998). To do so, Karg and colleagues (2015b) used a competitive setup very similar to Melis and co-workers (2006). Chimpanzees attempted to steal food from a human experimenter who could not see them until they reached their hand into the tubes where the food was located. From a distance, both tubes appeared to block the human's view but one was actually see-through. Before the test phase, subjects were shown peanuts behind each surface so that they could experience the occlusive properties of the tubes. In the test phase, they avoided the see-through tube, even though from afar both looked identical. These results suggest that chimpanzees were able to infer what their competitor could see based on their own self-experience with these novel occluders.

Attributing knowledge, ignorance and false belief to others

In addition to reasoning about what others can perceive in the present, experimental evidence suggests that chimpanzees are capable of understanding what others *have perceived* in the past. In a competitive paradigm, subordinate chimpanzees were more likely to approach hidden food when their competitor was uninformed or misinformed about its location than when he was knowledgeable (Hare et al., 2001). Similarly, chimpanzees were reluctant to approach food when their competitor had witnessed it being hidden but not if their competitor was switched out for a new dominant individual who had not seen the baiting process (Hare et al., 2001). In the 'chimp chess' paradigm in which food was baited in several cups and two chimpanzees took turns searching for that food—each blind to the other's choices—subjects adjusted their behavior depending on whether their competitor was knowledgeable or ignorant of the food's location (Kaminski et al., 2008). When the subject chose second, she avoided food that her competitor knew about (inferring that the competitor would have already retrieved this item). MacLean and Hare (2012) presented chimpanzees with a series of experiments in which subjects witnessed an experimenter become aware of or remain ignorant to an object that was placed in his vicinity (by visually orienting to it or not). The experimenter later looked in the direction of the object and mimicked surprise.

Subjects recognized when the experimenter already knew about the object, and in these cases were more likely to follow the experimenter's line of sight beyond the object in search of an alternate target of his attention. Finally, in a field experiment, Crockford and colleagues (2012) showed that chimpanzees in the presence of a snake were more likely to alert an incoming groupmate to the snake if she had not yet seen it than if she was already knowledgeable. Together, these studies show that chimpanzees can discriminate between knowledgeable and ignorant individuals, they have different expectations of how these agents will behave, and they use this information both to outcompete ignorant groupmates and potentially even to inform them of danger.

To-date, however, no experimental demonstration is consistent with chimpanzees explicitly attributing false beliefs to others (Call and Tomasello, 1999; Hare et al., 2001; Kaminski et al., 2008; Krachun et al., 2009). A false belief exists when an individual believes something that is different from reality. Unlike knowledge or ignorance, understanding false beliefs requires representing not only whether or not the individual is aware of the true state of the world but also specifically the way in which she construes the world to be different than it is. In humans, this ability appears to be tightly linked to language development (Milligan et al., 2007). In one study with chimpanzees, an experimenter baited food in one of two containers out of view of the subject but within view of a human competitor (Krachun et al., 2009). In the critical condition, the competitor then either turned around or left the room and the experimenter switched the locations of the cups. When the competitor returned she reached effortfully for the now incorrect location. While in one experiment chimpanzees tended to look at the baited container in false belief trials, they still followed the competitor's reaching gesture and selected the incorrect cup. Although their looking behavior may reflect some implicit understanding of belief, unlike human children, chimpanzees were unable to act on this understanding.

Recognizing others' goals and intentions

A variety of experiments suggest that chimpanzees and other apes have some understanding of others' goals and intentions. Chimpanzees treat the actions of a human, but not a mechanical claw, as goal-directed and anticipate the outcomes of these goal-directed actions (Kano and Call, 2014b; Myowa-Yamakoshi et al., 2012). They correctly identify an actor's goal based on the context of his actions (Buttelmann et al., 2012). Chimpanzees are also able to complete others' failed actions (Tomasello and Carpenter, 2005). They provide targeted help to conspecific and human partners based on a specific understanding of their partner's needs (Melis and Tomasello, 2013; Warneken et al., 2007; Yamamoto et al., 2012). Chimpanzees also discriminate intentional from accidental actions (Call and Tomasello, 1998; Tomasello and Carpenter, 2005) and respond differently to an experimenter when he is unwilling to help them versus unable to do so, even though the experimenter performs very similar actions in each case (Call et al., 2004). Taken together, these studies suggest that chimpanzees have a robust understanding of the goals and intentions that govern others' actions.

Given the cooperative nature of human society, humans are also known to exhibit shared intentionality (Tomasello et al., 2005). Shared intentionality describes our ability to hold, and attribute to others, shared goals and intentions that structure collaborative or joint activities. For example, when performing a collaborative activity, all collaborators are aware of their shared goal as well as each other's complementary roles required to accomplish the goal. Although chimpanzees engaged in triadic turn-taking activities with humans and even elicited further interaction after the activities were terminated, it remains unknown if these activities are supported by an understanding of shared intentions (MacLean and Hare, 2013; Warneken et al., 2006). Some evidence suggest that chimpanzees are aware of their partners' role in collaborative activities and even the means that are necessary for the partner to accomplish that role, suggesting that differences in motivation rather than cognition may underlie species differences in shared intentionality (Melis and Tomasello, 2013).

Summary

Chimpanzees have at least a basic understanding of others' perspectives: they know *whether* or not an agent can see or hear a stimulus. They also know whether or not an agent is knowledgeable or ignorant about that stimulus, and they can reason about the goals and intentions that underlie agents' actions. However, they may not possess explicit comprehension of others' false beliefs, or the cooperative motivations that are necessary for shared intentionality. This cognitive profile may also be shared by the last common ancestor of humans and chimpanzees. However, such an evolutionary inference requires either assuming commonality between bonobo and chimpanzee ToM or that the last common ancestor was more similar cognitively to chimpanzees than bonobos.

Bonobo cognitive evolution

The singular focus on chimpanzees as a model for human cognitive evolution is problematic because, like humans, chimpanzees too have experienced 5–7 million years of evolution since our lineages diverged. Likewise, bonobos and chimpanzees have experienced around a million years of independent evolution. Although the general assumption has often been that bonobos will perform similarly to chimpanzees on most cognitive tasks, this is not always the case (Herrmann et al., 2010). Only through studying the three species can we confidently infer which aspects of our cognition are shared evolutionarily, and which are unique to each species. Perhaps even more importantly, there is much reason to believe that bonobos will offer some very different clues than chimpanzees about human evolutionary history (Hare, 2011). Below, we outline a framework that guides our analysis of existing comparative studies and our suggested directions for future work. We describe the self-domestication and empathizing–systemizing hypotheses, which complement one another and provide testable ultimate and proximate explanations for differences in chimpanzee and bonobo social cognition (Figure 6.2).

The self-domestication hypothesis

The self-domestication hypothesis is based on the observation that domestication—which frequently involves artificial selection against aggression—consistently produces a suite of changes in morphology, physiology, behavior and cognition relative to a species' wild counterpart (Hare et al.,

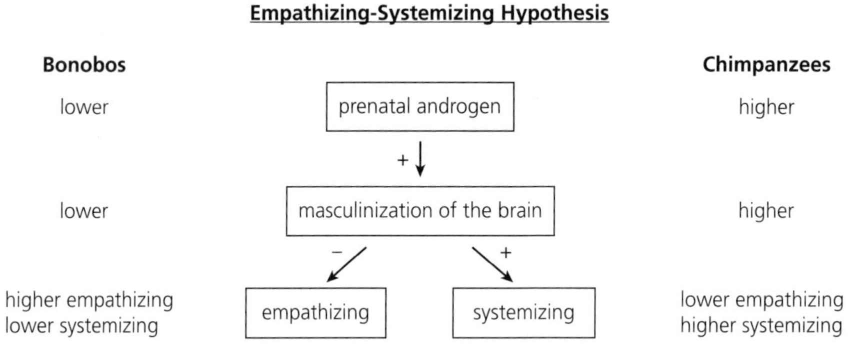

Empathizing-Systemizing Hypothesis

Bonobos		Chimpanzees
lower	prenatal androgen	higher
lower	masculinization of the brain	higher
higher empathizing lower systemizing	empathizing systemizing	lower empathizing higher systemizing

Bonobos	Predictions	Chimpanzees
higher	Social Tolerance	lower
higher	Theory of Mind Skills	lower
lower	Casual Reasoning and Tool Use	higher
lower	Signs of Masculinization	higher

Figure 6.2 Diagram and predictions of the Empathizing–Systemizing hypothesis.
(Diagramme et prédictions de l'hypothèse Empathizing-Systemizing.)

2012). Importantly, in many cases of domestication, including that of dogs from wolves, a phase of self-domestication has been proposed to have preceded domestication by artificial selection. During self-domestication of dogs, natural selection most likely favoured individuals who were more socially tolerant and less fearful of humans, as these individuals could live in closer proximity to humans and exploit new resources such as human refuse. In this way, selection against aggression is thought to have exerted pressures on canine evolution through natural selection even before the intervention of artificial selection.

Relative to chimpanzees, bonobos exhibit many of the characteristic changes associated with the domestication syndrome, and are hypothesized to have been self-domesticated via natural selection for prosociality and against extreme forms of aggression (Hare et al., 2012). Bonobos exhibit higher levels of tolerance while co-feeding than chimpanzees and consequently are more flexible cooperators in instrumental tasks (Hare et al., 2007). Whereas chimpanzees are highly xenophobic, bonobos are known to be prosocial even with strangers (Tan and Hare, 2013). Researchers have also documented species differences in apes' neuroanatomy and hormonal profiles and a delayed pattern of social development in bonobos relative to chimpanzees, which most likely constitute the proximate underpinnings of bonobos' greater social tolerance and prosocial flexibility (Rilling et al., 2012; Stimpson et al., 2015; Wobber et al., 2010a; Wobber et al., 2010b). Importantly, although both species live in similar large multi-male, multi-female, promiscuous, fission–fusion societies, bonobos are believed to have experienced more relaxed feeding competition than chimpanzees throughout their evolutionary history (Wrangham, 1993). Reduced competition may have allowed selection to favour increased social tolerance and ultimately produce the other correlated by-products of the self-domestication process.

From the self-domestication hypothesis, we can derive a key prediction about the social cognitive abilities of chimpanzees and bonobos. The self-domestication hypothesis posits heterochonic shifts—changes in the timing or pattern of development—as a major proximate mechanism underlying species differences (Hare et al., 2012). Specifically, domesticates show juvenilization or retention of paedomorphic traits into adulthood that in non-domesticates are only present in infants and juveniles. They also often exhibit delayed patterns of development. Since relaxed feeding competition is believed to have facilitated selection against aggression in bonobos, delays are expected to impact behaviors related to feeding competition. Such delays have been shown relative to chimpanzees in the ontogeny of social intolerance, social inhibition and spatial memory (Rosati and Hare, 2012; Wobber et al., 2010b).

Therefore, the chief prediction that we can draw from the self-domestication hypothesis is that there will be a relationship between delayed patterns of development and ToM abilities in bonobos. However, there are several competing hypotheses about the specific influence that delayed development will have on adult ToM. Additionally, it is not clear whether all ToM skills will be impacted in the same way.

One possibility is that social cognitive traits themselves will be delayed. Nevertheless, despite the delay, bonobos may eventually demonstrate comparable performance to chimpanzees on ToM tasks (e.g. as is the case with social inhibition, Wobber et al., 2010b). Alternatively, developmental delay may result in adult bonobos failing to show comparable social cognitive flexibility to chimpanzees (e.g. as is the case with spatial memory, Rosati and Hare, 2012). It is also possible that delayed development may result in a longer period of developmental plasticity and permit bonobos to develop skills that chimpanzees do not. Indeed, humans have evolved an extended period of juvenility and adolescence that is associated with greater cognitive flexibility (Kaplan et al., 2000). Consistent with this hypothesis, there is evidence that bonobos reach certain developmental milestones beyond the typical period of chimpanzee development. For example, chimpanzees experience changes in TT3 thyroid hormone levels at the end of their somatic growth, and bonobos exhibit juvenile levels of these hormones ten years longer than chimpanzees do (Behringer et al., 2014).

Another alternative is that social cognition may not have been directly shaped by self-domestication; however, traits that were shaped by selection against aggression may still influence the ontogeny of social cognition. For example, retention of juvenile levels of social tolerance into adulthood (Wobber et al., 2010b) may provide more opportunities for bonobos to learn about the behavior and mental states of their conspecifics, ultimately resulting in more flexible ToM skills in bonobos. Research in human children has already linked temperament to ToM, showing in particular that children who are less aggressive, more shy and more observant at age three perform better on false belief tasks at age five (Wellman et al., 2011). These findings also align with the empathizing–systemizing hypothesis' prediction that bonobos will exhibit more attuned ToM skills as a result of reduced masculinization, a trait which most likely arose through self-domestication (see later in this chapter).

The empathizing–systemizing hypothesis

A potentially complementary hypothesis—this one relevant to the proximate origins of chimpanzee and bonobo species differences—comes from the literature on autism spectrum disorders. This hypothesis, the empathizing–systemizing hypothesis, links degree of masculinization of the brain during prenatal development to differences in adult ToM skills. Variation in masculinization has been documented between chimpanzees and bonobos, perhaps resulting from self-domestication. According to the empathizing–systemizing hypothesis, this variation may lead to differences in ToM abilities between the two species (Figure 2).

The empathizing–systemizing hypothesis was originally developed to provide an explanation for why people with autism spectrum disorders are characterized by deficits in empathizing and average or above-average abilities in systemizing (Baron-Cohen, 2009). Empathizing in this context refers to the ability to understand others' mental states and to respond appropriately to their emotions. Systemizing refers to a propensity for creating and analysing predictable, rule-governed systems. Men tend to exhibit higher levels of systemizing and lower levels of empathizing than women, and the even more extreme pattern of high systemizing and low empathizing seen in autism spectrum disorder is thought to result from hyper-masculinization of the brain during prenatal development (Baron-Cohen, 2009).

Bonobos have more human-like 2D:4D ratios than chimpanzees, suggesting lower prenatal androgen exposure and less masculinization of the brain during development (McIntyre et al., 2009). If non-human apes adhere to the pattern of individual differences seen in humans, bonobos should exhibit greater skill in empathizing but less skill in systemizing than chimpanzees (MacLean, 2016). Neurological evidence further supports this claim: relative to chimpanzees, bonobos have more grey matter in the right dorsal amygdala and the right anterior insula—areas that function to perceive distress (Rilling et al., 2012). They also have stronger connections between the amygdala and the ventral anterior cingulate cortex, a pathway that contributes to aversion to others' harm, as well as greater serotonergic innervation of the amygdala than chimpanzees (Rilling et al., 2012; Stimpson et al., 2015). Relative to other apes, humans and bonobos also have a larger and more diversified posterior orbitofrontal cortex, which has been implicated in emotional responsiveness to social stimuli (Semendeferi et al., 1998). Thus, both physiological and neurological evidence support the empathizing–systemizing hypothesis and suggest that bonobos may be more sensitive to others' mental states than chimpanzees.

Recent human evolution has been characterized by a reduction in androgen activity, as reflected in the feminization of craniofacial morphology (Cieri et al., 2014). This shift is believed to partially explain our unusual levels of within-group social tolerance (Cieri et al., 2014). It may therefore be that the same forces involved in bonobo evolution have shaped the cognitive and behavioral phenotype of our species. If the predictions of the self-domestication and empathizing–systemizing hypothesis are supported in bonobos and chimpanzees, there is reason to believe that humans' unique social cognitive abilities also evolved, at least in part, through a process of feminization in response to selection for prosociality and against aggression.

Theory of mind in bonobos

The self-domestication hypothesis predicts developmental delays that may impact the emergence of bonobos' ToM abilities. Consistent with this prediction, MacLean and Hare (2012) found that chimpanzees succeeded in three related perspective-taking tasks involving reasoning about what an actor has previously seen whereas bonobos succeeded in only one. MacLean and Hare (2012) suggested that this differential performance might reflect a developmental delay in bonobos relative to chimpanzees (Wobber et al., 2010b). However, this potential species difference must be interpreted with extreme caution for several reasons. First, the bonobos tested in the task were, on average, three years younger than the chimpanzees and roughly half were non-adults. Second, in a control pre-test, bonobos did not respond as strongly as chimpanzees to human vocalizations, which were a key component of test stimuli, making it unclear whether the methods were simply more ecologically valid for chimpanzees. Beyond this study, which was not directly aimed at exploring developmental differences in chimpanzee and bonobo social cognition, there is no published work that directly addresses this question.

The principal prediction of the empathizing–systemizing hypothesis is that, due to reduced masculinization perhaps as a result of self-domestication, bonobos will show more flexibility than chimpanzees on ToM tasks and less on systemizing tasks. Herrmann and colleagues (2010) administered a cognitive test battery of 16 tasks to a large sample of bonobos and chimpanzees. As predicted by the empathizing–systemizing hypothesis, the results showed that chimpanzees are more skilled than bonobos on tasks requiring understanding of physical causality and tool use (traits associated with systemizing), whereas bonobos performed more skilfully than chimpanzees on tasks related to ToM, especially gaze following (traits linked to empathizing). However, a much wider range of tasks than presented in Herrmann and co-workers (2010) is needed to characterize more fully the differences in ToM abilities between chimpanzees and bonobos.

For example, bonobos consistently appear to be more sensitive than chimpanzees to certain types of social information. They gaze follow more reliably and more flexibly than chimpanzees (Herrmann et al., 2010), including in response to allospecific models (Kano and Call, 2014a). They also make more eye contact than chimpanzees (Kano et al., 2015), which in humans may be associated with greater sensitivity to others' mental states (e.g. Phillips et al., 1992) and in other species appears to be a product of domestication (Miklosi et al., 2003).

However, in most other ToM skills for which both species have been tested, their performance has been comparable (except MacLean and Hare, 2012, reviewed earlier). For example, in basic understanding of attention and agency, it appears that there are no differences between species. Both species, as well as the other great apes, interpret others' actions as goal-directed (Buttelmann et al., 2012; Kano and Call, 2014b). They also follow gaze geometrically and check back with the actor when they cannot identify the target of her gaze (Brauer et al., 2005; Povinelli and Eddy, 1996; Tomasello et al., 1999). In addition, all great apes will walk to the front of an agent before attempting to communicate with her (Liebal et al., 2004), and use more gestural communication when the recipient of their gestures is oriented towards them rather than away (Kaminski et al., 2004).

In object choice tasks where subjects must follow a human's gesture to locate hidden food, bonobos have always performed similarly to chimpanzees (Herrmann et al., 2010). The one study that tested bonobos on their explicit understanding of others' beliefs also did not find positive evidence in either bonobos or chimpanzees (Krachun et al., 2009).

In addition to controlled cognitive experiments, several studies have investigated in each species behaviors that may be related to empathy. Though no studies have quantitatively compared the two species in these behaviors, bonobos and chimpanzees both exhibit consolation behavior, affiliative contact towards victims of aggression, as well as contagious yawning (Campbell and de Waal, 2011; Demuru and Palagi, 2012; Fraser et al., 2008; Palagi and Norscia, 2013). These behaviors are generally directed at bond partners and close kin, and have been argued to reflect rudimentary forms of empathy.

Taken together, the existing qualitative and quantitative work on bonobo and chimpanzee ToM suggests comparable skills in both species. The major exception is that bonobos have shown greater sensitivity than chimpanzees to eye contact and gaze—foundational traits that are used for gathering social information that informs inferences about others' psychological states. Eye contact is also associated with lower levels of prenatal testosterone in human infants (Lutchmaya et al., 2002). Meanwhile, chimpanzees exhibit more skill in tasks relating to causal reasoning or reasoning about unobservable properties of the physical world. These data provide preliminary support for the empathizing–systemizing hypothesis that, within the *Pan* clade, chimpanzees are more the systemizers and bonobos the empathizers. These findings are also consistent with the self-domestication hypothesis. However, developmental comparisons (as well as comparisons between adults of each species) thus far suggest, for the most part, that if self-domestication has shaped bonobo ToM abilities, any developmental delay that bonobos experience may be overcome by adulthood through the developmental plasticity that such a delay affords. Future work will be essential in further testing the predictions of these hypotheses.

Moving forward

ToM abilities are among the defining features of our species yet there remain many unanswered questions about their evolutionary origins. Comparative research involving humans' two closest relatives, chimpanzees and bonobos, provides the most powerful tool for understanding the evolution of these skills in the lineage leading to humans. While work in chimpanzees has demonstrated that they possess many, but certainly not all, of the ToM skills that are exhibited by humans, many more gaps exist in our knowledge of these abilities in bonobos (see Table 6.1). These gaps hinder inferences about the cognitive phenotype of the last common ancestor of chimpanzees, bonobos and humans as well as the selective forces that shaped the cognitive abilities of each species.

We have outlined two hypotheses—the self-domestication hypothesis and the empathizing–systemizing hypothesis—that together provide testable predictions that will hopefully inspire future comparisons. The self-domestication hypothesis suggests that selection against aggression and for prosociality may have resulted in heterochronic changes that impact the emergence of ToM skills. These same selective pressures may

Table 6.1 Theory of mind skills in chimpanzees and bonobos. (Capacités de la théorie de l'esprit chez les chimpanzés et les bonobos.)

Understanding of others'	Chimpanzees	Bonobos
Gaze direction[1]	✓	✓
Attention[2]	✓	✓
Seeing[3]	✓	?
Hearing[4]	✓?	?
Knowledge and Ignorance[5]	✓	✓?
False beliefs (Implicit)[6]	✓?	✓?
False beliefs (Explicit)[7]	X	X
Goals[8]	✓	✓
Intentions[9]	✓	✓?
Shared goals[10]	?	?

1 (Brauer et al., 2005; Povinelli and Eddy, 1996; Tomasello et al., 1999);
2 (Hostetter et al., 2001; Kaminski et al., 2004; Leavens et al., 2004; Liebal et al., 2004);
3 (Hare et al., 2000; Hare et al., 2006; Karg et al., 2015a; Karg et al., 2015b; Melis et al., 2006);
4 (Brauer et al., 2008; Melis et al., 2006);
5 (Crockford et al., 2012; Hare et al., 2001; Kaminski et al., 2008; MacLean and Hare, 2012);
6 (Krupenye et al., 2016);
7 (Call and Tomasello, 1999; Hare et al., 2001; Kaminski et al., 2008; Krachun et al., 2009);
8 (Buttelmann et al., 2012; Kano and Call, 2014b; Melis and Tomasello, 2013; Myowa-Yamakoshi et al., 2012; Tomasello and Carpenter, 2005; Warneken et al., 2007; Yamamoto et al., 2012);
9 (Call et al., 2004; Call and Tomasello, 1998; Krupenye and Hare, in preparation; Tomasello and Carpenter, 2005);
10 (MacLean and Hare, 2013; Melis and Tomasello, 2013; Warneken et al., 2006).

also be responsible for reduced masculinization of the brain during prenatal development in bonobos relative to chimpanzees. The empathizing–systemizing hypothesis, in turn, suggests that these differences in masculinization may be responsible for greater flexibility in empathizing (i.e. ToM abilities) in bonobos and in systemizing (i.e. physical causal understanding) in chimpanzees.

There are several key areas that should be explored in order to test these hypotheses and provide the most direct insights into the evolutionary history of humans and their closest relatives. First, if the social cognitive abilities of these species have been shaped by a self-domestication process, delayed development relative to chimpanzees may impact the ToM skills of bonobos. Therefore, a critical test of this hypothesis requires a comparative developmental investigation of chimpanzee and bonobo ToM. Second, as a result of reduced masculinization (predicted by the self-domestication hypothesis), the empathizing–systemizing hypothesis predicts that bonobos will exhibit greater competence than chimpanzees in employment of some, if not all, ToM abilities. Thus, future work must address the existing gaps in our understanding of ToM in both species.

To address these gaps, several concurrent strategies are necessary. On the one hand, it will be important to replicate with bonobos ToM studies that have been completed with chimpanzees. For example, almost no work has investigated their perspective-taking skills, including their understanding of what others perceive and what others know. In other areas, such as understanding of others' goals and intentions, bonobos have only been tested on a minority of paradigms. It will also be critical to develop new paradigms that maximize ecological validity for bonobos. Previous experiments that have been designed to study social cognition in chimpanzees may not hold the same ecological validity for bonobos, and poor performance by bonobos in such tasks could reflect species differences in experimental motivation more than cognitive ability (with the reverse applying to chimpanzees for more bonobo-centric paradigms). Thus, a combination of existing paradigms and new approaches adapted for the unique characteristics of each species will be essential to fully capture the nuanced cognitive and motivational differences between them.

Since motivation and temperament interact with ToM abilities, many of the most successful studies of chimpanzee ToM have capitalized on competitive settings that maximize chimpanzees' motivation to demonstrate their cognitive skill (Hare et al., 2000; Hare et al., 2001). However, competitive contexts may not motivate bonobos in the same way that they do chimpanzees. Research has shown that whereas male chimpanzees experience a spike in testosterone preceding competition over food, male bonobos experience a spike in cortisol (Wobber et al., 2010a). These differential responses indicate that while chimpanzees are emboldened in anticipation of conflict, bonobos become stressed and are probably motivated to avoid, rather than engage in, competition. In contrast, bonobos have been shown to perform a handful of cooperative behaviors more flexibly than chimpanzees, owing to their heightened social tolerance (Hare et al., 2007; Tan and Hare, 2013). Therefore, it may be in noncompetitive contexts that bonobos are most motivated to demonstrate their cognitive skill. Future work should take advantage of such contexts to investigate social cognition in bonobos. For example, studies of targeted helping and sharing could shed light on bonobos' understanding of other's goals and intentions. Krupenye and colleagues (in preparation) have found that bonobos actively transfer food, even doing so proactively (without request by the recipient)—prosocial behaviors not exhibited by chimpanzees—although, unlike chimpanzees, bonobos do not transfer tools or other objects. These findings suggest that while bonobos and chimpanzees have a similar understanding of others' needs, motivational differences impact their tendency to demonstrate that understanding.

Human infants discriminate between agents based on their prosocial and antisocial actions and intentions (Hamlin, 2013), and these abilities may also be particularly pronounced in bonobos. Indeed, Krupenye and Hare (in preparation-a; in preparation-b) recently found evidence that bonobos are able to discriminate between unfamiliar social partners based on both their actions and intentions. Future research should continue to examine social cognition in cooperative settings, including perspective-taking abilities, cognitive and affective empathy and shared intentionality.

Finally, many areas of ToM are understudied in general. For example, little is known about apes' understanding of others' emotions, and with their heightened sensitivity to faces and gaze bonobos may be likely to excel in this area. Moreover, to-date no study has provided positive evidence of explicit false belief understanding in nonhumans, and researchers are just beginning to investigate implicit understanding of false belief. This new direction relies on the kinds of spontaneous anticipatory looking or violation of expectation measures that have proven fruitful in human infants (Baillargeon et al., 2010). For example, using an anticipatory looking paradigm, Krupenye, Kano and colleagues (2016) recently discovered that—just like human infants—bonobos, chimpanzees and orangutans successfully predicted that an agent would search for a hidden object where he last saw it, even if the object had been moved while the agent wasn't looking. While further work is necessary to confirm the mechanism underlying these results and identify its limits, Krupenye, Kano and colleagues (2016) data provide the first evidence that great apes may have at least an implicit understanding of others' false beliefs.

Future research on social cognition in bonobos and chimpanzees promises to yield exciting developments. Examining these traits will clarify whether social cognition is largely similar between the two species—that is, whether both species have a *Pan*-typical ToM. Alternatively, future work may demonstrate that the species' independent evolutionary histories have produced key differences in their social cognitive abilities. In many cases, this work will permit more accurate reconstructions of the cognitive phenotype of the last common ancestor by clarifying which social cognitive abilities are shared by bonobos, chimpanzees and humans. Additionally, by identifying cognitive differences between the three species, we will gain critical insights into the selective forces that have shaped cognitive evolution not only in our closest relatives but in humans as well.

References

Baillargeon, R., Scott, R.M., and He, Z. (2010). False-belief understanding in infants. *Trends in Cognitive Sciences*, 14, 110–18.

Baron-Cohen, S. (2009). Autism: the empathizing–systemizing (E–S) theory. *Annals of the New York Academy of Science*, 1156, 68–80.

Behringer, V., Deschner, T., Murtagh, R., Stevens, J.M., and Hohmann, G. (2014). Age-related changes in thyroid hormone levels of bonobos and chimpanzees indicate heterochrony in development. *Journal of Human Evolution*, 66, 83–8.

Brauer, J., Call, J., and Tomasello, M. (2005). All great ape species follow gaze to distant locations and around barriers. *Journal of Comparative Psychology*, 119, 145–54.

Brauer, J., Call, J., and Tomasello, M. (2008). Chimpanzees do not take into account what others can hear in a competitive situation. *Animal Cognition*, 11, 175–8.

Buttelmann, D., Schütte, S., Carpenter, M., Call, J., and Tomasello, M. (2012). Great apes infer others' goals based on context. *Animal Cognition*, 15, 1037–53.

Call, J., Hare, B., Carpenter, M., and Tomasello, M. (2004). 'Unwilling' versus 'unable': chimpanzees' understanding of human intentional action. *Developmental Science*, 7, 488–98.

Call, J. and Tomasello, M. (1998). Distinguishing intentional from accidental actions in orangutans (*Pongo pygmaeus*), chimpanzees (*Pan troglodytes*), and human children (*Homo sapiens*). *Journal of Comparative Psychology*, 112, 192–206.

Call, J. and Tomasello, M. (1999). A nonverbal false belief task: The performance of children and great apes. *Child Development*, 70, 381–95.

Campbell, M.W. and De Waal, F.B. (2011). Ingroup–outgroup bias in contagious yawning by chimpanzees supports link to empathy. *PLoS One*, 6, e18283.

Cieri, R.L., Churchill, S.E., Franciscus, R.G., Tan, J., and Hare, B. (2014). Craniofacial feminization, social tolerance, and the origins of behavioral modernity. *Current Anthropology*, 55, 419–43.

Crockford, C., Wittig, R.M., Mundry, R., and Zuberbuhler, K. (2012). Wild chimpanzees inform ignorant group members of danger. *Current Biology*, 22, 142–6.

Demuru, E. and Palagi, E. (2012). In bonobos yawn contagion is higher among kin and friends. *PLoS One*, 7, e49613.

Fraser, O.N., Stahl, D., and Aureli, F. (2008). Stress reduction through consolation in chimpanzees. *Proceedings of the National Academy of Sciences*, 105, 8557–62.

Hamlin, J.K. (2013). Failed attempts to help and harm: intention versus outcome in preverbal infants' social evaluations. *Cognition*, 128, 451–74.

Hare, B. (2001). Can competitive paradigms increase the validity of experiments on primate social cognition? *Animal Cognition*, 4, 269–80.

Hare, B. (2011). From hominoid to hominid mind: what changed and why? *Annual Review of Anthropology*, 40, 293–309.

Hare, B., Call, J., Agnetta, B., and Tomasello, M. (2000). Chimpanzees know what conspecifics do and do not see. *Animal Behaviour*, 59, 771–85.

Hare, B., Call, J., and Tomasello, M. (2001). Do chimpanzees know what conspecifics know? *Animal Behaviour*, 61, 139–51.

Hare, B., Call, J., and Tomasello, M. (2006). Chimpanzees deceive a human competitor by hiding. *Cognition*, 101, 495–514.

Hare, B., Melis, A.P., Woods, V., Hastings, S., and Wrangham, R. (2007). Tolerance allows bonobos to outperform chimpanzees on a cooperative task. *Current Biology*, 17, 619–23.

Hare, B., Wobber, V., and Wrangham, R. (2012). The self-domestication hypothesis: evolution of bonobo psychology is due to selection against aggression. *Animal Behaviour*, 83, 573–85.

Herrmann, E., Hare, B., Call, J., and Tomasello, M. (2010). Differences in the cognitive skills of bonobos and chimpanzees. *PloS One*, 5, e12438.

Heyes, C.M. (1993). Anecdotes, training, trapping and triangulating: do animals attribute mental states? *Animal Behaviour*, 46, 177–88.

Heyes, C.M. (1998). Theory of mind in nonhuman primates. *Behavioral and Brain Sciences*, 21, 101–14; discussion 115–48.

Hostetter, A.B., Cantero, M., and Hopkins, W.D. (2001). Differential use of vocal and gestural communication by chimpanzees (*Pan troglodytes*) in response to the attentional status of a human (*Homo sapiens*). *Journal of Comparative Psychology*, 115, 337–43.

Kaminski, J., Call, J., and Tomasello, M. (2004). Body orientation and face orientation: two factors controlling apes' begging behavior from humans. *Animal Cognition*, 7, 216–23.

Kaminski, J., Call, J., and Tomasello, M. (2008). Chimpanzees know what others know, but not what they believe. *Cognition*, 109, 224–34.

Kano, F. and Call, J. (2014a). Cross-species variation in gaze following and conspecific preference among great apes, human infants and adults. *Animal Behaviour*, 91, 137–50.

Kano, F. and Call, J. (2014b). Great apes generate goal-based action predictions: an eye-tracking study. *Psychological Science*, 25, 1691–8.

Kano, F., Hirata, S., and Call, J. (2015). Social attention in the two species of *Pan*: bonobos make more eye contact than chimpanzees. *PLoS One*, 10, e0129684.

Kaplan, H., Hill, K., Lancaster, J., and Hurtado, A. M. (2000). A theory of human life history evolution: diet, intelligence, and longevity. *Evolutionary Anthropology*, 9, 156–85.

Karg, K., Schmelz, M., Call, J., and Tomasello, M. (2015a). Chimpanzees strategically manipulate what others can see. *Animal Cognition*, 18, 1069–76.

Karg, K., Schmelz, M., Call, J., and Tomasello, M. (2015b). The goggles experiment: can chimpanzees use self-experience to infer what a competitor can see? *Animal Behaviour*, 105, 211–21.

Krachun, C., Carpenter, M., Call, J., and Tomasello, M. (2009). A competitive nonverbal false belief task for children and apes. *Developmental Science*, 12, 521–35.

Krupenye, B. and Hare, B. in preparation-a. Bonobos prefer individuals that hinder third parties over those that help.

Krupenye, C. and Hare, B. in preparation-b. Intentions outweigh outcomes in bonobos' social evaluations.

Krupenye, C., Kano, F., Hirata, S., Call, J., and Tomasello, M. (2016). Great apes anticipate that other individuals will act according to false beliefs. *Science*, 354, 110–14.

Krupenye, C., Tan, J., and Hare, B. in preparation. Bonobos directly transfer food but not their toys or tools.

Leavens, D.A., Hostetter, A.B., Wesley, M.J., and Hopkins, W.D. (2004). Tactical use of unimodal and bimodal communication by chimpanzees, *Pan troglodytes*. *Animal Behaviour*, 67, 467–76.

Liebal, K., Call, J., Tomasello, M., and Pika, S. (2004). To move or not to move: How apes adjust to the attentional state of others. *Interaction Studies*, 5, 199–219.

Lutchmaya, S., Baron-Cohen, S., and Raggatt, P. (2002). Foetal testosterone and eye contact in 12-month-old human infants. *Infant Behavior and Development*, 25, 327–35.

Maclean, E.L. (2016) . Unraveling the evolution of uniquely human cognition. *Proceedings of the National Academy of Sciences*, 113, 6348–54.

Maclean, E.L. and Hare, B. (2012). Bonobos and chimpanzees infer the target of another's attention. *Animal Behaviour*, 83, 345–53.

Maclean, E. and Hare, B. (2013). Spontaneous triadic engagement in bonobos (*Pan paniscus*) and chimpanzees (*Pan troglodytes*). *Journal of Comparative Psychology*, 127, 245–55.

Mcintyre, M.H., Herrmann, E., Wobber, V., Halbwax, M., Mohamba, C., De Sousa, N., Atencia, R., Cox, D., and Hare, B. (2009). Bonobos have a more human-like second-to-fourth finger length ratio (2D:4D) than chimpanzees: a hypothesized indication of lower prenatal androgens. *Journal of Human Evolution*, 56, 361–5.

Melis, A. P., Call, J., and Tomasello, M. (2006). Chimpanzees (*Pan troglodytes*) conceal visual and auditory information from others. *Journal of Comparative Psychology*, 120, 154–62.

Melis, A. P., and Tomasello, M. (2013). Chimpanzees' (*Pan troglodytes*) strategic helping in a collaborative task. *Biology Letters*, 9, 20130009.

Miklosi, A., Kubinyi, E., Topal, J., Gacsi, M., Viranyi, Z., and Csanyi, V. (2003). A simple reason for a big difference: Wolves do not look back at humans, but dogs do. *Current Biology*, 13, 763–6.

Milligan, K., Astington, J.W., and Dack, L.A. (2007). Language and theory of mind: Meta-analysis of the relation between language ability and false-belief understanding. *Childhood Development*, 78, 622–46.

Myowa-Yamakoshi, M., Scola, C., and Hirata, S. (2012). Humans and chimpanzees attend differently to goal-directed actions. *Nature Communications*, 3, 693.

Palagi, E. and Norscia, I. (2013). Bonobos protect and console friends and kin. *PLoS One*, 8, e79290.

Penn, D. and Povinelli, D. (2007). On the lack of evidence that non-human animals possess anything remotely resembling a 'theory of mind'. *Philosophical Transactions of the Royal Society of London B*, 362, 731–44.

Phillips, W., Baron-Cohen, S., and Rutter, M. (1992). The role of eye-contact in the detection of goals: Evidence from normal toddlers, and children with autism or mental handicap. *Development and Psychopathology*, 4, 375–83.

Povinelli, D.J. and Eddy, T.J. (1996). Chimpanzees: Join visual attention. *Psychological Science*, 7, 129–35.

Povinelli, D.J., Nelson, K.E., and Boysen, S.T. (1990). Inferences about guessing and knowing by chimpanzees (Pan troglodytes). *Journal of Comparative Psychology*, 104, 203–10.

Premack, D. and Woodruff, G. (1978). Does the chimpanzee have a theory of mind. *Behavioral and Brain Sciences*, 1, 515–26.

Rilling, J.K., Scholz, J., Preuss, T.M., Glasser, M.F., Errangi, B.K., and Behrens, T.E. (2012). Differences between chimpanzees and bonobos in neural systems supporting social cognition. *Social Cognitive and Affective Neuroscience*, 7, 369–79.

Rosati, A.G. and Hare, B. (2012). Chimpanzees and bonobos exhibit divergent spatial memory development. *Developmental Science*, 15, 840–53.

Semendeferi, K., Armstrong, E., Schleicher, A., Zilles, K., and Van Hoesen, G.W. (1998). Limbic frontal cortex in hominoids: A comparative study of area 13. *American Journal of Physical Anthropology*, 106, 129–55.

Stimpson, C.D., Barger, N., Taglialatella, J.P., Gendron-Fitzpatrick, A., Hof, P.R., Hopkins, W.D., and Sherwood, C.C. (2015). Differential serotonergic innervation of the amygdala in bonobos and chimpanzees. *Social Cognitive and Affective Neuroscience*, 11(3): 413–422.

Tan, J. and Hare, B. (2013). Bonobos share with strangers. *PLoS One*, 8, e51922.

Tomasello, M. *and* Call, J. (1997. *Primate Cognition*, New York, NY: Oxford University Press.

Tomasello, M., Call, J., and Hare, B. (1998). Five primate species follow the visual gaze of conspecifics. *Animal Behaviour*, 55, 1063–9.

Tomasello, M., Call, J., and Hare, B. (2003). Chimpanzees understand psychological states: the question is which ones and to what extent. *Trends in Cognitive Sciences*, 7, 153–6.

Tomasello, M. and Carpenter, M. (2005). The emergence of social cognition in three young chimpanzees. *Monographs of the Society for Research in Child Development*, 70.

Tomasello, M., Carpenter, M., Call, J., Behne, T., and Moll, H. (2005). Understanding and sharing intentions: The origins of cultural cognition. *Behavioral and Brain Sciences*, 28, 675–691.

Tomasello, M., Hare, B., and Agnetta, B. (1999). Chimpanzees, *Pan troglodytes*, follow gaze direction geometrically. *Animal Behaviour*, 58, 769–77.

Warneken, F., Chen, F., and Tomasello, M. (2006). Cooperative activities in young children and chimpanzees. *Child Development*, 77, 640–63.

Warneken, F., Hare, B., Melis, A.P., Hanus, D., and Tomasello, M. (2007). Spontaneous altruism by chimpanzees and young children. *PLoS Biology*, 5, e184.

Wellman, H.M., Lane, J.D., Labounty, J., and Olson, S.L. (2011). Observant, nonaggressive temperament predicts theory of mind development. *Developmental Science*, 14, 319–26.

Wobber, V., Hare, B., Maboto, J., Lipson, S., Wrangham, R., and Ellison, P.T. (2010a). Differential changes in steroid hormones before competition in bonobos and chimpanzees. *Proceedings of the National Academy of Sciences*, 107, 12457–62.

Wobber, V., Wrangham, R., and Hare, B. (2010b). Bonobos exhibit delayed development of social behavior and cognition relative to chimpanzees. *Current Biology*, 20, 226–30.

Wrangham, R.W. (1993). The evolution of sexuality in chimpanzees and bonobos. *Human Nature*, 4, 47–9.

Yamamoto, S., Humle, T., and Tanaka, M. (2012). Chimpanzees' flexible targeted helping based on an understanding of conspecifics' goals. *Proceedings of the National Academy of Sciences*, 109, 3588–92.

CHAPTER 7

What did we learn from the ape language studies?

Michael Tomasello

Abstract The 'ape language' studies have come and gone, with wildly divergent claims about what they have shown. Without question, the most sophisticated skills have been displayed by Kanzi, a male bonobo exposed from youth to a human-like communicative system. This chapter attempts to assess, in an objective a manner as possible, the nature of the communicative skills that Kanzi and other great apes acquired during the various ape language projects. The overall conclusion is that bonobos and other apes possess most of the requisite cognitive skills for something like a human language, including such things as basic symbol learning, categorization, sequential (statistical) learning, etc. What they lack are the skills and motivations of shared intentionality—such things as joint attention, perspective-taking and cooperative motives—for adjusting their communicative acts for others pragmatically, or for learning symbols whose main function is pragmatic.

Il y a eu beaucoup d'études sur la langue des singes avec des résultats très divergents. Sans question, on a vu les compétences les plus avancées chez Kanzi, un bonobo mâle qui a été exposé dès la jeunesse à un système de communication humain. Ici j'essaye d'évaluer le plus objectivement possible l'origine des compétences de communication que Kanzi et d'autres Grands singes ont appris pendant les différents projets linguistiques. Je conclue que les bonobos et les autres grands singes possèdent la plupart des compétences cognitives nécessaires à un langage humain, inclut les bases d'apprentissage de symboles, catégorisation, apprentissage séquentiel statistique, etc. Ils manquent les compétences et motivations d'intentionnalité commune—comme attention commune, prendre une perspective différente, motifs coopératifs—pour qu'ils améliorent leurs actes communicatives pragmatiquement.

Most animal communication takes place through evolved communicative signals, either physical or behavioral, that are more or less fixed in expression. In addition, however, great apes sometimes employ their cognitive skills to communicate via learned signals that are used flexibly and strategically, that is, intentionally. This happens most often and most readily not in the vocal domain—where fixed signals are the norm—but rather in the gestural domain (Tomasello, 2008).

In all four great ape species individuals ritualize with one another gestural signals that they use to flexibly regulate their social interactions, from grooming to sex to play (Call and Tomasello, 2007). Included in these gestural signals are so-called attention-getters—things such as slapping the ground or holding out an object (or even pointing for humans)—that work by directing the recipient's attention either to the self or to some external entity or event. These special signals suggest that great apes understand something about how the recipient processes and comprehends their signals; indeed in some situations apes can even inhibit a fixed communicative signal so that others cannot process it at all (Tanner and Byrne, 1993).

In this chapter I tell the story of how a handful of behavioral researchers over the past several decades have attempted to exploit great apes' sophisticated skills of intentional communication to teach them a human-like system of 'linguistic' communication. The essence of linguistic communication is (1) the use of learned symbols to refer the attention of others to external entities and events (for various purposes), and (2) combining such symbols to communicate things that would be difficult or impossible to communicate with single symbols.

Tomasello, M., *What did we learn from the ape language studies?* In: *Bonobos: Unique in Mind, Brain, and Behavior*.
Edited by Brian Hare and Shinya Yamamoto: Oxford University Press (2017).
© Oxford University Press. DOI: 10.1093/oso/9780198728511.003.0007

Mastering these two aspects of linguistic competence in a human-like way relies not only on sophisticated cognitive skills but also on a sophisticated understanding of the pragmatics of communication; that is, an understanding of how psychological agents manage to manipulate one another's mental states in desired ways. The point of these so-called ape language studies was both to investigate the potential and the limitations of great apes' communicative skills and to shed some light on the evolutionary roots of human linguistic communication. Of special relevance to the current volume, the main character in this story will be the world's most famous bonobo.

A brief history

Kellogg and Kellogg (1933) and Hayes and Hayes (1951) attempted to teach some English words to their infant chimpanzees, Gua and Viki, respectively, but mostly failed because the chimpanzee vocal apparatus is not capable of making many sounds important in human speech. Gardner and Gardner (1969) attempted to overcome apes' difficulties with speech by teaching a chimpanzee, Washoe, some parts of American Sign Language. Their apparent success led to similar projects with a gorilla, Koko (Patterson, 1978), and an orangutan, Chantek (Miles, 1990), both of whom were at least as proficient as Washoe. Other attempts to circumvent the speech barrier were made by Premack (1976) who used plastic tokens to stand for spoken words in communicating with the chimpanzee Sarah, and by Rumbaugh and colleagues (1977), who created for the chimpanzee Lana a visual language based on graphic symbols (lexigrams) depicted on a computerized keyboard.

The early work on ape language was put to a severe test by the critiques of Terrace and colleagues who raised their own chimpanzee, Nim, with American Sign language (e.g. Terrace et al., 1979). On the basis of their experience with Nim, as well as a re-analysis of the data of other investigators (including videotapes of Washoe), Terrace and colleagues argued that the linguistic apes were not really linguistic at all. What the apes were doing was learning operant responses to get food and other human-controlled rewards. These responses just happened to correspond to human sign language signs, but they should not be considered linguistic symbols in any sense of the term. Almost all of Nim and Washoe's productions, at least, were requests for things, mostly food. Terrace and co-workers also found that the apes engaged in a lot of mimicking and repetition of these responses from the humans interacting with them that were then taken to be 'sentences' and 'conversations'. Thus, in response to the question 'Do you want to eat an apple?' an ape might produce 'Apple eat eat apple eat apple hurry apple hurry hurry', which would have been reported as 'Eat apple hurry' and which would appear to an outsider as a conversationally appropriate response in the form of a sentence. Terrace and his team also noted that almost all of these kinds of 'sentences' were highly redundant within themselves, with longer productions conveying no more information than shorter productions. To prove all of these points in one study, the team simply trained a pigeon to peck the keys of a linear array in particular sequences in the presence of particular visual stimuli (meant to be analogous to the visually based systems of Premack and Rumbaugh) and argued that the language of the apes could in no way be discriminated from the productions of the pigeon (Terrace et al., 1977).

Part of what Terrace and colleagues argued was true, but only of their own chimpanzee, Nim, who had been taught its 'language' by a fairly rigid set of behavioristic training regimes. Nim was indeed tied fairly tightly to producing signs for food rewards, and he did mimic and repeat a lot of human signs simply because he was rewarded for doing so. This does not seem to be true of all of the linguistic apes; for example, some other linguistic apes do not mimic and repeat any more than human children, and they do so under very similar pragmatic circumstances (Greenfield and Savage-Rumbaugh, 1991). In answering the critique presented by Terrace and colleagues, one set of investigators in particular has documented that apes can use something resembling human linguistic symbols and can produce and understand combinations of those symbols in creative ways. Savage-Rumbaugh and colleagues took up the challenge of Terrace and colleagues' critique and set out to work with apes in different ways to see if they could learn more than operant responses.

Symbols

In response to the critique presented by Terrace and colleagues (1977, 1979), Savage-Rumbaugh and colleagues decided that the primary issue in ape language research was the issue of whether apes could learn to use something resembling human linguistic symbols. In working with the two chimpanzees, Sherman and Austin, they concentrated on individual lexigrams (created so as not to resemble their referents) and their symbolic status. Like Lana before them, these two chimpanzees were taught to use a specially constructed computer keyboard filled with lexigrams. Based on their analyses of what constitutes a human symbol, however, Savage-Rumbaugh and colleagues attempted to make sure that Sherman and Austin learned: (1) to comprehend and produce lexigrams; (2) to use their lexigrams in a wide array of situational contexts, even in the absence of their referents; and (3) to use their lexigrams spontaneously to communicate with each other. They had some success in this effort but also some failures. The nature of their efforts and successes taught them a great deal about what it means to use a symbol (Savage-Rumbaugh, 1986; Savage-Rumbaugh et al., 1980).

The most important of their subsequent experiences was their work with a wild-caught bonobo named Matata and her adopted son, Kanzi. Savage-Rumbaugh and his team attempted unsuccessfully to teach Matata lexigrams. However, her son Kanzi, who had spent his first two-and-a-half years holding onto and observing his mother while she was interacting with humans around the keyboard (and also interacting with humans in other ways), learned many of the lexigrams his mother had not (Figure 7.1, Plate 7). Since he could only have acquired these observationally—he had not been directly trained—the investigators decided to continue allowing Kanzi to acquire his signs incidentally rather than through direct training with external reinforcement. The subsequent procedure used with Kanzi was based on child language research showing that human children acquire most of their early linguistic symbols in highly predictable routine interactions with adults, without training (Bruner, 1983). Kanzi was thus exposed to the lexigram language in as natural a manner as possible as he and his caretakers went about their daily activities. His only reward was communication itself and the things it helped directly to accomplish.

Figure 7.1 Savage-Rumbaugh communicating with Kanzi using the lexigram keyboard (see Plate 7). (Savage-Rumbaugh communiquant avec Kanzi à l'aide du clavier à lexigrammes.) Picture from wikicommons: File:Kanzi,_conversing.jpg.

Kanzi quickly learned a wide array of lexigrams. His early lexigrams were generalized to novel referents without specific training, including for perceptually absent referents (Savage-Rumbaugh et al., 1986). In many ways, his early vocabulary matched the early vocabularies of human children, including proper names for individuals; labels for common objects; words for actions, locations and properties; and even a few function words such as *no* and *yes* (Nelson, 1987). Using this same general teaching method, two other bonobos and a common chimpanzee showed similar competencies, demonstrating that Kanzi was not a unique genius among bonobos and bonobos were not uniquely equipped for this task among apes. Kanzi and those following him were thus clearly capable, from the beginning, of associating arbitrary lexigrams flexibly with a wide array of different kinds of novel referents.

Domestic dogs, parrots, dolphins and some other animals are able to learn to recognize various kinds of arbitrary signs as well, and so the question becomes how Kanzi understands what he is doing. One obvious difference is that, unlike dogs and dolphins, from the beginning Kanzi produced what he comprehended (and vice versa) without specific training of this correspondence (and so unlike Sherman and Austin), perhaps suggesting an understanding of symbols as communicatively bidirectional. Alex the African grey parrot—hundreds of millions of years away in evolutionary distance—does this as well in the vocal modality (Pepperberg, 1999). It may be helpful in this regard that the medium of communication with Kanzi is pointing to a lexigram on a keyboard because all he has to do is touch the same place that others have previously touched in association with the same referent rather than imitating sounds or actions (and all parrots have to do is mimic sounds, which comes quite naturally to them). In any case, it would seem that Kanzi—perhaps because he acquired his lexigrams spontaneously by observing and comprehending their use by his caretakers—does seem to have an understanding of the way that his lexigrams work bidirectionally in a way that previously language trained apes (and animals other than parrots) do not.

Perhaps the clearest limitation of Kanzi's communication with lexigrams is that virtually all of his productions are imperatives: requests for objects or actions. In one of the few systematic analyses of a fixed corpus, Greenfield and Savage-Rumbaugh (1990) found that fully 96 per cent of all of Kanzi's productions were imperatives, and many of the others were responses to human demands to name things (for similar results for Washoe and her associates, see Rivas, 2005). When great apes 'point' for humans, they also do so almost exclusively for imperative purposes, and indeed they do not even comprehend informative pointing gestures (see Tomasello, 2006, for a review). In contrast, even prelinguistic human infants also communicate to inform others of things helpfully and to share attention with them to exciting objects and events (Tomasello et al., 2007). This raises the question of whether Kanzi's pointing to a lexigram—even if it is in some senses both referential and bidirectional—is a communicative act designed to influence the psychological states of others, or something more in the direction of an effective procedure for getting what he wants.

Evidence from the natural gestural communication of bonobos and other great apes (e.g. Call and Tomasello, 2007) suggest that they do understand something of the process of referential communication, as they use attention-getters aimed at drawing the attention of others to things—typically themselves or a part of their body, but with humans to external objects—which then should induce those others to do what they want them to. To my knowledge, no species of mammal other than great apes uses attention-getters of this type in its natural communication, and so it is likely that Kanzi and other language-learning apes are doing more than just producing actions that induce others to do what they want; they know that they must go through the other's attention for this to happen. To really nail this down would require some experiments that have yet to be conducted (the closest is Lyn and Savage-Rumbaugh, 2000). For example, Baldwin and colleagues (1996) played a word through a loudspeaker as children were looking at an object. They did not learn the word under these conditions but they did quite readily when it was used by an adult in a social context. Similarly, Baldwin (1991) waited until a child was looking at one object and then she looked at a different object and labelled it; children learned the word not for the object they were looking at but for the one the adult wants or is attempting to draw their attention to through her use of the novel word (see Tomasello, 2001, for other experiments that would also help to determine exactly what word learners understand about the use of language). Absent studies such as these with apes, we will have to be content with concluding that Kanzi does understand something of the way the referential communication works psychologically, though precisely how much we cannot say.

One interesting issue concerns a particularly important type of word that is found in all human languages and that works in a special way: pronouns (and other pro-forms). It is not that an ape could not learn a pronoun such as *he* or *it*; they probably could be trained to use a single form for all men or all objects, for example, but this would not be a pronoun. Pronouns are used when the communicator and recipient share common ground about what the communicator wishes to refer the recipient's attention to. Then, instead of saying 'the chair' or 'the chair next to the window', he may say 'it' if the partner can guess that 'it' refers in this context to the chair. Appropriate use of a pronoun thus requires an assessment of one's common ground with a communicative partner and how readily the partner can access it, and to-date none of the linguistic apes has demonstrated an ability to do this. (N.B. for a child to learn when to use a pronoun rather than a full noun phrase, she must first note that mature speakers are assessing their common ground with their listeners and choosing a pronoun or noun phrase as a result; then she must be able to do the same thing herself.) Again, the only way to know for certain would be with future experiments. The fact that great apes only give visually based gestures when their partner is looking at them (see Call and Tomasello, 2007, for a review) suggests that they do have at least some skills in assessing the state of their partner before communicating, but this is based simply on visual orientation and not on the basis of the knowledge states of the communicative partner.

So we may say that Kanzi and very likely other linguistic apes are comprehending and producing their communicative devices in some sense referentially; that is, they know that they are achieving their end by directing the attention of the recipient. However, the pragmatics of communication of the human variety also includes the ability to use communicative devices for many different social ends—including to inform others of things and to share attention with them to interesting events—and these other motivations (i.e. other than the imperative) just do not seem to come naturally to great apes. Choosing a formulation based on the knowledge states of a partner or one's common ground with her also does not seem to be something the great apes do as a natural matter of course. So it is not so much the cognitive task of learning referential devices that is the ape–human difference in using symbols but rather the pragmatic dimension of using them for varied non-imperative purposes and adjusting them for the communicative needs of the partner.

Combinations of symbols

To look at the 'syntax' and 'grammar' of linguistic apes, we need to distinguish two levels of grammatical competence. First, an individual may produce combinations of symbols in novel ways that create new meanings. The most well-known example is Washoe calling a duck 'water bird', a combination that surprised even his caretakers. There are many other well-documented cases as well involving more mundane action–object combinations and the like, for example, by Kanzi and other apes as reported in Greenfield and Savage-Rumbaugh (1990; 1991). Second, a subject may produce combinations that have specific symbols to serve as grammatical markers: that is, although 'water bird' is a creative combination there are no symbolic indications in this utterance of how these two signs are to be related to one another; the relating is left up to the listener. In adult human language, on the other hand, sentences have second-order grammatical symbols that indicate how the individual signs are interrelated. For example, in the English sentence 'John called Mary', who is doing the calling and who is being called is indicated by the order in which the symbols are produced. The fact that word order is indeed being used as a communicatively significant symbol is demonstrated by the fact that the opposite ordering of elements would make Mary the caller and John the callee (so-called contrastive usage). In other languages, word order is not used for this purpose but rather special word endings (case markers) indicate the roles being played by John and Mary (Bates and MacWhinney, 1979).

In the context of ape language research it is especially important to recognize that using a consistent word order does not mean that word order is being used as a contrastive symbolic device indicating the relations among words (Tomasello, 1992). Kanzi, as well as many other linguistic apes, has combined his lexigrams in some creative ways indicating some first-level grammatical competence (Greenfield and Savage-Rumbaugh, 1990; 1991). In these productions—which constitute about 10 per cent of Kanzi's total productions—he uses some consistent ordering patterns. For example, in his two-lexigram combinations Kanzi uses the action words *tickle* and *chase* about twice as often as the first element than as the second element. In general, Kanzi normally uses lexigrams for action as first element and lexigrams for the object of action as the second element, but this instead directly reflects the order in which he has experienced those used by caretakers. There is no evidence that a different ordering of lexigrams would produce a different meaning. Kanzi also combines his lexigrams with gestures in creative ways, mostly by using pointing to indicate who is to perform some action or the location toward which an action should be directed. He mostly uses the gesture after the lexigram, but the opposite ordering does not indicate a consistent change of meaning. To repeat: a consistent ordering pattern is not the same thing as the contrastive use of order as a grammatical device conveying a consistent second-order meaning.

One of the unique things about Kanzi is that he seems to comprehend large portions of spoken English, and this comprehension appears to go well beyond his productions. From the beginning, Kanzi's comprehension seemed to go well beyond just picking up some words, the way a dog might tune into the word 'dinner'. He seemed to be comprehending whole sentences in ways that suggested a creative synthesis of the words, not a rote learning of a recurrent sound pattern. After several years of working with Kanzi, Savage-Rumbaugh and colleagues systematically tested his comprehension of English commands such as, for example, 'take the telephone outdoors' and 'give Rose a shoe' (Savage-Rumbaugh et al., 1993). Over a period of several months, Kanzi and a human two-year-old were presented with several hundred sentences in similar ways. Almost all sentences were given as requests involving an action word and one or more object or person words, with the task of the subjects being to fulfil the request. For each trial for each subject multiple objects were present, affording many different types of actions. Several dozen different action verbs were used in the different sentences. Many sentences were constructed so that the actions named would not be deducible from knowledge of the objects involved (e.g. 'put the money on the mushrooms'), and some were such that the role of the two objects involved could easily be reversed (e.g. 'put some milk in the water').

Both Kanzi and the human two-year-old were very competent at carrying out requests of all types. Kanzi carried out 59 per cent of all blindly presented sentences promptly and correctly, the child 54 per cent. When generic prompting and encouragement by the experimenter were allowed, Kanzi was correct on 74 per cent of the blind trials, the child on 65 per cent. In attempting to determine what knowledge of language underlay Kanzi's performance in this study, Tomasello (1994) re-analysed the data to discriminate among a number of different hypotheses. He found that Kanzi's performance indicated that he was doing more than simply responding to the words in the sentences in some plausible way; he understood the use of word order as a contrastive symbolic device. For example, Kanzi responded correctly to the two requests: 'put the *hat* on your *ball*' and 'put the *ball* on the *hat*'. Overall, he understood both members of 57 per cent of pairs using the 'put X on Y' formula (to the child's 39 per cent), which is significantly different from the chance probability of acting out both sentences correctly (25 per cent). Tomasello's conclusion was thus that Kanzi knows, for example, that the object being put (e.g. 'hat') is named first and the place it is being put (e.g. 'on the ball') is named only afterwards, even when the object and location could easily have been reversed. He knows that the person biting or chasing is mentioned before *bite* or *chase*, and the one being bitten or chased is only mentioned after. He knows these things presumably because he has seen one order paired with one version of the key event (e.g. X chasing Y) and another order paired with another version of the same event (e.g. Y chasing X).

The most plausible hypothesis is thus that, like 24-month old children, Kanzi understands actions and events and the words that designate them. He understands that actions and events have participants and that the order in which these participants are mentioned for a particular event determines what role each plays in that particular event. He has something like a grammatical category of noun allowing him to comprehend newly learned object labels in the same sentence frames in which previously learned object labels have been used. However, he is still limited. Specifically, he has yet to form any sentence schemas that would license grammatical categories and other generalizations across events. This means that his grammatical categories are of the type 'hitter' and 'thing hit with', 'putter' and 'thing put', not generalized categories such as 'agent' and 'patient' or 'subject' and 'direct object'. This is basically the level of grammatical competence of young two-year-old children, who also work with 'verb islands' and other item-based constructions of this type (Tomasello, 2003).

Nevertheless, the fact that Kanzi does not also do these things in production gives pause, as does the fact that a number of other non-primate species, such as dolphins and parrots (Herman, 2005; Pepperberg, 2000), have also shown essentially the same ability to recognize correlations between sign orders and particular types of requested actions. For none of these animals is there is a corresponding competence with sign order in the *production* of communicative acts, and so their comprehension skills might be based on many different kinds of cognitive and/or learning skills, some of them having very little to do with communication in particular. A simple explanation for why 'linguistic' apes do not use syntactic devices in their productive communication with humans (even if they may comprehend the contrastive use of order when signed or spoken to) is that all of their communication is designed for the requestive function. This exclusive focus on requests in the here and now of our current social interaction means that there are almost no functional demands in the gestural or sign production of these apes for syntactically marking the roles that different actors play in an event, or for identifying more explicitly the different actors involved (as in noun phrases), or for designating the time of an event (as with tense markers), or for marking a topic (as with topic markers), or for designating speech act function (as with intonation or special constructions), or for doing any of the myriad other things that go into human grammatical competence beyond simply the backbone of combining symbols in creative ways.

The upshot is this. Many of the communicative acts produced by Kanzi and friends are clearly complex, structured by a kind of event–participant structure, reflecting a partitioning of situations into the participants involved and the relations or actions among them. Despite this complexity, however, some key devices that are essential to human linguistic communication are missing: all those aspects of human grammar that conceptually structure constructions for others given their particular knowledge, expectations and perspective. The apes have learned to distinguish and combine appropriately items for events and participants (and perhaps locations). What is missing is all of those aspects of human grammar that are geared at making the utterance comprehensible to the recipient—a key part of the cooperative motive in which we work together toward communicative success. Fundamentally:

- Linguistic apes do not 'ground' their acts of reference for the listener to help them identify the referent. That is to say, they do not have noun phrases with things like articles and adjectives that help to specify which ball or cheese is wanted, for example. Nor do they have any kind of markers of tense that would suggest which event, as shown by when it occurred, they intend to indicate.
- They do not produce second-order symbols such as case markers or word order to mark semantic roles and so to indicate who is doing what to whom in the utterance. Communicators do not need this information; they know what they want to say. It is provided to make sure that the listener understands the role of each participant in the larger situation or event being communicated about.
- They do not have constructions or other devices for indicating for listeners what is old versus new versus contrasting information. For example, if you adamantly expressed that Bill broke the window, I probably would correct you by using a cleft construction and say 'no, it was FRED that broke the window'. Apes do not have any such constructions.

- They do not choose constructions based on perspective. For example, I might describe the same event either as 'I broke the vase' or 'the vase broke', based on your knowledge and expectations and my communicative intentions, whereas linguistic apes have not learned constructional alternatives of this type.
- They do not specifically indicate in their utterances their communicative motive (why should they, since it is always a request), nor anything of their epistemic or modal attitudes toward the referential situation (e.g. that one *must* or *can* do it).

Importantly in all of this, however, is question of whether and to what extent apes have had the opportunity to acquire such linguistic constructions, as we do not know for any language-learning ape precisely what kinds of linguistic experience they have had, and with what frequency in what communicative contexts.

In any case, the key theoretical point is that beyond just supplying ordering preferences for utterances, human linguistic constructions are created with adaptations for the recipients' knowledge, expectations and perspective in mind. In addition, even very simple constructions like noun phrases require adaptations to the recipient's knowledge, expectations and perspective. Humans also use conventional expressions of motives and epistemic and modal attitudes in their constructions, such things as *might, ought* and *should*. In terms of conversation, even the analysis of Greenfield and Savage-Rumbaugh (1991) is not convincing that Kanzi or any other linguistic ape knows how to produce a fully fledged conversational turn in which the topic of conversation is acknowledged and specifically indicated, and simultaneously something new is predicated of that topic (e.g. 'now *that* [indication of shared topic] was a surprising turn of events'). Call all of this the pragmatic dimension of grammar, and call it uniquely human.

Conclusion

It is fair to say that before the ape language studies, very few psychologists and linguists would have predicted the great apes have the 'linguistic' skills that they apparently do. In an analysis of apes' natural gestural communication, Tomasello (2008) calls their attention-getting gestures 'the missing link' in the evolution of language because they do their work not directly on the behavior of others but indirectly on the behavior of others through their attention. The fact that linguistic apes can even learn arbitrary devices for directing the attention of others to external referents, with at least some degree of understanding, is remarkable. Their use of creative combinations beyond what they are specifically taught is astonishing on top of that. The cognitive dimensions of language acquisition—acquiring symbols and combining them in creative ways—is at least to some significant extent shared between humans and other apes.

With the pragmatic dimensions of the process, the differences are stark. Humans communicate cooperatively: they work together to achieve successful communication, and each adjusts to the other toward this end. They communicate not only to get others to do things but also simply to inform them of things helpfully and to share attention of mutually interesting events. Their grammatical constructions are specifically designed for situations that differ in what kinds of knowledge, expectations and perspective are shared by communicator and recipient, and what kinds are not. In the account of Tomasello (2008; 2009), this difference has roots that go much deeper than communication and language per se, and they have to do with the overall more cooperative lifestyles that humans have adopted in the 6 million years since they separated from chimpanzees and bonobos, resulting in a species-unique suite of skills and motivations for shared intentionality (Tomasello, 2014). It is therefore not likely a difference that could be eliminated via specific rearing environments or training for apes. But who knows? The scientific community has been surprised before and it could be again.

It is not clear that there will be other linguistic apes in the future, as this research was in many ways a product of its time both scientifically and ethically. The research could have been improved in many different ways, especially with a more experimental focus more attuned to issues of the pragmatics of communication, but it is a good thing that this research was undertaken as it provides us with important information about what it takes to create and learn a human language. As for the uniqueness of bonobos, although there is no doubt that Kanzi is the most accomplished of all apes in comprehending and using a human-like language, the kinds of systematic comparisons needed to make such a determination have not been done.

References

Baldwin, D.A. (1991). Infants' contribution to the achievement of joint reference. *Child Development*, 62(5), 875–90.

Baldwin, D.A., Markman, E.M., Bill, B., Desjardins, R.N., Irwin, J., and Tidball, G. (1996). Infants' reliance on a social criterion for establishing word-object relations. *Child Development*, 67(6), 3135–53.

Bates, E. and MacWhinney, B. (1979). A functionalist approach to the acquisition of grammar. In: Ochs, E. and Schieffelin, B. (eds). *Developmental Pragmatics*. New York, NY: Academic Press, pp. 167–211.

Bruner, J. (1983). *Child's Talk: Learning to Use Language*. New York, NY: Norton.

Call, J. and Tomasello, M. (eds). (2007). *The Gestural Communication of Apes and Monkeys*. New York, NY: Erlbaum.

Gardner, R.A. and Gardner, B.T. (1969). Teaching sign language to a chimpanzee. *Science*, 165(3894), 664–72.

Greenfield, P.M. and Savage-Rumbaugh, E.S. (1990). Grammatical combination in *Pan paniscus*: Processes of learning and invention in the evolution and development of language. In: Parker, S.T. and Gibson, K.R. (eds) *'Language' and Intelligence in Monkeys and Apes: Comparative Developmental Perspectives*. Cambridge: Cambridge University Press, pp. 540–78.

Greenfield, P.M. and Savage-Rumbaugh, E.S. (1991). Imitation, grammatical development, and the invention of protogrammer. In: Krasnegor, N., Rumbaugh, D., Studdert-Kennedy, M., and Schiefelbusch, R. (eds). *Biological and Behaviorial Determinants of Language Development*. Hillsdale, NJ: Erlbaum, pp. 235–58.

Hayes, K.J. and Hayes, C. (1951). The intellectual development of a home-raised chimpanzee. *Proceedings of the American Philosophical Society*, 95(2), 105–9.

Herman, L. (2005). Intelligence and rational behavior in the bottle-nosed dolphin. In: Hurley, S. and Nudds, M. (eds). *Rational Animals?* Oxford: Oxford University Press.

Kellogg, W.N. and Kellogg, L.A. (1933). *The Ape and the Child*. New York, NY: Hafner Publishing Company.

Lyn, H. and Savage-Rumbaugh, E.S. (2000). Observational word learning in two bonobos (*Pan paniscus*): Ostensive and non-ostensive contexts. *Language and Communication*, 20(3), 255–73.

Miles, L. (1990).The cognitive foundations for reference in a signing orangutan. In Parker, S.T. and Gibson, K.R. (eds). *'Language' and Intelligence in Monkeys and Apes: Comparative Developmental Perspectives*. Cambridge: Cambridge University Press, pp. 511–39.

Nelson, K. (1987). What's in a name? Reply to Seidenberg and Petitto. *Journal of Experimental Psychology*, 116, 293–6.

Patterson, F.G. (1978). The gestures of a gorilla: Language acquisition in another pongid. *Brain and Language*, 5, 72–97.

Pepperberg, I.M. (1999). *The Alex Studies: Cognitive and Communicative Abilities of Grey Parrots*. Cambridge, MA: Harvard University Press.

Premack, D. (1976). Language and intelligence in ape and man. *American Scientist*, 64(6), 674–83.

Rivas, E. (2005). Recent use of signs by chimpanzees (*Pan troglodytes*) in interactions with humans. *Journal of Comparative Psychology*, 119(4), 404–17.

Rumbaugh, D.M. (ed.) (1977): *Language Learning by a Chimpanzee*. New York, NY: Academic Press.

Savage-Rumbaugh, E.S. (1986): *Ape Language*. New York, NY: Columbia University Press.

Savage-Rumbaugh, S., MacDonald, K., Sevcik, R.A., Hopkins, W. D., and Rubert, E. (1986). Spontaneous symbol acquisition and communication use by pygmy chimpanzees (*Pan paniscus*). *Journal of Experimental Psychology: General*, 115(3), 211–35.

Savage-Rumbaugh, E.S., Murphy, J., Sevcik, R., Brakke, K.E., Williams, S.L., and Rumbaugh, D.M. (1993). Language comprehension in ape and child: with commentary by Elizabeth Bates. *Monographs of the Society for Research in Child Development*, 58(3–4).

Savage-Rumbaugh, E.S., Rumbaugh, D.M., and Boysen, S. (1980): Do apes use language? *American Scientist*, 68(1), 49–61.

Tanner, J.E. and Byrne, R.W. (1993). Concealing facial evidence of mood: Perspective-taking in a captive gorilla? *Primates*, 34, 451–7.

Terrace, H.S., Petitto, L.A., Sanders, R.J., and Bever, T.G. (1979). Can an ape create a sentence? *Science*, 206(4421), 891–902.

Terrace, H.S., Straub, R.O., Bever. T.G., and Seidenberg, M.S. (1977). Representation of a sequence by a pigeon. *Bulletin of the Psychonomic Society*, 10, 269.

Tomasello, M. (1992). *First Verbs: A Case Study of Early Grammatical Development*. New York, NY: Cambridge University Press.

Tomasello, M. (1994). Can an ape understand a sentence? A review of Language Comprehension in Ape and Child by E.S. Savage-Rumbaugh et al. *Language and Communication*, 14(4), 377–90.

Tomasello, M. (2001). Perceiving intentions and learning words in the second year of life. In: Bowerman, M. and Levinson, S. (eds) *Language Acquisition and Conceptual* features, such as language, must include insights from both species (de Waal, 1997). Compared to an impressive body of research on chimpanzees, much less is known about bonobos, probably due to their late discovery, more remote geographic range and lower representation in captivity (de Waal, 1997). Moreover, while a substantial body of research exists on their grasp of artificial and human language systems (Savage-Rumbaugh et al., 1993; see also Chapter 7 in this book), research on their natural communicative behavior has only recently gained momentum.

Here, we address this by reviewing recent research that has been conducted on bonobo natural communication, focusing specifically on what insights can be made about the evolution of language. Human language is based on a multitude of capacities that includes a sophisticated social awareness and specialized social skills in detecting, understanding and influencing the mental states of others. A key question that we therefore aim to explore is the interaction between communication and social cognition, and the kinds of inferences that bonobos can make about their social world through attending to each other's communicative signals (Clay and Zuberbühler, 2014).

Vocal communication in the forest habitat

As with other great apes, bonobos are typically tropical forest dwellers. Thus, due to the dense nature of their habitat, individuals are regularly are out of visual access to one another. Their low-visibility habitat, combined with the fission–fusion dynamics that characterize their social organization, mean that vocalizations are likely to play a critical role in coordinating the activities and behaviors between and within communities.

To-date, the first and only major description of the bonobo vocal repertoire was made by de Waal in the late 1980s (de Waal, 1988), comparing the vocal repertoire of captive bonobos living in San Diego Zoo with that of chimpanzees (Marler and Tenaza, 1977; van Hooff, 1973). Since this study was undertaken, only one preliminary analysis of the repertoire of wild bonobos has been conducted (Bermejo and Omedes, 1999). While Bermejo and Omedes' research was promising, the study individuals were not fully habituated, nor identified and no quantitative acoustic analyses were performed. Therefore, a more detailed analysis of the wild bonobo vocal repertoire remains outstanding.

As with chimpanzees (Marler and Tenaza, 1977), the bonobo vocal system is highly graded (de Waal, 1988). The graded nature of an animal vocal system refers to the scaling of acoustic similarity between call types (Marler and Tenaza, 1977). Although more difficult to describe systematically, the acoustic variation present in graded signals has the potential for considerable communicative complexity and variability, especially if receivers are able to perceive graded signals categorically.

Perhaps the most striking feature of bonobo vocalizations is their remarkably high pitch, especially compared to those of chimpanzees (de Waal, 1988; Mitani and Gros-Louis, 1995). For example, the mean frequency of chimpanzee screams was shown to be 1275 Hz compared to 2846 Hz for bonobos (Mitani and Gros-Louis, 1995). The reasons underlying this high pitch are not well understood, although anatomical and social differences, related to the presence of neotenous characteristics, may go some way to explain it. As well as being smaller in body size, bonobos show juvenilized features in their cranio-morphology and in regions surrounding the basicranium (Laitman and Heimbuch, 1984). These anatomical differences are likely to give rise to related variations in laryngeal mechanisms and vocal tract length (Mitani and Gros-Louis, 1995). From an evolutionary perspective, it has been suggested that bonobos tendency to forage for terrestrial herbaceous vegetation may have selected for greater dependency on the vocal modality (Taglialatela et al., in review), which may have resulted in greater vocal control, a prerequisite for the production of high-frequency vocalizations. An alternative hypothesis is that their higher vocal pitch is a by-product of selection for the retention of more juvenilized traits and lower aggression, a process that has been suggested to be a form of 'self-domestication' in this species (Hare et al., 2012). This would be consistent with similar patterns observed in artificially domesticated Siberian red foxes, who, unlike their control counterparts, retained a juvenile communicative repertoire into adulthood (Gogoleva et al., 2008).

De Waal (1988) described the bonobo vocal repertoire as being composed of 12 principal call types, which includes three hoots, ('high hoot', 'contest hoot', 'low hoot'), three peeps ('food peep', 'alarm peep', 'peep-yelp'), two barks ('wieew bark', 'whistle bark') as well as grunts, pant laughs, pout moans and screams. However, it is likely, as was suggested by de Waal, that greater variation exists within and beyond these call categories. The repertoire documented for wild bonobos (Bermejo and Omedes, 1999) largely confirmed de Waal's findings. The authors stressed the important role that combinatorial vocal sequences seem to play in bonobo vocal communication. The flexible use of heterogeneous vocal sequences highlights a further potential for the calls to be combined in different ways to provide different meanings (Clay and Zuberbühler, 2009; 2011a).

Although bonobo vocalizations are generally higher pitched compared to those of chimpanzees (Mitani and Gros-Louis, 1995), there are numerous parallels in their acoustic form and contextual usage, which is perhaps unsurprising considering their recent phylogenetic divergence (Pruefer et al., 2012). For example, bonobo pant laughs, pout moans, low hoots, and wieew barks overlap in both acoustic structure and contextual usage with those of chimpanzees. Bonobos, like chimpanzees, use hoot vocalizations for long-distance communication, as well as in response to food discovery and other relevant events or disturbances (de Waal, 1988; Marler and Tenaza, 1977; Schamberg et al., 2016, 2017; van Hooff, 1973;). Structurally, however, the chimpanzee 'pant hoot' is a composite vocalization, composed of four distinct phases (introduction, build up, climax and downward phase), while the homologous bonobo 'high hoot', is a 'whooping' call that can be composed of rapid sequences of both staccato and legato elements which, although they may increase in speed and amplitude, do not possess the phrase-like form of chimpanzee pant-hoots (Hohmann and Fruth, 1994) (see Figure 8.1).

Despite parallels between the vocal repertoires, some notable differences do exist (de Waal, 1988) which provide relevant insights into underlying differences in their social systems. For instance, bonobo males produce an acoustically distinct 'contest-hoot' vocalization, which appears to function as a ritualized form of vocal provocation towards specific targets in order to test and explore social relationships (de Waal, 1988; Genty et al., 2014; see Figure 8.2). Chimpanzee males do not show an equivalent vocalization, probably because males signal dimensions of their strength and competitiveness through more pronounced visual gestural signals, including unintended cues such as pilo-erection as well as via physically aggressive behaviors and elaborate displays that combine both visual and acoustic components (Goodall, 1986).

Chimpanzees produce a formalized greeting signal of subordinance to more dominant individuals, termed the 'pant grunt' (Goodall, 1986; Newton-Fisher, 2004). Pant grunts occur reliably up the dominance hierarchy to the extent that they are sometimes used as a direct proxy of chimpanzee dominance relationships (e.g. Newton-Fisher, 2004; Slocombe and Zuberbühler, 2005a; Townsend et al., 2008). In contrast, numerous captive and wild studies have concluded that bonobos lack a formalized signal of subordinance (Furuichi and Ihobe, 1994; Vervaecke et al., 2000), which has been suggested to reflect their more relaxed and less linear dominance style and more peaceful disposition (Vervaecke et al., 2000). Nevertheless, recent analyses have indicated that wild bonobos from Lui Kotale and sanctuary-living bononbos from Lola ya Bonobo do, in fact, produce a submissive grunt vocalization equivalent to a pant-grunt although its usage appears to be less pronounced and less formalized (Clay et al., in review). While more infrequently than in chimpanzees, bonobos from these two sites similarly produced pant grunts up the hierarchy. However, despite the central position of females in bonobo society (Furuichi, 2011), the bonobo pant grunt vocalizations were more or less exclusively produced by males towards other males, rarely being produced by or towards females (Clay et al., in review). These intriguing sex- and species-based differences may reflect the fact that female dominance is less linear and less driven by aggressive displays of force in bonobos, as it is for male chimpanzees, thus may be less dependent on formalized signals of subordinance. In this way, the less aggressive nature of bonobo society, combined with the elevated roles of females, may have resulted in a shift away from vocalizations which primarily expresses linear and despotic dominance relationships that are associated with displays of physical aggression.

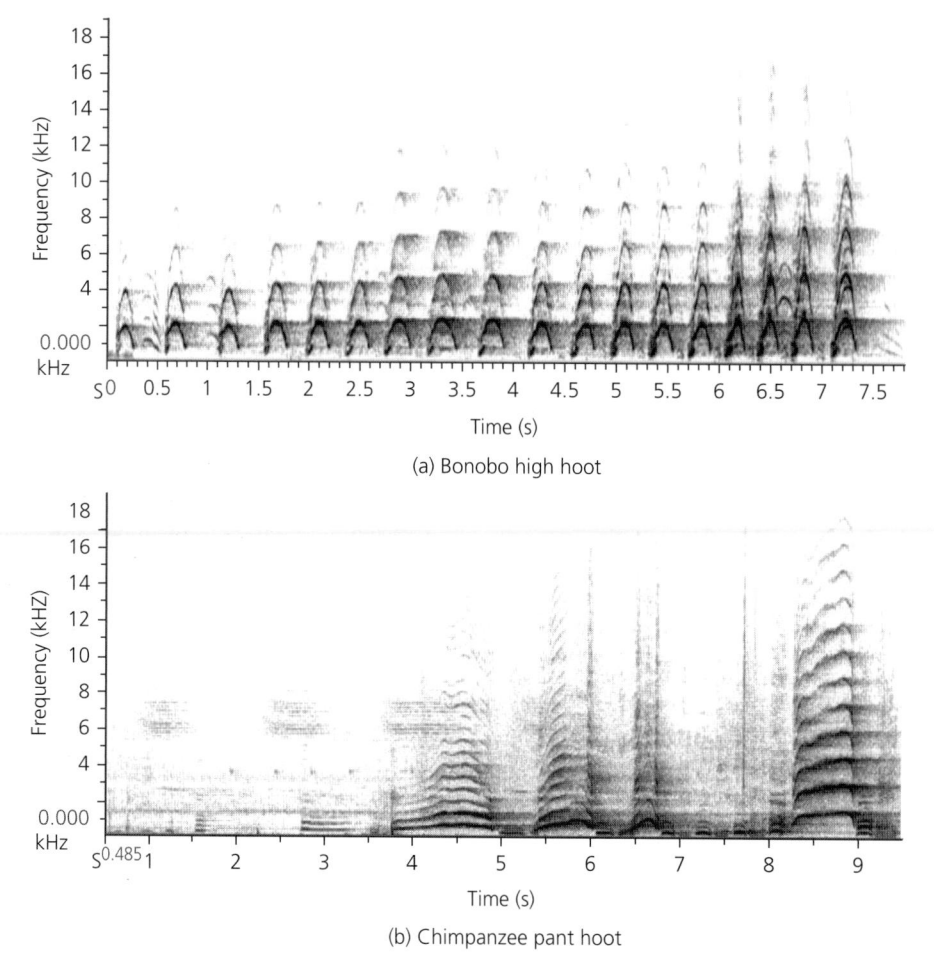

Figure 8.1 Time-frequency spectrograms showing representative examples of bonobo high hoots chimpanzee pant-hoots. These individually distinct and partially socially learned call types are used in long-distance communication in both species.
(Les spectrogrammes de fréquence et durée montrant des exemples représentatives des cri des bonobos et ceux des chimpanzés. Ces appels individuellement distincts et socialement appris partiellement sont utilisés pour communications à longues distances chez les deux espèces.)

Bonobo vocal communication and social awareness

How individuals produce and respond to vocalizations provides a useful window into underlying cognitive mechanisms and social awareness as well as offering insights into the evolution of the social skills required for language (Seyfarth and Cheney, 2003; 2014).

Compared to humans and animals capable of vocal learning, primates have considerably less control over their vocal production. Some have thus taken the view that primate vocalizations are highly canalized, involuntary expressions of emotions, which lack flexibility and are broadcast indiscriminately (Tomasello, 2008). Nevertheless, recent research has demonstrated a greater level of vocal control, flexibility (Lemasson, 2011) and even intentionality (Crockford et al., 2012) that goes beyond these previous assumptions. For instance, there is evidence that great apes can sometimes control, modify and target their vocalizations depending on their audience (Crockford et al., 2012; Genty et al., 2014), as well as to reshape existing vocalizations through vocal learning (Watson et al., 2015). Recent studies have also demonstrated vocal flexibility via voluntary control of the

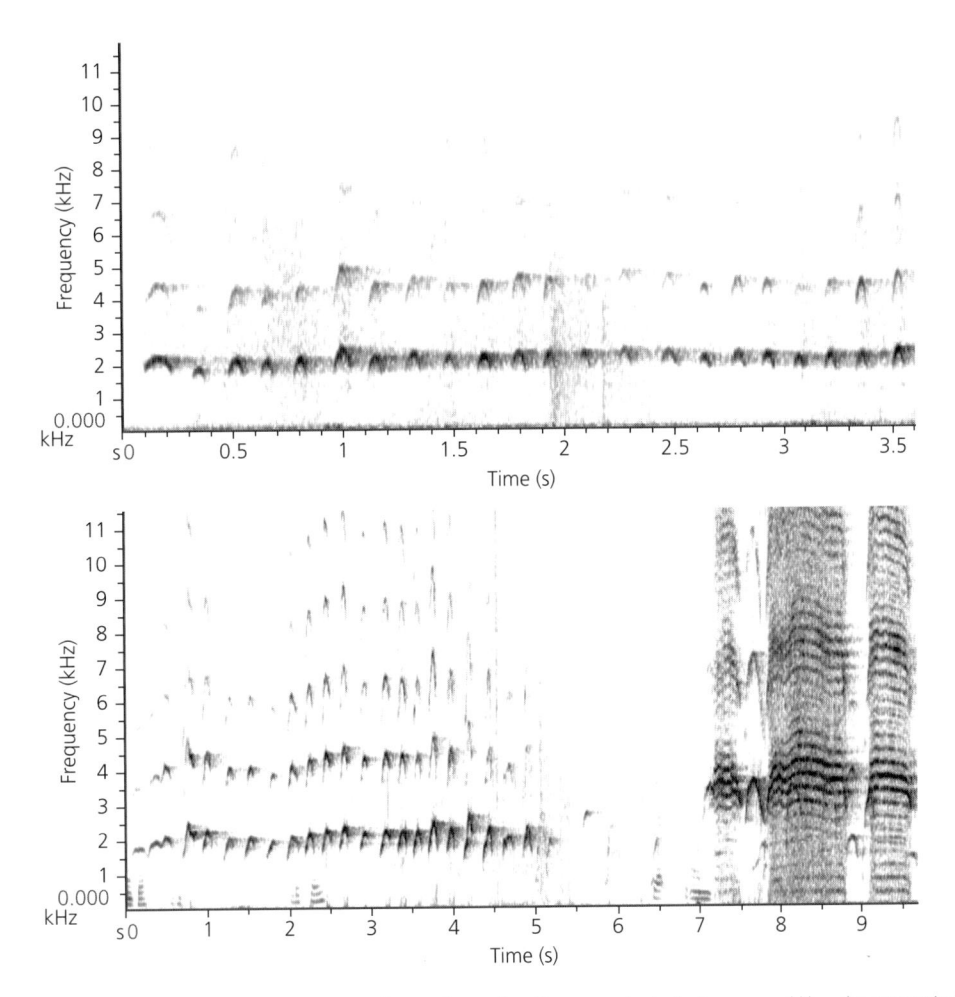

Figure 8.2 Time-frequency spectrograms showing contest hoot series produced by two males in the Bompusa wild bonobo community in Lui Kotale. The bottom pane shows the submissive scream response of the signaller's social target, a lower-ranked male compared to the target. (Les spectrogrammes de fréquence et durée montrant les cris de compétition produits par deux mâles au communauté naturel Bompusa de bonobos à Lui Kotale. Le cadre en-dessous montre le cri submissive en réponse de la cible sociale du signaleur, un mâle de rang inférieur à celui de la cible.)

breathing and articulator apparatus; the spontaneous production of unvoiced, novel calls by captive great apes, including bonobos, being one relevant example. Individuals across a number of captive facilities have been shown to voluntarily produce atypical, voiceless calls (e.g. 'raspberrys') in order to gain the attention of human caregiver/other conspecifics (e.g. *Pongo pygmaeus* (Lameira et al., 2015) *P. troglodytes* (Hopkins et al., 2007); *P. paniscus* (Taglialatela et al., 2003)), some of which have been acquired through social learning (Taglialatela et al., 2012).

Currently, the extent to which free-living bonobos are able to control their vocalizations requires further investigation. However, a recent study revealed that wild bonobos are able to use their peep vocalizations in functionally flexible ways to express a full range of emotional valence (positive–neutral–negative) across much of their daily lives (Clay et al., 2016). Functional flexibility refers to calls whose functions vary based on context rather than being tied to a specific context. With peeps apparently less tied to a predefined function, recipients may need to take pragmatic information into account to make inferences about call meaning. In this way, evidence of functional flexibility is indicative of an evolutionary early transition away from fixed vocal signalling towards more functionally flexible signals on the pathway towards language.

Compared to primate vocal production, greater flexibility and cognitive sophistication have been demonstrated in the context of signal comprehension (Seyfarth and Cheney, 2003; 2014). For instance, there is good evidence that numerous primate species can draw inferences about external events by attending to vocalizations (Seyfarth and Cheney, 2003; 2014), and that their interpretation of signal meaning varies based on surrounding contextual information (Scarantino and Clay, 2014). In addition to responding to ecological events, primates can make assessments about their social worlds by attending to vocalizations (Clay and Zuberbühler, 2014; Seyfarth and Cheney, 2003). In bonobos, one relevant context for studying social awareness is their socio-sexual behavior (Clay et al., 2011b; Clay and Zuberbühler, 2012; 2014). Compared to other primates, bonobos exhibit a heightened socio-sexuality, with individuals frequently engaging in sexual interactions across all age and sex combinations and in a multitude of positions (Hohmann and Fruth, 2000; Kano, 1992). In bonobos, sex does not only serve a reproductive function but is also used as a kind of social tool to reduce social tension as well as to express, establish, repair and consolidate social relationships (de Waal, 1987; Hohmann and Fruth, 2000). For instance, socio-sexual interactions appear to facilitate integration of newly immigrating females into the community and their subsequent development of affiliative relationships with non-related group members, especially established females (Furuichi, 2011; Kano, 1992).

During same-sex interactions, females sometimes produce 'copulation calls', as when copulating with males (Clay and Zuberbühler, 2011b). Despite obvious differences in the physical nature of the interaction, calls produced while interacting with male and female partners share the same acoustic morphology and cannot be statistically discriminated (Clay and Zuberbühler, 2011b). Nevertheless, these calls convey an array of social information, such as cues to caller identity and subtle differences in call delivery that are indicative of partner sex and dominance rank. Although females are more likely to call with male partners, females demonstrate considerable awareness of dyadic dominance ranks, with low-ranking females more likely to vocalize when interacting with high-ranking partners, regardless of partner sex (Clay et al., 2011). As shown in Figure 8.3, this rank effect is consistent with what is observed in female chimpanzees, who are more likely to produce copulation call with high-ranking males as compared to low-ranking males (Townsend et al., 2008). Low-ranking female bonobos were more likely to call when invited to have sex by a high-ranking female rather than when soliciting her. One interpretation of these rank and invitation effects is that female bonobos produce copulation calls to express or even advertise their socio-positive sexual with socially important group members, a sign of social recognition. This would be compatible with recent data from wild bonobos in Wamba that demonstrate the important role that high-ranking, older females play in supporting low-ranking females in coalitions against males (Tokuyama and Furuichi, 2016).

Interestingly, unlike these social variables, markers of the effects of physical stimulation, genital contact duration and spatial position showed no effect on call production (Clay and Zuberbühler, 2012). Similar to their sexual behavior, bonobo copulation calls thus appear to have become ritualized away from their original reproductive function to be used in social ways to express relationships with same-sex coalition partners. This makes an interesting contrast with chimpanzees, where copulation calls seem to have retained their original reproductive function of inciting male–male sexual competition (Townsend et al., 2008).

During female–female interactions, bonobo females are sensitive to the composition of their audience, being more likely to call when the alpha female is bystander (Clay and Zuberbühler, 2012). Interestingly, female chimpanzees are more likely to produce copulation calls in the presence of high-ranking audiences, although this effect is demonstrated for high-ranking males in the audience, not females (Townsend et al., 2008). The unusual shift towards sensitivity to female audiences is likely to relate to the central position that females play in bonobo society (Furuichi, 2011). In bonobos, establishing social relationships with dominant females is crucial for establishing a female's own social position as well as receiving female coalitionary support against males (Furuichi, 2011; Tokuyama and Furuichi, 2016).

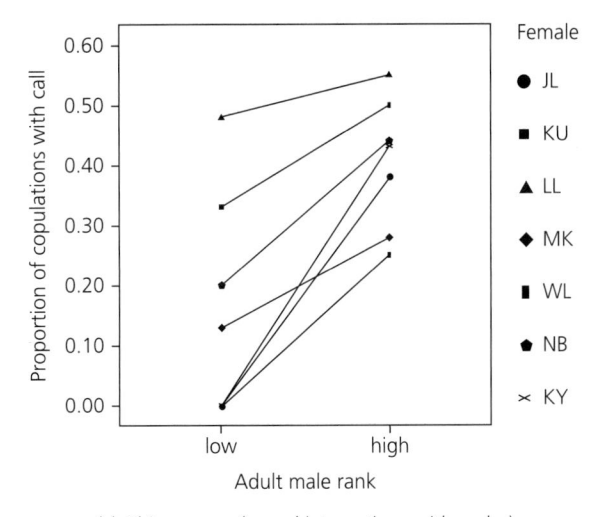

(a) Chimpanzee (sexual interactions with males)

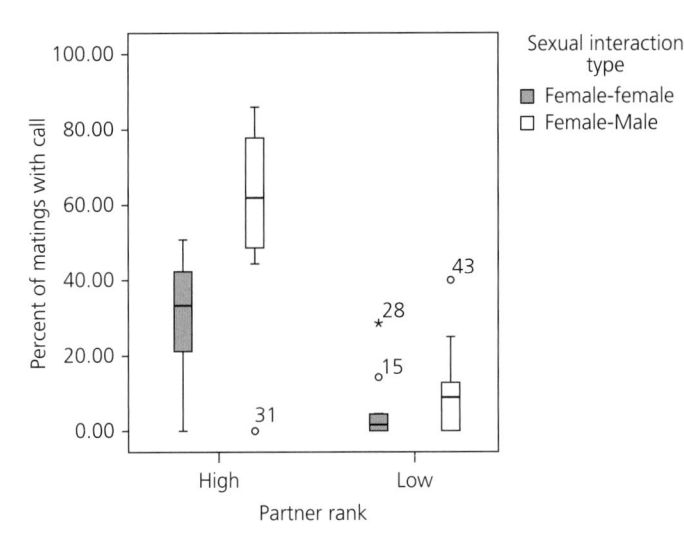

(b) Bonobo (sexual interactions with females and males)

Figure 8.3 Effects of social dominance and partner sex on copulation calling behavior in (a) female chimpanzees and (b) bonobos. Figures reproduced from (a) Townsend et al., 2008 and (b) Clay et al., 2011.
(Les effets de la domination sociale et sex sur le le comportement d'appel de copulation chez (a). Les chimpanzés femelles et (b) les bonobos femelles. Figures reproduites de (a) Townsend et al., 2008 et (b) Clay et al., 2011.)

As with sexual interactions, social conflicts represent another important event for any socially living animal. Social conflicts can have serious consequences for the broader social structure and dominance hierarchies, therefore being able to use vocalizations to understand the outcomes of third-party interactions can allow listeners to predict the interactions of group members and to better manage their own relationships. Recent research has shown that numerous primates are sensitive to vocalizations given during agonistic interactions and can draw inferences about resulting social relationships. For instance, in chimpanzees, playback studies have demonstrated that chimpanzees hearing different scream variants can infer something about the social roles of the conflicting individuals (i.e. victim or aggressor), the resulting direction of aggression (Slocombe et al., 2009) as well as the conflict severity (Slocombe et al., 2010).

In bonobos, Clay and co-workers examined the acoustic structure of screams produced by victims of aggression in response to socially expected or unexpected aggression (Clay et al., 2016). Expected aggression included conflicts over a contested resource and conflicts that were provoked by the victim, while unexpected aggression was any spontaneous, unprovoked hostility towards the victim. While bonobo victim screams carried strong acoustic cues to caller identity, the degree to which the conflict could be socially predicted by the victim predicted the scream acoustic structure better than conflict severity. The impact of social expectation on scream structure is consistent with previous research showing that great apes possess certain psychological expectations about social regularities in their groups and will protest when their personal expectations about a social situation are violated (Brosnan and de Waal, 2014). The fact that perceived violations of expectations more strongly discriminated scream structure than conflict severity suggests that the underlying psychological components of a conflict may be more important to bonobos than only the physical experience alone. The finding that bonobos vocally signal this to others suggests that audience members may be sensitive to it as well. In chimpanzees and other primates, some individuals intervene and police the social interactions of others (Goodall, 1986). Thus, one explanation is that these protesting screams may facilitate such interventions to benefit the signaller.

As well as during sex and social conflicts, bonobos also appear to be sensitive to social factors during feeding events and may be able to control vocal production in order to pursue certain goals. For instance, a captive study indicated that bonobos may be able to control the production of food-associated calls as a means promote reproductive and social strategies (van Krunkelsven et al., 1996). In this study, high-preference food items were hidden in the enclosure and individuals were subsequently released to forage, either alone or with others. Food-associated calls produced by males often resulted in female approach, which frequently resulted in copulation, suggesting that despite the fitness costs of attracting a feeding competitor, males may receive sexual benefits by attracting females to a feeding source. Females may also accrue benefits by attracting female coalition partners, who will ultimately enhance their status and enable them to monopolize feeding over males.

Bonobo call combinations

The production of food-associated vocalizations in bonobos represents a useful bridge to another aspect of primate vocal behavior relevant to the evolution of language, the production and comprehension of call combinations. Recent studies have indicated that bonobos can produce and understand call sequences that convey meaningful information about something in the external world (Clay and Zuberbühler, 2009; 2011a see also Schamberg et al., 2016, 2017).

Animal call combinations have received increasing attention from those interested in the evolution of syntax and higher-order linguistic structures (see Collier et al., 2014, for discussion). Currently, there is growing evidence that some primate species combine different calls in context-specific ways that are meaningful to recipients (Zuberbuhler and Lemasson, 2014). Combining calls has been interpreted as a way for primates to escape the limitations of poor vocal control and limited vocal repertoires. Relevant questions are whether component calls have their own independent meanings and, if so, whether they are combined in an additive or combinatorial way, to generate meanings beyond those of component calls (Collier et al., 2014).

Clay and Zuberbühler (2009; 2011a) examined bonobo call combinations during foraging. Unlike chimpanzees, who produce a single graded vocal type during feeding, the rough grunt (Slocombe and Zuberbühler, 2005), bonobos produce five distinct classes of food-associated calls during the discovery and consumption of food which they regularly combine into longer sequences (Clay and Zuberbühler, 2009). While these call types were produced in response to food of differing perceived qualities, the structure of the call sequences probably related to the quality of food encountered (Clay and Zuberbühler, 2009). In a subsequent playback study, receivers heard different combinations of food-associated call types which they used to guide their foraging decisions. By integrating information across calls individuals were able to extract meaning regarding the food quality encountered by the caller in order to search at the correct food location (Clay and Zuberbühler, 2011a).

While there appears to be some degree of complexity in bonobo call sequences in this context, it is essential to note that the manner in which bonobos appear to combine calls differs markedly from the hierarchical organization and complexity seen in human syntax. Nevertheless, the fact that bonobos were able to extract specific information from the calls suggests that, at least, receivers may possess some of the cognitive processes required for the comprehension of rudimentary syntactic structures, pointing to shared ancestry. This is consistent with findings that Kanzi, a language-competent bonobo, appears to understand syntactically complex spoken human sentences used in request contexts to a level equivalent of a human child (Savage-Rumbaugh et al., 1993; Tomasello, this chapter). Further research is needed to explore the production and perception of call combinations in bonobos, and the extent to which combinations of calls can be used to alter signal meaning.

Long-distance communication: vocal flexibility and call convergence

Vocalizations play a critical role in coordinating party movements across long distances, with individuals frequently depending on long-distance vocalizations to make decisions about where to go, when and with whom. One study suggested that, similar to what has been found in chimpanzees, wild bonobos are able to modify their long-distance 'high-hoot' vocalizations in order to converge and synchronize with other group members (Hohmann and Fruth, 1994). This study found high degrees of behavioral synchronization, with individuals producing high hoots in distinct alternating sequences with out-of-sight group members. Individuals also shifted their hoot pitch to correspond with those of responding group members. This surprising degree of vocal flexibility and synchronization suggests that bonobos may be able to control and modify this signal in response to social variables.

In addition to high hoots, bonobos produce tonally flat vocalizations, termed 'whistles', that occur in a wide variety of contexts, mostly preceding other calls (Bermejo and Omedes, 1999, Schamberg et al., 2016). While neither whistles nor high hoots appear to exhibit a high degree of context specificity, a sequential combination of the whistle-high hoot vocalization reliably signals a vocalizer's imminent travel to join a different party (Schamberg et al., 2016). Interestingly, high hoots can also be combined with another call, the 'low hoot' in the context of inter-party communication (Schamberg et al., 2017). Low hoots are an acoustically noisy, low-pitched vocalization in which the caller produces sound through both inspirations and expirations (de Waal, 1988; Bermejo and Omedes, 1999). A recent study showed that, in contrast to the whistle-high hoot combination, which signals the vocaliser's imminent travel to another party, the low hoot-high hoot combination is more likely to result in inter-party recruitment, that is individuals from other parties are more likely to approach the caller following this call combination (Schamberg et al., 2017). The production of distinct call combinations by bonobos to facilitate inter-party decision-making suggests a sophisticated degree of vocal flexibility. Current analyses are focused on understanding the context in which whistle-high hoot and low hoot-high hoot combinations are given and how this maps on to the bonobos' social relationships (Schamberg et al., 2016, 2017).

Bonobo gestural communication

Ape gestural communication and the evolution of language

Along with vocalizations, gestural communication in great apes provides critical insights into the evolution of language and communication, yet it has received relatively little research attention compared to other types of social behavior. Moreover, until recently, most studies of gesture concerned artificial gestures produced by human-trained apes, such as those taught American Sign Language (e.g. Patterson, 1978), or gestures produced during experimental tasks to ask questions about imitation (e.g. Custance et al., 1995), intentionality and perspective taking (Hopkins and Leavens, 1998). Less research has focused on how gestures are used to mediate social life in a naturalistic environment. Nevertheless, great apes use gestures extensively in daily life across a multitude of contexts; for instance, to request and share food (de Waal, 1997; Kano, 1992), to request grooming (de Waal, 1988; Pika and Mitani, 2006; van Hooff 1973) and frequently during play (e.g. Genty et al., 2009; Pika and Zuberbühler, 2008). These studies have already demonstrated that the production of great ape gestures is under considerable voluntarily control, show signs of flexibility and can be used in a socially targeted, intentional manner in order to alter a recipient's behavior in a desired way (Call and Tomasello, 2007; Genty et al. 2009). It has also been demonstrated that apes are able to generate novel gestures (Call and Tomasello, 2007; Genty et al., 2009; Goodall, 1986). These features observed in great ape gestures—context-independence, voluntary control, intentionality and generativity—are important components of human language, suggesting that they evolved before humans separated from the last common ancestor with modern great apes.

Definition of gestures

Before discussing advances made in the study of bonobo gesture research it is first relevant to define the term 'gesture'. A gesture is defined as a movement of the limbs and head that qualifies as intentional, in that it shows signs of (a) being socially directed, (b) being goal-directed, (c) being mechanically ineffective (i.e. the movement is insufficient to cause a result by physical force) and (d) having received a voluntary response from the targeted recipient. For a signal to be considered *socially directed*, the signaller has to produce the signal only in the presence of an audience while orienting its body and/or gaze to a specific recipient. It also has to show signs of audience checking (the subject looks at the targeted recipient before or during signalling and/or alternates gaze between the recipient and an event or object. For a signal to be considered *goal-directed*, there has to be evidence that the signaller pursues a specific goal. This criterion is fulfilled if two main sub-criteria are met. The first sub-criterion is that the signaller adjusts the signal modality according to the recipient's attentional state (uses silent gestures when recipient is fully attending, produces attention getters to inattentive recipients, i.e. audible or tactile gestures, or changes its location to face the recipient). The second sub-criterion is that the signaller shows response waiting (pauses and maintains visual contact with recipient after signalling) if a desired goal is not met immediately. In case of failing to get a response, the signaller is then expected to show persistence in communication, either by repeating the same signal or through elaboration by switching to other signals. In the case of getting a desired response, the signaller is expected to stop signalling (e.g. Call and Tomasello, 2007; Cartmill and Byrne 2007; Liebal et al., 2013).

Bonobo gestural communication

The first report of wild bonobo gestural communication came from the pioneering work of Kano (1980), among others. For example, Kano (1980) described that bonobos regularly use hand-begging gestures to request food from conspecifics. Nevertheless, as with the first bonobo vocal repertoire, the most extensive descriptions of bonobo gestures come from captive studies (e.g. Savage-Raumbaugh et al. 1977). Savage-Rumbaugh and co-workers (1977) described the gestures deployed during sexual solicitations to position the partner and demonstrated that specific gestures resulted in specific postures. De Waal (1988) produced the first extensive gestural repertoire of bonobos, describing in detail the form and context of production of 15 gestures. For example a 'stretch over' gesture is defined as an arm, stretched and raised until about head level, with the palm facing downward. This gesture is produced in the context of sexual solicitation to invite a partner to come to body contact. Pika and co-workers (2005) have also provided an inventory of sub-adult bonobo gestural repertoire and use.

In recent years, the systematic investigation of great ape gestural communication has revealed the presence of numerous core abilities required for language. Individuals can address others intentionally, usually to alter a recipient's behavior in a goal-directed way (Call and Tomasello 2007; Genty et al., 2009; 2014; Hobaiter and Byrne, 2011) or to convey information referentially (Genty and Zuberbühler, 2014; Leavens et al., 2004; Pika and Mitani, 2006). They also seem to understand the communicative function of their gestures in that they require the presence and attention of a suitable recipient, as documented by a large number of studies (e.g. Call and Tomasello, 2007; Genty et al., 2009; Pika et al., 2005). Bonobos and chimpanzees can also communicate cooperatively by using gestural turn-taking sequences (Fröhlich et al., 2016). A general conclusion from this literature is that great ape gestural communication is intentional, that they act as if taking into account basic mental states of their recipients, and that key cognitive abilities necessary for language were already present in the common ancestor of modern humans and great apes.

Flexibility of gesture production

a) Pragmatics: adjustment of gestures to signaller's goal

Beyond gesturing flexibly and understanding their communicative function (e.g. Call and Tomasello, 2007; Cartmill and Byrne, 2007; Genty et al., 2009; Pika et al., 2005), great apes gesture in variety of different contexts to convey different intentions or requests. For example, the typical 'hand reach' can be used to request food or to solicit an approach or a sexual interaction. Similarly, several gestures can be used to convey a single request. For example, in bonobos a range of gestures, postures and vocalizations are used to initiate sexual interactions (de Waal, 1988; Genty and Zuberbühler, 2014; Savage-Rumbaugh et al., 1977). A recent study by Genty and co-workers (2015) demonstrates that, in the context of sexual initiation, the signalling behavior of bonobos varies systematically depending on the initiator's social goals and sex. Five gestures were used frequently in the context of sexual solicitations: 'arm up', 'arm reach', 'hand reach', 'stretch over' and 'touch'. For example, in females 'hand reach' was specifically linked to sex initiations for appeasement, suggesting that primate gestures can sometimes be as functionally specific as vocalizations, in line with previous work showing that primate gestures can be used to clarify the signaller's intention in ambiguous situations (Genty et al., 2014). These results indicate that gestures do not only function to persuade a sexual partner but also to clarify the goal of the signaller. These findings also reflect that during the course of evolution, complex and flexible communication can evolve from basic and evolutionary old biological functions, such as reproduction, by increasing access to higher cognitive processes.

b) Shared knowledge: adjustment of gestures to familiarity of recipient

Much of human communication is based on knowledge and conventions shared between interlocutors. The history of shared experiences between a signaller and a recipient determines what inferences can be made about mutual knowledge. For instance, when infants are misunderstood they are more likely to elaborate linguistic requests to an unfamiliar than a familiar adult, indicating that they have some understanding of their shared interaction history (Tomasello et al., 1984). When communicating, great apes act as if taking into account their recipient's attention (Call and Tomasello, 2007) and comprehension (Cartmill and Byrne, 2007). However, it remains unknown whether they take into account familiarity, what others know based on shared interactions history, and whether they use that knowledge to communicate efficiently. Results from experimental studies indicate that bonobos and chimpanzees can infer the target of an experimenter's attention by remembering what he had previously seen or not seen (e.g. Maclean and Hare, 2012), which suggest that bonobos and chimpanzees can adjust their behavior efficiently to obtain food based on their knowledge of what one knows.

In a recent study, Genty and co-workers (2015) experimentally documented precursors of the ability to communicate through shared knowledge in bonobos. Their subjects, raised at the Lola Ya Bonobo sanctuary, Democratic Republic of Congo, usually requested food from their keepers by producing either common hand-begging gestures (de Waal, 1997; Kano, 1980) and, for some individuals, a number of idiosyncratic signals, like 'hand-claps + rasp grunts' or 'arm-up while standing bipedally'. The authors analysed how subjects adapted their communicative signal production to request food from a human partner depending on its familiarity ('familiar' versus 'unfamiliar'). The results showed that subjects behaved similarly to human infants interacting with adults (Tomasello et al., 1984) and language-trained bonobos interacting with their caretakers (Pedersen and Fields, 2009). Bonobos adapted signal production according to whether they knew the recipient or not. In case of communicative failure, they used more repetitions with a familiar recipient but elaborated more, by using new signals with an unfamiliar one, suggesting an apparent attempt to rectify an unsuccessful communication event and misunderstanding. The subjects seemed to understand which signals were more likely to persuade the familiar recipient to deliver the food and they seemingly quickly understood that the same signals are not necessarily equally effective when interacting with an unfamiliar recipient with whom they have no shared history. These results seem to indicate that bonobos take into account recipient familiarity and possibly consider the knowledge they share with them, acquired jointly during a common history of interactions, to adjust their communicative effort efficiently. Further research is necessary to highlight the underlying mechanisms responsible for the distinction between familiar and unfamiliar recipients and to what degree mental state attribution is involved.

Meaning in bonobo gestures

Despite evidence for shared features with language in great apes gestures, in particular their flexible and goal-directed use, it has been surprisingly difficult to identify their semantic content, or 'meaning'. Great ape gestures appear to be given to initiate, maintain on-going or terminate social interactions. Several gestures can be used to achieve the same goal and each gesture can be used to convey different intentions. Therefore, although some gestures seem to convey a more specific meaning than others (Hobaiter and Byrne, 2014), they generally seem to convey a broad range of meanings that depend on the context of production rather than possessing semantic or referential content, suggesting that meaning resides more in the pragmatic context than in the morphological form of the signal.

Humans regularly use *deictic* gestures, such as pointing, to direct a recipient's attention to a particular object or location, but also *iconic* gestures that recreate an aspect of the shape or movement of an object or event (Tomasello, 2008). Both types of gestures are defined as referential in that they either direct attention to a present referent or generate a mental representation of an absent referent. So far, there has been very little evidence for the presence of such referential gestures in the natural repertoire of great apes (e.g. Pika and Mitani, 2006) and we are not aware of any systematic analyses. This is surprising because language-trained apes can learn to communicate this way with humans (e.g. Savage-Rumbaugh et al., 1977). For example, Savage-Rumbaugh and colleagues (1986) reported that Kanzi and Mulika made hitting motions towards nuts they wanted a human observer to crack open for them. For Tomasello (2008), the iconic nature of ape gestures described in previous studies lies purely in the eye of the observer and is caused by 'garden-variety ritualized behaviors' of animals trying to move others in desired ways.

In a recent paper Genty and Zuberbühler (2014) revisit this claim with a study on a previously overlooked human-like gesture in bonobos that has both deictic and iconic character: beckoning in the context of initiating a sexual interaction (Figure 8.4). The gesture is produced with the apparent goal of persuading the recipient to approach and jointly retreat to a different location for sex. The default way to initiate sexual interactions in bonobos is to approach actively and solicit sex *in situ*, but sometimes initiators attempt to attract recipients to a different location, for example because of social competition, spatial inconvenience or other reasons. The beckoning gesture fulfils key criteria of deixis and iconicity (physical resemblance between the gesture and its referent) in that it communicates to a distant recipient the desired direction and travel path in relation to a specific social intention, to have sex at another location.

This study is, to our knowledge, the only case in the literature that goes beyond anecdotes about single individuals and represents the first significant evidence of a gesture within the natural repertoire of great apes that is deployed intentionally and to which recipients respond by mapping onto the signal's form, thus qualifying as iconic. The authors also found evidence that the signallers were making sure that the signal had been understood and responded to appropriately by gazing back to check that the recipient was following and by repeating the signal when unsuccessful in provoking the recipient's approach.

These findings are relevant in that they provide evidence that great apes can naturally use spatial reference as part of a communicative effort, with recipients responding to such signals appropriately. The ability to produce gestures that depict some spatial features of a desired action was therefore probably already present in the common ancestor of humans and apes. Although these results suggest that apes understand beckoning without problem, it is not clear how aware signallers really are when producing their beckoning gesture. For example, it

Figure 8.4 Illustration of the most frequent beckoning sequence to persuade a distant partner to approach and jointly move to a different location. Illustrations depict the sexual initiation posture (A) followed by a beckoning gesture: arm stretch towards recipient (B), sideways arm sweep towards self (C–E) and wrist twirl (F), then a body pivot (G–H) before walking away and regularly gazing back to check if recipient follows (I). (Illustration d'un bonobo démontrant la séquence d'appel la plus fréquente pour persuader une partenaire à distance à se rapprocher et se à déplacer ensemble. Illustrations montrant la posture d'initiation sexuelle (A) suivie par une gesture d'appel: bras étendue vers la recipiente (B), l'étendue du bras du côté vers soi (C–E) et la rotation du poignet (F), ensuite une rotation du corps (G–H) avant de partir et regarder le partenaire en s'éloignant pour voir si elle suivra (I). (Genty et Zuberbühler, 2014).)

is possible that the physical resemblance between the gesture (beckoning) and its referent (desired travel path) was entirely incidental or part of a biological signalling predisposition. A potentially fruitful line of research would be to test apes in their comprehension of novel iconic signals that depict other aspects of their natural behavior or even physical objects that are relevant to them.

What remains unknown is how this human-like beckoning gesture was acquired in this population. A long-standing debate in the ape gesture literature is whether gestures are learned across ontogenetic development (Call and Tomasello, 2007) or part of a species-specific repertoire (Genty et al., 2009; Hobaiter and Byrne, 2011). Most likely, gestures are acquired via a combination of both mechanisms.

In this study, the beckoning gesture was produced by several individuals from two different social groups and the authors found a significant positive correlation between the number of observed sexual initiations and the frequency of beckoning across individuals, suggesting that with more observations, the behavior would probably be observed in most or all mature individuals and could therefore be considered a species characteristic of bonobos. Ontogenetic ritualization (Call and Tomasello, 2007), therefore, can probably be discarded as an acquisition method in this particular case. A systematic study of this gesture in more captive and wild populations is required to answer this question.

It is, however, puzzling that great apes do not take more regular advantage of referential gestures in their natural communication to refer to absent entities, a fundamentally human capacity. More research is needed to reveal the cognitive and psychological limitations responsible for the rarity of these signals in great apes.

Combining vocalizations and gestures: multimodal communication in bonobos

While considerable progress has been made in our understanding of the vocal and gestural components of natural bonobo communication, a promising new direction in studies related to the evolution of language is the manner in which these and other modalities interact multi-modally (Liebel et al., 2013; Pollick and de Waal, 2007; Slocombe et al., 2011; Taglialatela et al., 2011). While this research area for primates is still young, multimodal signals have been demonstrated to play an important role in communication behavior of various animals (Partan and Marler, 2005) and represent an undisputedly core component of human language and communication. In human communication, for example, speech signals are routinely combined with vocal and visual signals to convey and modify the speaker's intended meaning (Tomasello, 2008).

While primates regularly combine signals multi-modally, an unresolved question is whether the integration of signal modalities functions to increase signal amplitude (i.e. to generate redundancy) or whether it can actually change the semantic meaning of the signal set (Partan and Marler, 2005).

Pollick and co-workers (2007) compared the multi-modal communication of bonobos and chimpanzees and found that bonobos showed greater flexibility than chimpanzees in their combination of signals. Furthermore, when bonobos combine gestures with facial and/or vocal signals, the gestures are more effective at eliciting a response than when used alone (Pollick et al., 2007).

Genty and co-workers (2014) recently found that the multi-modal integration of vocalizations and gestures during bonobo contest hoot displays appeared to disambiguate their intended meaning. As well as socially to provoke a target, 'contest hoots' were sometimes also used during friendly play (Genty et al., 2014). While the calls' acoustic structure remained the same across play and agonistic contexts, there was a significant difference in the choice of associated gestures between the two calling contexts. Gestures were classified into two categories: 'rough' and 'soft' gestures. Rough gestures were performed with force (i.e. flap) whereas soft gestures were silent and performed without force (i.e. hand reach). During play, contest hoots were significantly more often combined with soft than rough gestures compared to agonistic challenges. The authors concluded that contest hoots indicate the signaller's intention to interact socially with important group members, while the gestures provide additional cues concerning the nature of the desired interaction. These data reveal that multi-modal signals may function to help convey the signaller's apparent goal, or as in the case of play, maintain an ongoing activity.

Conclusion

Although research on bonobo communication is still in its infancy, the studies documented here already highlight the considerable flexibility, diversity, complexity and, moreover, the frequent subtlety of bonobo communication across multiple modalities. These studies indicate a sophisticated social awareness, which bonobos use to modify their communicative output in response to specific targets in order to pursue different social goals. In the vocal domain, this includes, for example, using vocalizations to advertise socially relevant sexual interactions to important group members (Clay and Zuberbühler, 2012), to protest against violated social expectations (Clay et al., 2015), as well as to combine calls together in order to communicate travel intentions (Schamberg et al., 2016, 2017) or to convey information about food (Clay and Zuberbuhler, 2009; 2011a). In the gestural domain, bonobos are able to use gestures in intentional (Pika and Zuberbühler, 2008) and iconic ways to depict a desired intention and travel direction of their recipient (Genty and Zuberbühler, 2014). Bonobos also seem to possess a rudimentary understanding of shared conventions that allow them to adjust their communication signals when misunderstood (Genty et al., 2015). Moreover, when bonobos combine gestures with vocal signals, they are more effective at eliciting a response than when they used alone (Pollick et al., 2007), and appear to assist in conveying the signaller's apparent goal (Genty et al., 2014). Overall, studies of natural communication in bonobos, our closest living relatives along with chimpanzees, provide crucial insights into underlying cognitive mechanisms and social awareness that most likely provided the critical socio-cognitive scaffolding that supported the evolution of human language.

Acknowledgements

We sincerely thank Klaus Zuberbühler for his ongoing support and involvement in much of the research presented here. We are very grateful to Brian Hare and Shinya Yamamoto for organizing the 2012 Bonobo Symposium at the International Primatology Society Congress in Mexico and for inviting us to take part in this book which emerged from this meeting. We also thank Brian Hare, William Hopkins and Katie Slocombe for their insightful comments on earlier versions of this manuscript. Much of the empirical research reviewed in this chapter has been funded with grants by the Leverhulme Trust, the BBSRC, the LSB Leakey Foundation, the National Geographic Society: Committee for Research and Exploration Grant, the British Academy Small Research Grant, the European Union Seventh Framework Programme for research, technological development, and demonstration under grant agreement 283871, private donors associated with the British Academy and the Leakey Foundation and the Emory College of Arts and Sciences. We thank the Ministries of Research and Environment in the Democratic Republic of Congo for their support (permit no. MIN. RS/SG/004/2009). We thank Claudine André, Fanny Mehl, Dominique Morel, Valery Dhanani, Pierrot Mbonzo and the Lola ya Bonobo staff for their support. We thank Frans de Waal, Gottfried Hohmann, Barbara Fruth, Isaac Schamberg, the village of Lompole, all staff of the Lui Kotale Bonobo research Project in the Democratic Republic of Congo and the Max Planck Institute of Evolutionary Anthropology for their collaboration.

References

Bermejo, M. and Omedes, A. (1999). Preliminary vocal repertoire and vocal communication of wild bonobos (*Pan paniscus*) at Lilungu (Democratic Republic of Congo). *Folia Primatologica*, 70(6), 328–57.

Brosnan, S.F. and de Waal, F.B.M. (2014). Evolution of responses to (un)fairness. *Science*, 346, 6207.

Call, J. and Tomasello, M. (2007). *The Gestural Communication of Apes and Monkeys*. New York, NY: Taylor and Francis Group/Lawrence Erlbaum Associates.

Cartmill, E.A. and Byrne, R.W. (2007). Orangutans modify their gestural signaling according to their audience's comprehension. *Current Biology*, 17, 1345–48.

Clay, Z., Archbold, J., and Zuberbühler, K. (2015). Functional flexibility in wild bonobo vocal behaviour. *Peer Journal*, 3, p.e1124.

Clay, Z., Pika, S., Gruber, T., and Zuberbühler, K. (2011). Female bonobos use copulation calls as social signals. *Biology Letters*, p.rsbl20101227.

Clay, Z., Ravaux, L., de Waal, F.B.M., and Zuberbühler, K. (2016). Bonobos vocally protest to violations of social expectations. *Journal of Comparative Psychology*, 130(1), 44.

Clay, Z. and Zuberbuhler, K. (2014). Social awareness in chimpanzees and bonobos. In: Dor, D., Knight, C., and Lewis, J. (eds). *The Social Origins of Language*, Vol. 19 pp. 141–156. Oxford: Oxford University Press.

Clay, Z. and Zuberbühler, K. (2009). Food-associated calling sequences in bonobos. *Animal Behaviour*, 77(6), 1387–96.

Clay, Z. and Zuberbühler, K. (2011a). Bonobos extract meaning from call sequences. *PloS One*, 6(4), e18786.

Clay, Z. and Zuberbühler, K. (2011b). The structure of bonobo copulation calls during reproductive and non-reproductive sex. *Ethology*, 117(12), 1158–69.

Clay, Z. and Zuberbühler, K. (2012). Communication during sex among female bonobos: Effects of dominance, solicitation and audience. *Scientific Reports*, 2, 291.

Collier, K., Bickel, B., van Schaik, C.P., Manser, M.B., and Townsend, S.W. (2014). Language evolution: Syntax sharing contributed to the establishment of highly cooperative human societies.

Bonobos and chimpanzees are ideal study species because they frequently share food between non-kin. Sharing with kin is relatively prevalent in primates, and can be explained by kin selection (Hamilton, 1964); in contrast, sharing with non-kin is largely limited to apes and New World monkeys (Jaeggi and van Schaik, 2011). Many researchers have struggled to explain the mechanism and evolution of non-kin sharing since it cannot be explained by kin selection theory. The prevailing hypotheses are reciprocity and sharing-under-pressure. The reciprocity hypothesis posits that animals share in exchange for a past or future benefit (e.g. receiving the food items or a different currency, such as coalition membership or mating opportunities). In contrast, the harassment hypothesis explains that there is an immediate benefit for the owner. Thus, when the beggar negatively affects the owner's feeding rate, the owner may give up some food but retain the majority; sharing to avoid further harassment from the beggar ('sharing-under-pressure', Wrangham, 1975; termed 'tolerated theft' for human sharing, Blurton Jones, 1984).

These previously postulated hypotheses are attractive and plausible for some aspects of food sharing; however, many other aspects could be missed because these hypotheses were predominantly based on the chimpanzee sharing behavior while largely ignoring its sister species, the bonobo. Previous studies have pointed out many conspicuous differences between chimpanzees and bonobos (review: Hare and Yamamoto, 2015), which also are seen in food sharing (e.g. Yamamoto, 2015). Thus, bonobo food sharing might be explained by other mechanisms, which could further our understanding of how humans evolved various forms of food sharing. The comparison between chimpanzees and bonobos is one of the main purposes of this chapter, and it suggests a fundamental question: did cooperation evolve in severe situations, which drove animals to share out of necessity, or in rich environments where resource abundance allowed animal generosity? Previous theories employing evaluation of nutrition or fitness cost–benefit were proposed based on the former assumption; that is, food sharing should increase nutritional intake and would be essential for survival in severe environments. In humans, however, it has been suggested that hunter–gatherer sharing in most cases may result from surplus food being obtained and shared during central place foraging (Berbesque et al., 2016). In addition, some forms of human food sharing seem to be purely social and ritualized with benefits beyond their immediate nutritional significance (Mauss, 1925). A Japanese proverb says that the poor cannot afford manners; only when basic needs for living are met can people spare the effort to be polite. Research is needed to determine if non-human animals demonstrate such forms of social sharing.

In this chapter, we first discuss some important characteristics of food sharing among wild bonobos in the Wamba forest. Wamba is the longest-lasting research site, established in 1973, and the bonobos have been well habituated to humans such that we could observe their food sharing and social interactions in excellent observational conditions (for basic information regarding Wamba, see Furuichi et al., 2012). We refer primarily to the most recent study (Yamamoto, 2015; for food sharing in Wamba, see also: Kano, 1980; Kuroda, 1984; Hirata et al., 2011) as well as new reports of significant events and data. We make comparisons between bonobos and chimpanzees as well as within-bonobo comparisons for those in Wamba and other research sites. We also discuss annual differences in sharing frequency in Wamba, sex differences and developmental changes in bonobo food sharing. Through these comparisons we propose an explanatory hypothesis, that is, bonobos demonstrate courtesy food sharing characterized by begging for a social bond.

Food sharing in Wamba

Food type

The main food type that is shared among wild bonobos of the E1 group in Wamba is plant-based (Kano, 1980; Kuroda, 1984; Yamamoto, 2015). Prey animals (*Anomalurus* sp.) are also consumed but this is infrequent and only four cases involving sharing of prey have been reported so far (one single case involved multiple sharing events between multiple pairs) (Hirata et al., 2010; Ihobe, 1992; Ingmanson and Ihobe, 1992; Yamamoto, 2015). Among the plants eaten, large fruits of *Anonidium mannii* and *Treculia africana* (as large as, or sometimes lager than, a rugby football and basketball, respectively) are the most frequently shared items in the natural environment (Figure 9.1; see Table 9.1 for food types shared among Wamba bonobo reported in Yamamoto, 2015).

One of the most important points is that the plant food can be obtained by independent individuals without cooperation or specialized skills, which is a prominent difference from the acquisition of prey. Although the large fruits are often available in small numbers compared to smaller fruits, Yamamoto (2015) reported that individuals of any age–sex class, except dependent infants, could obtain and consume *Anonidium* fruits (also known as junglesop) independently (i.e. not through sharing from others). The data showed that not only individuals who were unable to obtain junglesop fruit independently but also skilled foragers begged for the fruit from others. The bonobos also shared smaller fruits and other plant food, which were abundant and available for any individual at a site. We will later describe some cases of the sharing of abundant food in detail.

Table 9.1 Food items shared among independent individuals in Wamba observed during a seven-month study. (Data republished from Table 1 in Yamamoto 2015, *Behavior*).
(Aliments partagés parmis des Bonobos Wamba durant une étude de sept mois. (vu dans Yamamoto 2015, *Behavior*).)

Species	Local name	Food type	Size of food	No. of events
Anonidium mannii	Bolingo	Fruit	Big	150
Treculia Africana	Boimbo	Fruit	Big	6
Anomalurus spp.	Itere	Fruit	Big	3
Saba florida	Bossenda	Fruit	Big	2
Brachystegia laurentii	Langa	Fruit (seed)	Big (small)	2
Cola chlamydantha	Bokotikoti	Fruit	Small	2
?	Botete	Fruit	Small	2
Meliponinae spp.	Liutsu	Honey	Big	2
Isolona congolana	Bofiningo	Fruit	Small	1
Musanga cecropioides	Bombambo	Fruit	Small	1
Dialium pachyphyllum	Elimilimi	Fruit	Small	1
Dacryodes edulis	Bosou	Fruit	Small	1
Parkia bicolor	Lilembe	Fruit	Small	1
Pancovia laurentii	Botende	Fruit	Small	1
Landolphia awariensis	Batofe	Fruit	Small	1
Raphia sp.	Bolilo	Pith	Big	1
Guarea laurentii	Litoku	Pith	Big	1
Total				178

Figure 9.1 (a) A fruit of *Anonidium mannii*, and (b) its remains after a bonobo ate the fruit. Some flesh remained around the seeds (Photo by Shinya Yamamoto; this figure is republished from Figure 1 in Yamamoto 2015, *Behavior*).
((a) Un fruit Anonidium mannii, et (b) ces déchets après qu'un bonobo l'a mangé. De la peau restante autour des pépins (Photo par Shinya Yamamoto; Cette figure est publiée dans Yamamoto 2015, *Behavior*).)

Meat sharing observed in Wamba is similar to fruit sharing in that females are predominantly involved in the events, which is different from chimpanzees' meat sharing (Sakamaki and Furuichi, unpublished data). However, the bonobos' meat sharing seems to differ from their fruit sharing in terms of food transfer frequency (Hirata et al., 2010; for Lomako data see also White, 1994). Meat owners often denied giving up their prey, although recipients begged for meat more intensively than for fruit. Meat can be considered a hard-to-obtain, precious food item, and in this sense, it is similar to meat sharing in chimpanzees. Thus, to highlight bonobo-specific points (i.e. courtesy food sharing) hereafter we will focus on their plant food sharing.

Transfer type

Almost all food transfers are initiated by recipients. Yamamoto (2015) categorized food transfer observed during 150 *Anonidium* sharing events into three types and reported the frequency for each: proactive transfer (found in one out of 150 sharing events; 0.7 per cent), which was initiated by the owner in the absence of begging by the recipient; reactive transfer (found in two events; 1.3 per cent), where the owner facilitated sharing upon the recipient's begging; and passive transfer (found in all 150 events; note that a single sharing event involved multiple transfers, which were sometimes categorized into different types of transfer), where the owner did not facilitate sharing but simply tolerated the recipient's food taking. Numerous earlier studies pointed out the importance of recipient begging to solicit food transfer (e.g. Hirata et al., 2010; Kuroda, 1984; Yamamoto, 2015; Figure 9.2, Plate 9). The most conspicuous form of begging is extending an arm towards the food in the possessor's hand, leg or mouth. Although close observation (Yamamoto, 2015; 'staring' in Kuroda, 1984; 'peering' in Furuchi, 1989) has been described as 'begging' in some previous studies (Fruth and Hohmann, 2002; Gilby, 2006; Goldstone et al., 2016), Wamba researchers normally exclude this behavior from the 'begging' category. This is because: 1) close observation has almost no effect on the solicitation of sharing among bonobos (Furuichi, 1989; Kuroda, 1984; Yamamoto, 2015); 2) the animals might only be interested in the item in another individual's hand without having any intention of taking it; and 3) close observation is also conspicuous in other contexts, particularly during social learning in apes (e.g. Yamamoto et al., 2013). Therefore, the function and motivation of close observation is ambiguous. Sharing events normally occurred peacefully and no physically agonistic interactions were observed during the 150 sharing events reported by Yamamoto (2015).

Figure 9.2 Food begging among wild bonobos in Wamba (Photo by Takeshi Furuichi). Bonobos often beg and obtain a share from a food-owner's mouth. This means that the food owner has already consumed much of an edible part before the food item is transferred to a recipient. (Suppli pour la nourriture chez les bonobos sauvages à Wamba. (Photo par Takeshi Furuichi).Les bonobos supplient fréquemment et obtiennent une portion de la nourriture de la bouche du propriétaire. Ceci signifie que le propriétaire de la nourriture a déjà consommé la majorité de la partie comestible d'un aliment avant l'avoir donné à un récipient.) (see Plate 9)

Social relationships

Adult females play a central role in food sharing (Table 9.2). Among the 150 *Anonidium* sharing events between independent individuals reported by Yamamoto (2015), adult females accounted for 92.7 per cent of the owners (adult male: 7.3 per cent; and no sharing from juvenile owners) and 50.7 per cent of the recipients (adult male: 3.3 per cent; juveniles: 46.0 per cent). When restricted to non-kin sharing, adult females accounted for 89.5 per cent of the owners and 80.0 per cent of the recipients. Although females have priority access to food (Furuichi, 1989, 1997; Kano, 1992; White and Wood, 2007), these figures are notable considering that all age–sex classes consume the fruits (adult females: 58.9 per cent; adult males: 32.2 per cent; juveniles: 8.8 per cent). Food was transferred predominantly from dominant to subordinate individuals. Among the 150 sharing events, 129 events (86.0 per cent) were from dominants to subordinates, 10 (6.7 per cent) were from subordinates to dominants and 11 (7.3 per cent) were between individuals where the dominance relationship was ambiguous. New analyses revealed that in 36 (50.0 per cent) of 72 sharing events between adult females, the recipients were newly (less than five years) immigrant young females who were subordinate to the residents.

We found no conspicuous evidence for reciprocity in bonobo food sharing. It appears that the bonobos did not establish food-for-food reciprocity. The food transfer was unidirectional from the dominant to the subordinate, and no significant relationship was observed between the number of times giving and the number of times receiving (Yamamoto, 2015). Considering that most sharing among adults was observed between same-sex individuals, the food-for-sex hypothesis is not applicable. Food-for-support is also not plausible. Tokuyama and Furuichi (2016) revealed that dominant females support subordinate females when they make alliances to combat male harassment, the direction of which is the same as that of food sharing (although this data were obtained in a neighbouring group in Wamba, we assume this pattern is the same in the E1 group). Thus, dominants are helpful towards subordinates in both food sharing and support. Food-for-grooming might be possible because dominants attract more groomers than subordinates (Ryu et al., 2015). However, while there has been no systematic examination of the relationship between food sharing and grooming in the E1 group of Wamba, several intensive studies have suggested no evidence for food-for-grooming in other research sites for wild bonobos (Lomako: Fruth and Hohmann, 2002; LuiKotale: Goldstone et al., 2016).

Begging for social bonds

Begging for abundant plant food

The characteristics of food sharing among wild bonobos suggest that bonobo sharing cannot be explained solely by the nutritional functions of food. In chimpanzees, highly nutritious meat and hard-to-obtain, delicious cultivated fruit is shared (e.g. Boesch, 1994; Hockings et al., 2007). Equal distribution is impossible and, in a typical case, only a single individual can possess a prey at the beginning. It is natural that other individuals surround the owner and beg for the precious meat and fruit, which otherwise they could never eat. In contrast, animals can find and obtain natural plant food by

Table 9.2 Sex and developmental stage of individuals that were included in 150 events of junglesop fruit sharing during a 7-month study in Wamba. The number of individuals in each age–sex class in the E1 group is represented by the median number throughout the study periods. 'Adult' in the present study includes adolescent and adult individuals in the typical categorization. (Data republished from Table 4 in Yamamoto 2015, *Behavior*).
(Les étages sexuelles et développementales des individus qui étaient inclus dans 150 événements de partage du fruit Anonidium mannii (junglesop) dans une étude de 7 mois à Wamba. (vu dans Yamamoto 2015, *Behavior*).)

Shared from	To				
	Adult Female	Adult Male	Juvenile Female	Juvenile Male	Total (%)
Adult Female (N = 10)	72	5	16	46	139 (92.7)
Adult Male (N = 9)	4	0	0	7	11 (7.3)
Juvenile Female (N = 1)	0	0	0	0	0 (0.0)
Juvenile Male (N = 2)	0	0	0	0	0 (0.0)
Total (%)	76 (50.7)	5 (3.3)	16 (10.7)	53 (35.3)	150 (100)

themselves without any cooperation or specialized skills. Actually, all the individuals in the E1 group of Wamba, even young juveniles at the age of four and elders who could travel on foot, could find and eat the junglesop fruit by themselves. Therefore, the bonobos do not beg solely for nutritional benefits and it is plausible that social interaction over food is itself important.

Here, we report two new, typical cases of this socialized food sharing.

Case 1 on 13 July 2011

Anna (low-ranking, recent immigrant, young female) and Yukiko (juvenile female) were obtaining parts of the fruit of *Treculia africana* (45 cm in diameter) from Nao (high-ranking female) sitting on a big branch 4 m above the ground. Nao did not provide a share proactively or reactively, but just tolerated their taking some small portions from her hand or mouth (i.e. passive transfer). Seemingly by accident, Nao dropped approximately half the fruit on the ground. Neither Anna nor Yukiko descended the tree to acquire the dropped fruit, although it was clearly available and they noticed it fell. Instead, they continued to beg for a share from the rest in Nao's hand. When another individual, Yuki (high-ranking female), obtained the dropped fruit and climbed up the tree with it, they clustered around Yuki to obtain a share.

Case 2 on 3 August 2011

Hoshi (middle-ranking female) begged Kiku (high-ranking female) to share food particles of pith from *Guarea laurentii*, which was abundant and easily available to them at the site. Kiku chewed the pith, like sugarcane, to drink its juice, and Hoshi took the wadge from Kiku's mouth. When Kiku left the site, Hoshi stayed in the same place and ate some fresh pith available within her arm's reach.

Both of these cases are characterized by the recipient begging for another individual's food item, which was readily available to them at the site. It seems clear that the bonobos did not beg solely for the nutritional benefit obtained by food sharing because there were easier ways to get the same food in larger quantities (and probably of better quality) at the site. Thus, it is likely that they were motivated by the social interaction occurring around food itself. In both cases, subordinate individuals begged for food from dominant owners. Furthermore, Anna was a young female who had immigrated into the E1 group five months prior to the sharing event (Case 1). This type of observation supports the idea that subordinate recipients seek to bond with elder and dominant group members in part through food sharing. Observational evidence of this type of socialized food sharing is accumulating, and further analysis should reveal its social aspect in more detail (Yamamoto et al., in preparation).

Courtesy food sharing

We have labelled this type of socialized food sharing 'courtesy food sharing'. In humans, we can easily find this type of sharing in daily life. In Japan, we have a custom called *osusowake* in which people share food with neighbours when they make dishes containing more than they or their family can eat. The goal of this is not predominantly to help poor people; rather, it functions to facilitate communication and develop social bonds with neighbours. Mauss (1925) introduced details of several cultures' sharing traditions, such as *kula* in the Trobriand Islands and *potlach* in north-western American hunter–gatherers. He showed that sharing (giving, receiving and reciprocating) is regulated by social norms, and that properties which are transferred between individuals (or social groups) are not restricted to material benefits but include social, normal, ethical and even religious and spiritual ones. People strengthen social relationships with other individuals (or social groups) through sharing.

Bonobos might be demonstrating a basic form of courtesy food sharing and might be the only non-human primate demonstrating such a highly socialized form of food sharing with a function beyond its nutritional benefits. Let us repeat the important point. The bonobos sometimes beg for food which is easily attainable without asking another individual to share. Although many researchers have suggested that wild chimpanzees share food to establish social relationships with others and/or promote their status (Gomes and Boesch, 2009; Hockings et al., 2007; Mitani and Watts, 2001), the food itself is always a valuable item, and thus it serves as a 'currency'. In contrast, in the bonobo sharing just described—although this is the case in some of their sharing but not necessarily all—the transferred food has little nutritional value itself but instead probably serves as a 'catalyst' for a positive social interaction that potentially strengthens social bonds.

Now, let us address the difference between bonobos and humans, which seem to be unique in primates in displaying courtesy food sharing in contexts other than mating. It is notable that proactive giving characterizes human sharing. Although some of the giving is regulated by the recipients' implicit request or social norm of reciprocity, unsolicited giving and helping is conspicuous in humans, even in early developmental stages (Mussen and Eisenberg-Berg, 1978; Warneken, 2016). In contrast, bonobo sharing is almost always initiated by the recipients' explicit request. Bonobos seem to make social bonds through receiving or begging rather than through giving.

We emphasize the importance of shifting our viewpoint from donors to recipients when examining bonobo sharing. Many researchers have focused on the perspective of donors and sought explanations for the reason why food owners relinquish their precious items, which is represented in many article titles containing phrases such as 'to give, or not to give' (e.g. Gurven, 2004). When observing bonobo food sharing from the donors' viewpoint, we can see that their food sharing is not an important phenomenon. Donors may share only the surplus or a non-important food item. Occasionally, a single individual cannot eat an entire junglesop fruit, and at other times, they can easily find plenty of the same food items in the same food patch. They often share just a small portion after they have consumed most of the best parts (see the example in Case 2). Therefore, donors may incur little cost from sharing. This may not attract much attention from evolutionary theorists because the evolution of no-cost cooperation is easily explained theoretically. However, this becomes interesting when we see the interaction from the recipients' viewpoint. There are three notable points we would like to repeat. First, the bonobo recipients begged for plant food, which they could have found and obtained by themselves and the same food was often clearly abundant in the site. Second, young and subordinate individuals begged for food from elder and dominant owners. Finally, recipients did not reciprocate in a clear manner.

In fact, in bonobos and chimpanzees, cooperative behavior is characterized by the importance of the recipients' request, not only in food sharing but also in helping interactions. Several experimental studies with captive chimpanzees have shown that they help others upon request, but proactive helping is infrequent (Melis et al., 2011; Warneken et al., 2007; Yamamoto et al., 2009, 2012). In experiments conducted in the Primate Research Institute, Kyoto University, Japan (for recent projects here, see Matsuzawa, 2013), chimpanzees handed a tool, predominantly upon request, to a conspecific partner who needed the tool in order to drink juice (Yamamoto et al., 2009). In subsequent tests, it was shown that the chimpanzees selected the appropriate tool from a set of seven objects, according to the partner's situation. Thus, they could understand the partner's need and desire; however, this perspective-taking ability did not automatically lead to prosocial behavior in chimpanzees (Yamamoto et al., 2012). Bonobos may have more proactive tendencies, because experimental studies with captive bonobos have shown they prefer to share food with a conspecific, even more with a stranger, rather than monopolizing a food item (Hare and Kwetuenda, 2010; Tan and Hare, 2013). However, other studies have found bonobos' reluctance to share with groupmates (Bullinger et al., 2013; Jaeggi et al., 2010a; Parish, 1994), and proactive food transfer was almost never observed in the wild (Yamamoto, 2015). Thus, it seems safe to assume that recipients play an important role in various contexts of ape cooperation.

Regarding courtesy food sharing, we propose the idea that the evolutionary origin of this highly socialized food sharing, which goes beyond its nutritional function, can be found in the 'begging for social bond' observed in bonobos. The human 'giving for social bond' should have evolved after humans acquired proactivity in prosociality. Recent intensive discussion on the evolution of proactive prosociality

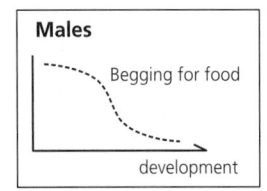

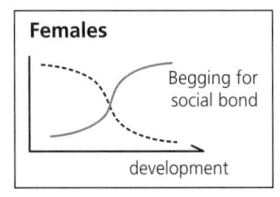

Figure 9.3 Hypothesized sex difference and developmental change of food begging, which consists of 'begging for food' (dashed line) and 'begging for social bond' (solid line). This hypothesis predicts that 'begging for food' declines with development, but 'begging for social bond' increases only in females after their immigration into a new non-natal group. This might be a mechanism to maintain the high-frequency food sharing in adult females, but not in males.
(L'hypothèse sur la différence sexuelle et le changement développemental du suppli pour la nourriture, qui est 'supplier pour la nourriture' (ligne pointillée) et 'supplier pour un lien social' (ligne solide). Cette hypothèse prédit que 'supplier pour la nourriture' diminue avec le développement, mais 'supplier pour un lien social' augmente seulement chez les femelles après leur immigration dans un nouveau groupe non natal. Ceci peut être le mécanisme qui maintient la haute fréquence du partage de nourriture chez les femelles adultes, mais pas chez les mâles.)

has led to some explanatory hypotheses. For example, allomaternal care is considered a candidate explanation. By systematic comparison of 24 groups belonging to 15 primate species, Burkart and her colleagues (2014) suggested that proactive prosocial motivation arose whenever selection favours the evolution of cooperative breeding. Yamamoto and Tanaka (2009) also discussed the evolution of proactive prosociality in relation to the evolution of social norm and reputation system in humans. In a society with indirect reciprocity networks, people would be driven to have proactive tendencies in cooperation because this would be evaluated by others and lead to a good reputation in a normative society. Given that the 'giving for social bond' seems to be restricted to humans, the socialized food sharing in humans may be highly developed in accordance with the emergence of the human norm and reputation system, albeit this socialized sharing has an evolutionary origin in non-human primates in its primitive form as 'begging for social bond'.

Comparisons

Development and sex differences

Here we discuss development and sex differences in bonobo's courtesy food sharing. The data shown in Table 9.2 show that, although juveniles of both sexes beg for food, in adults (including sub-adults), females, but not males, account for the majority of recipients in junglesop fruit sharing. It appears that females and males may have different developmental processes concerning food sharing.

We propose a hypothesis that explains the development and sex differences by conceptually separating 'begging for food' and 'begging for social bond' (Figure 9.3). In juveniles who are somewhat dependent on their mother, both females and males may beg for food from their mother for the sake of the food itself ('begging for food'). As they develop and become capable of finding and accessing food patches more easily, this type of food-begging behavior probably decreases. On the other hand, 'begging for social bond' may increase in females after their immigration into a new non-natal group. Bonobos are characterized by their strong non-relative, female–female social bonds, despite female-biased dispersal in their society (Furuichi, 1989; 2011). They must establish social relationships with other resident females that were previously unknown to them (Idani, 1991). In this process, the 'begging for social bond' may play an important role. Therefore, for females, the frequency of food sharing remains high even among adults, although the characteristics of begging likely switches from 'begging for food' to 'begging for social bond'. Food sharing does not occur in adult males who have much weaker social bonds with one another and other females, except for their mother (Furuichi, 1997; 2011; Surbeck et al., 2011). It is notable that the five events where sharing occurred from an adult female to an adult male reported by Yamamoto (2015) were all between a mother and her mature offspring. Our hypothesis emphasizes the importance of social aspects in food sharing in bonobos, especially among adult females.

Annual changes within Wamba

We also discovered annual changes in sharing frequency. Table 9.3 presents numbers of junglesop fruit eaten and shared by independent individuals during our field research between 2010 and 2014 (seasons of high junglesop fruit production, except for 2012). Interestingly, the sharing rate, as calculated by the number of shared fruit divided by the total number of eaten fruit, significantly dropped in 2013 (sharing rate of less than 10 per cent was observed only in 2013). Considering that more than 200 instances of fruit eating were observed during each field season, except 2012 (the numbers of fruit eaten per observation day were 9.6, 9.4, 10.8 and 12.5 in 2010, 2011, 2013 and 2014, respectively), the amount of available fruit was most likely not to be the best explanation for the significant change in food-sharing frequency.

Decreased food sharing in 2013 may have been related to the social instability during this period. In August 2013, when we were collecting sharing data, we observed the alpha female Kiku being attacked by other females in the same group (Furuichi et al., unpublished data). During this aggressive attack, Kiku was injured and her social rank decreased. In accordance with this coup among females, conflicts between Kiku and another dominant female, Jacky, as well as between their mature sons, Kitaro and Jiro, became apparent. During this period, they experienced a socially unstable situation and this may have influenced food sharing; that is, their sharing frequency significantly dropped, although they ate almost the same amount of fruit. In the next season (2014), the social situation was settling, as Kiku recovered her dominance status, and food-sharing frequency began to recover. Thus, the frequency of food sharing seems to be linked to social stability, which again suggests social aspects of bonobo food sharing. We acknowledge that it can be predicted that bonobos try to establish social bonds by courtesy food sharing during periods of social instability. However, they did the opposite. Young females might not know who to establish bonds with in such situations. At the moment, we cannot fully explain this phenomenon but courtesy food begging might be effective in a stable peaceful social situation, but not during periods of rank uncertainty, which should be examined further in future studies.

Currently, we cannot totally eliminate the possibility that ecological factors affected the frequency of junglesop fruit sharing. We do not know exactly how much edible fruit was available during each field season. The quality or taste of available fruit was also not evaluated. These ecological factors possibly influenced bonobo food sharing. When they had a large quantity of fruit they might share more, although the opposite result could also occur. The analysis of quantitative data, however, did not reveal significant differences in frequency of eating junglesop fruit (the total number of fruit eaten per observational day) across the years (except 2012). Nonetheless, further research is needed to evaluate the effects of ecological factors and their interaction with social factors on food sharing. To strengthen our hypothesis of socially mediated food sharing, long-term monitoring of social stability and its relationship with food sharing is also needed.

Comparison with other bonobo sites

Another important approach is the comparison between different bonobo sites. Quantitative data

Table 9.3 Numbers of junglesop (*Anonidium mannii*) fruits eaten and shared by independent individuals during each field season between 2010 and 2014 (data added to Table 2 of Yamamoto 2015, *Behavior*).
(Nombre de fruits Anonidium mannii (junglesop) mangés et partagés par des individues indépendants durant chaque saison de champ entre 2010 et 2014 (données ajoutées à la Table 2 de Yamamoto 2015, *Behavior*).)

	Observation days	No. of fruits eaten	No. of fruits shared	Shared/eaten (%)
2010 (Jul–Aug)	35	337	54	16.0
2011 (Jun–Aug)	37	347	56	16.1
2012 (Sep–Nov)	37	2	2	100
2013 (Aug–Sep)	20	215	13	6.1
2014 (Aug–Sep)	19	237	25	10.5
Total	148	1138	150	13.2

on food sharing have been obtained from three major research sites for wild bonobos: Wamba, Lomako and LuiKotale. There appear to be differences among the three study sites in terms of the types of shared food items and sharing frequency, which might have been influenced by possible differences in the fruit and tree composition among the three forests. Of the two large fruit species, *Anonidium mannii* and *Treculia africana*, which are common to the study sites and make up most of the shared plant food, *Anonidium* fruits are preferred and more often consumed and shared than *Treculia* fruits among Wamba bonobos (Table 9.1; Yamamoto 2015). However, in Lomako and LuiKotale, *Treculia* is more popular and the bonobos spend more time feeding and sharing this fruit (in Lomako, *Treculia*: 93 per cent, *Anonidium*: 7 per cent of eaten fruits: Fruth and Hohmann, 2002; see also White, 1994, which reported 21 instances of *Treculia* sharing, but no instance of *Anonidium* sharing in Lomako; in LuiKotale, *Treculia*: 63.6 per cent, *Anonidium*: 36.4 per cent of feeding time, see Goldstone et al., 2016). In the high-production season for junglesop fruits in Wamba, we observed a maximum of 9 individuals who ate *Anonidium* fruits simultaneously at a feeding site, and counted up to 45 individuals during a single observation day, although on average 0.49 *Treculia* and fewer *Anonidium* fruits were eaten per day in Lomako (Fruth and Hohmann, 2002). These ecological differences may affect bonobo food sharing.

Despite the differences in their shared food items and frequency, food sharing at the three sites seems to occur in similar ways. Previous studies have detected no evidence of reciprocity at any bonobo research site (Wamba: Yamamoto, 2015; Lomako: Fruth and Hohmann, 2002; LuiKotale: Goldstone et al., 2016). This fact contrasts markedly with previous reports suggesting reciprocal food sharing among chimpanzees (Gomes and Boesch, 2009; Hockings et al., 2007; Mitani and Watts, 2001). Interestingly, Goldstone and his colleagues (2016) reported that LuiKotale bonobos are similar to Wamba bonobos in that their food sharing has considerable social aspects, which cannot be explained by nutritional functions alone. As expected from our hypothesis, they shared easily accessible plant food, and individuals with a less established position in the hierarchy—younger, recently immigrated females—were more likely to beg. This suggests that 'courtesy food sharing' could occur not only in Wamba bonobos but also in bonobos in general. Although Goldstone and his colleagues (2016) considered this type of food begging as begging for assessing relationships, we should wait for further empirical studies to evaluate whether 'begging for social bond' or 'begging for assessing relationships' is more plausible. Nonetheless, both studies from Wamba and LuiKotal agreed on emphasizing the importance of social rather than nutritional aspects in bonobos' food sharing.

The three research sites were located deep in the tropical rainforest of the Congo basin with little environmental variation, although there could have been some ecological differences, such as the *Anonidium/Treculia* ratio. The rich ecological environment may differ from chimpanzee habitat and this difference may affect food sharing. Unfortunately, however, we do not currently possess systematic comparative data on ecological factors that might influence sharing cooperation. Further comparative research among different sites of wild bonobos (and chimpanzees) in various environments is needed.

Comparison between bonobos and chimpanzees

As mentioned earlier, bonobo food sharing is distinguished from that of chimpanzees. Here, we summarize the differences. First, bonobos share plant food, which can often be obtained without any cooperation or specialized skills and thus, the importance of the link between cooperative hunting and food sharing suggested by meat sharing in chimpanzees (Boesch, 1994) does not apply to bonobos' sharing. Second, adult females played important roles in bonobo food sharing. Adult bonobo males were infrequently involved in food-sharing events unlike adult male chimpanzees that often controlled food-sharing events. Third, bonobos shared food peacefully compared to the excitement involved in chimpanzee meat sharing. Direct aggression was seldom observed over food and pressure from subordinate recipients seems to be low. Thus, sharing-under-pressure is less plausible for bonobos' sharing, although it could explain chimpanzee meat sharing (Gilby, 2006). Fourth, there is no evidence suggesting reciprocal exchange for bonobos, which is often cited to explain the function of chimpanzee food sharing. Fifth, bonobos share food even with out-group members in the wild (Yamamoto et al., unpublished data). Sixth, bonobos share abundant plant food, whereas in chimpanzees only high-quality, difficult-to-obtain food items (e.g. meat) are begged for and transferred, which suggests that nutritional acquisition is likely the primary function of soliciting food in chimpanzees.

Concerning social aspects of food sharing, the link between social relationships and sharing is important in both species; however, the direction of effects might be different between the two. In some chimpanzees' sharing events, food owners, normally dominant males, seem to select recipients such as allies and supporters in cooperative hunting (Boesch, 1994) and reproductively cycling females (Hockings et al., 2007). The reciprocal loop between sharing and social relationships is suggested in some chimpanzee communities. In these cases, existing social relationships seem to determine sharing occurrence, and food sharing strengthens the relationship as a positive feedback. In contrast, in bonobos, as mentioned in the explanation of their courtesy food sharing, newly immigrated young females often beg food from the same-sex senior dominants. They seem to want to establish social relationships with these unknown individuals through interaction with food. Thus, even between individuals without any established relationships, food sharing could take place, perhaps more easily than between individuals with established relationships. In bonobo food sharing, social relationships are likely an effect rather than a cause, which is worth further investigation.

Bonobo uniqueness and its implications

This chapter has illuminated the uniqueness of bonobo food sharing compared with that of chimpanzees. Subsequent investigation should be related to what formed these characteristics and the effects of this unique 'courtesy food sharing'. In this final section we discuss the relationship between food sharing and other unique traits. This will provide deep insights into the evolution of cooperative society from comparative viewpoints, and propose further research topics to advance our understanding of the evolution of *Homo* and *Pan*.

Tolerance, greetings and strong female–female bonding

The most plausible key element that facilitates bonobo food sharing might be their tolerance over food. Bonobos have a highly tolerant society that has often been cited as one of their remarkable traits during field observations (e.g. Furuichi, 1997; Kano, 1992) and experimental works (e.g. Hare et al., 2007). Hare and his colleagues (2007) found that bonobo dyads, compared with chimpanzee dyads, were more tolerant with each other over food, and were more likely to share monopolizable food. It is certain that high tolerance plays a major role in opportunities for them to share food with other group members, especially with newly immigrated subordinate females who have no established relationships with others.

It is also notable that female bonobos do not demonstrate explicit submissive greetings similar to chimpanzee pant-grunts, which are directed from subordinates to dominants (de Waal, 1988). Although male bonobos may form a linear hierarchy derived from agonistic interactions (Surbeck et al., 2011), females normally avoid conflict with each other (Tokuyama and Furuichi, 2016). Some studies suggest that female hierarchies are not linear and more complicated than those of males (Stevens et al., 2007). As reported by Goldstone and his colleagues (2016), female bonobos may need additional measures for assessing social relationships. For example, food begging may play a greeting role.

This may strengthen bonobo female–female bonds, which is exceptional within species with female-biased dispersal. Joint activity and/or mimicry of another's behavior often facilitates the smoothness of interactions and increases positive bonds between interacting partners in humans and some non-human animals (Chartrand and Barg, 1999; Paukner et al., 2009), which is considered closely related to empathy. A Japanese proverb says that 'people who eat from the same rice bowl become tied to each other' (English translation: 'to drink from the same cup'). Bonobo food sharing may have a similar function.

Environment for the evolution of sharing cooperation

Bonobo food sharing goes beyond nutritional functions, and thus nutrition does not seem to be the primary evolutionary driver of their courtesy sharing behavior. Differing from chimpanzee meat sharing, which might be driven by the difficulty of access to the prey, bonobo fruit sharing, especially courtesy sharing, might be facilitated by the ease with which food resource are obtained in their rich environment. When we reflect on the evolution of food sharing, it is plausible that food sharing originally emerged because of its nutritional benefit for immature individuals. This is supported by the fact that food sharing among adults only evolved in species already sharing with offspring (Jaeggi and van Schaik, 2011). Thus, kin selection is likely to be the evolutionary explanation for the emergence of sharing cooperation at the first stage. Next, a reciprocal system could allow animals to share with non-relatives. Phylogenetic analyses with 68 primate species suggested that sharing is traded for matings and/or cooperative support in the sense that these services are statistically associated (Jaeggi and van Schaik, 2011). We think this is reasonable and fits the prevailing evolutionary theories; however, bonobo food sharing seems to be exceptional and goes beyond the range of this explanation, which always assumes that the shared food is nutritionally precious. In the rich environment of the tropical rainforest with an abundance of plant food, the food may lose its nutritional importance and serve as a social lubricant.

A recent study suggested that ancestors of bonobos crossed the Congo River into the Congo Basin in the Pleistocene, approximately 1.0 Ma, where no ape species had been resident before (Takemoto et al., 2015; Chapter 18 in this volume). This may have created the separation between the two clades of *Pan*, and bonobos have evolved from the small population in the comparatively stable tropical rainforests. Evolution through this bottleneck in a rich environment might have formed the distinctive style of food sharing in bonobos.

Although the bonobo courtesy food sharing seems exceptional in non-human primates, it provides insights into the evolution of divergent human sharing behaviors and development of cooperative societies. Previous studies have mainly focused on hunting–sharing co-occurrence, emphasizing the nutritional importance of meat for hunter–gatherers. However, people share not only meat but also plant foods. They may have developed different sharing systems with different food items. Although chimpanzee meat sharing could be a model for sharing meat by early human hunter–gatherers, the bonobos' frequent fruit sharing may also shed light on food sharing in their society. They might have shared food for social bonding beyond its nutritional function when they could obtain food more than necessity, which might have played a substantial role in establishing a cooperative relationship for survival in an unstable environment. This is one of the challenges for future investigations.

Future directions

We have accumulated much qualitative and quantitative behavioral data on bonobos, as well as chimpanzees (Hare and Yamamoto, 2015); therefore, there is now sufficient background for systematic comparison between species. However, there are several challenges. For example, exact comparison of food-sharing patterns across species and/or across study sites within species are currently hampered by a lack of a standardized methodology for data collection. Many researchers have struggled with existing data to identify common factors that influence ape behavior but this is not an easy task. We should consider future collaborative work for systematic data collection across species and across study sites. We also need precise data for ecological factors, such as vegetation, seasonal change and animal foraging preference. To understand variation within a species, especially the bonobo, it is important to open up new research sites to investigate a wider range of populations.

We have stated the importance of the difference in sharing cooperation between chimpanzees and bonobos because the variation between the two is surprising considering they are phylogenetically very close to each other, diverging within the last million years or so (Prüfer et al., 2012; Takemoto et al., 2015; Chapter 18 in this volume). The difference between chimpanzees and bonobos is often attributed to difference in the environment of their habitat. For example, Furuichi (2011), who discussed female contributions to the peaceful nature of bonobo society, considered the ecologically rich environment in bonobo habitats as one of the primary factors that enhanced females' high status in bonobo society. The higher density of food patches in bonobo habitats, compared with those of chimpanzees, may reduce the travelling cost for the slower-moving females, which makes their gregariousness possible and enables them to form coalitions against males. Such rich ecological conditions may also have a strong influence on their tolerant nature and its subsequent effect on sharing cooperation.

It is true that many bonobos inhabit comparatively rich and stable environments; the main research sites for wild bonobos—Wamba, Lomako, and LuiKotale—are all located in the middle of extensive rainforests. Recently, however, wild bonobos have been found even in much drier forest–savannah mosaic areas (Inogwabini et al., 2007). Studies in these areas are promising, although bonobo societies and behaviors, as well as the composition and distribution of their edible plants are not yet fully investigated. A couple of research teams, including a Japanese one, have launched intensive projects in this area (e.g. Narat et al., 2015; Serckx et al., 2015). Comparison of bonobo behavior among such different environments would be key to understanding the environment–behavior interaction in bonobos. This will provide us with novel insights into the evolution of *Homo* and *Pan* societies, especially based on the environment in which cooperative societies evolved.

Experimental studies with captive bonobos are also useful for the examination of detailed factors and mechanisms. Ecological and social environments, such as food quality and quantity, combination of dyads interacting, and group composition, can be experimentally controlled in captivities. For food-sharing experiments, it is possible for researchers to decide the timing and individuals to provide the food resource. We expect that, if they demonstrate courtesy food sharing as hypothesized in this article, captive bonobos may share food more frequently when they need to develop or reconstruct their social relationships with others; for example, after they have conflicts and/or when they reunite after some separation because of changes in group composition that artificially simulate the fission–fusion system. Captive studies also have more advantages in that they can comparatively analyse the animals' physiological data. With these merits, it would be fruitful if researchers developed experiments and interpreted their results considering ecological validity. Thus, interdisciplinary collaboration is essential for understanding the evolution and mechanisms of cooperative society.

References

Berbesque, J.C., Wood, B.M., Crittenden, A.N., Mabulla, A., and Marlowe, F.W. (2016). Eat first, share later: Hadza hunter–gatherer men consume more while foraging than in central places. *Evolution and Human Behavior*, 37, 281–6.

Blurton Jones, N.G. (1984). A selfish origin for human food sharing: tolerated theft. *Ethology and Sociobiology*, 5, 1–3.

Boesch, C. (1994). Cooperative hunting in wild chimpanzees. *Animal Behaviour*, 48, 653–67.

Bullinger, A.F., Burkart J.M., Melis A.P., and Tomasello M. (2013). Bonobos, *Pan paniscus*, chimpanzees, *Pan troglodytes*, and marmosets, *Callithrix jacchus*, prefer to feed alone. *Animal Behaviour*, 85(1), 51–60. DOI:10.1016/j.anbehav.2012.10.006.

Burkart, J. M., Allon, O., Amici, F., Fichtel, C., Finkenwirth, A., Huber, J., Isler, K., Kosonen, Z. K., Martins, E., Meulman, E. J., Rueth, K., Spillmann, B., Wiesendanger, S., and van Schaik, C. P. (2014). The evolutionary origin of human hyper-cooperation. *Nature Communications*, 5, 4747.

Chartrand, T.L. and Bargh, J.A. (1999). The chameleon effect: the perception–behavior link and social interaction. *Journal of Personality & Social Psychology*, 76, 893–910.

de Waal F.B.M. (1988). The communicative repertoire of captive bonobos (*Pan paniscus*), compared to that of chimpanzees. *Behaviour*, 106 (3), 183–251. DOI: 10.1163/156853988X00269.

Dunfield K., Kuhlmeier V.A., O'Connell L., and Kelley E. (2011). Examining the diversity of prosocial behaviour: helping, sharing, and comforting in infancy. *Infancy*, 16, 227–47.

Feistner, A.T.C. and McGrew, W.C. (1989). Food-sharing in primates: a critical review. In: Seth, P.K. and Seth, S. (eds). *Perspectives in Primate Biology*, Vol. 3. New Delhi: Today and Tomorrow's, pp. 21–36.

Fruth, B. and Hohmann, G. (2002). How bonobos handle hunts and harvests: Why share food? In: Boesch, C.,

Hohmann, G., and Marchant, L. (eds). *Behavioral Diversity in Chimpanzees and Bonobos*. Cambridge: Cambridge University Press, pp. 231–43.

Furuichi, T. (1989). Social interactions and the life history of female *Pan paniscus* in Wamba, Zaire. *International Journal of Primatology*, 10, 173–97.

Furuichi, T. (1997). Agonistic interactions and matrifocal dominance rank of wild bonobos (*Pan paniscus*) at Wamba. *International Journal of Primatology*, 18(6), 855–75.

Furuichi, T. (2011). Female contributions to the peaceful nature of bonobo society. *Evolutionary Anthropology*, 20, 131–42.

Furuichi, T., Idani, G., Ihobe, H., Hashimoto, C., Tashiro, Y., Sakamaki, T., Mulavwa, M. N., Yangozene, K., and Kuroda, S. (2012). Long-term studies on wild bonobos at Wamba, Luo Scientific Reserve, D.R. Congo: towards the understanding of female life history in a male-philopatric species. In: Kappeler, P.M. and Watts, D.P. (eds). *Long-Term Field Studies of Primates*. Berlin: Springer-Verlag, pp. 143–433.

Gilby, I.C. (2006). Meat sharing among the Gombe chimpanzees: harassment and reciprocal exchange. *Animal Behaviour*, 71, 953–63.

Goldstone L.G., Sommer V., Nurmi N., Stephens C., and Fruth B. (2016). Food begging and sharing in wild bonobos (*Pan paniscus*): assessing relationship quality? *Primates*, 57(3), 367–76. DOI 10.1007/s10329-016-0522-6.

Gomes, C.M. and Boesch, C. (2009). Wild chimpanzees exchange meat for sex on a long-term basis. *PLoS One*, 4(4), e5116. DOI:10.1371/journal.pone. 0005116.

Hamann K., Warneken F., Greenberg J.R., and Tomasello M. (2011). Collaboration encourages equal sharing in children but not in chimpanzees. *Nature*, 476, 328–31. DOI:10.1038/nature10278.

Hamilton, W.D. (1964). Genetical evolution of social behaviour I. *Journal of Theoretical Biology*, 7, 1–16.

Hare, B., Melis, A.P., Woods, V., Hastings, S., and Wrangham, R. (2007). Tolerance allows bonobos to outperform chimpanzees on a cooperative task. *Current Biology*, 17, 619–23.

Hare, B. and Kwetuenda, S. (2010). Bonobos voluntarily share their own food with others. *Current Biology*, 20, R230–R231.

Hare B. and Yamamoto S. (2015). Moving bonobos off the scientifically endangered list. *Behaviour*, 152, 247–58.

Hirata, S., Yamamoto, S., Takemoto, H., and Matsuzawa, T. (2010). A case report of meat and fruit sharing in a pair of wild bonobos. *Pan Africa News*, 17(2), 21–3.

Hockings, K.J., Humle, T., Anderson, J.R., Biro, D., Sousa, C., Ohashi, G., and Matsuzawa, T. (2007). Chimpanzees share forbidden fruit. *PLoS One*, 2(9), e886. DOI:10.1371/journal.pone. 0000886.

Idani, G. (1991). Social relationships between immigrant and resident bonobo (*Pan paniscus*) females at Wamba. *Folia Primatologica*, 57, 83–95.

Idani, G. Kuroda, S., Kano, T., and Asato, R. (1994). Flora and vegetation of Wamba forest, Central Zaire with reference to bonobo (*Pan paniscus*) foods. *Tropics*, 3, 309–32.

Ihobe, H. (1992). Observations on the meat-eating behavior of wild bonobos (*Pan paniscus*) at Wamba, Republic of Zaire. *Primates*, 33, 247–50.

Ingmanson, E. and Ihobe, H. (1992). Predation and meat eating by *Pan paniscus* at Wamba, Zaire. *American Journal of Physical Anthropology*, Suppl. 14, 93.

Inogwabini, B.-I., Matungila, B., Mbende, L., Abokome, M., and wa Tshimanga, T. (2007). Great apes in the Lake Tumba landscape, Democratic Republic of Congo: newly described populations. *Oryx*, 41, 532–8.

Isaac, G. (1978). The food-sharing behavior of protohuman hominids. *Scientific American*, 238, 90–108.

Jaeggi, A.V., Stevens, J.M.G., and van Schaik, C.P. (2010a). Tolerant food sharing and reciprocity is precluded by despotism in bonobos but not chimpanzees. *American Journal of Physical Anthropology*, 143, 41–51.

Jaeggi. A.V. and van Schaik, C.P. (2011). The evolution of food sharing in primates. *Behavioral Ecology and Sociobiology*, 65, 2125–40.

Kano, T. (1980). Social behavior of wild pygmy chimpanzees (*Pan paniscus*) of Wamba: a preliminary report. *Journal of Human Evolution*, 9, 243–60.

Kano, T. (1992). *The Last Ape: Pygmy Chimpanzee Behavior and Ecology*. Stanford, CA: Stanford University Press.

Kuroda, S. (1984). Interaction over food among pygmy chimpanzees. In: Susman, R.L. (ed.). *The Pygmy Chimpanzee: Evolutionary Biology and Behavior*. New York, NY: Plenum Press, pp. 301–24.

Matsuzawa, T. (2013). Evolution of the brain and social behavior in chimpanzees. *Current Opinion in Neurobiology*, 23, 443–9

Mauss, M. (1925). *The Gift*. Trans. published in 2010, New York, NY: W.W. Norton & Co Inc.

Melis, A.P., Warneken, F., Jensen, K., Schneider, A.C., Call, J., and Tomasello, M. (2011). Chimpanzees help conspecifics obtain food and non-food items. *Proceedings of the Royal Society B*, 278, 1405–13.

Mitani, J.C. and Watts, D.P. (2001). Why do chimpanzees hunt and share meat? *Animal Behaviour*, 61, 915–24.

Mussen, P. and Eisenberg-Berg, N. (1978). *Roots of Caring, Sharing, and Helping: The Development of Pro-Social Behavior in Children*. Oxford: W.H. Freeman.

Narat, V., Pennec, F., Simmen, B., Ngawolo, J.C.B., and Krief, S (2015). Bonobo habituation in a forest–savanna mosaic habitat: influence of ape species, habitat type, and sociocultural context. *Primates*, 56, 339–49.

Parish, A.R. (1994). Sex and food control in the 'uncommon chimpanzee': how bonobo females overcome a phylogenetic legacy of male dominance. *Ethology and Sociobiology*, 15, 157–79.

Paukner, A., Suomi, S.J., Visalberghi, E., and Ferrari, P.F. (2009). Capuchin monkeys display affiliation toward humans who imitate them. *Science*, 325, 880–3.

Prüfer, K., Munch, K., Hellmann, I., Akagi, K., Miller, J. R., Walenz, B., Koren, S., Sutton, G., Kodira, C., Winer, R., Knight, J. R., Mullikin, J. C., Meader, S. J., Ponting, C. P., Lunter, G., Higashino, S., Hobolth, A., Dutheil, J., Karakoç, E., Alkan, C., Sajjadian, S., Catacchio, C. R., Ventura, M., Marques-Bonet, T., Eichler, E. E., André, C., Atencia, R., Mugisha, L., Junhold, J., Patterson, N., Siebauer, M., Good, J. M., Fischer, A., Ptak, S. E., Lachmann, M., Symer, D. E., Mailund, T., Schierup, M. H., Andrés, A. M., Kelso, J., and Pääbo, S. (2012). The bonobo genome compared with the chimpanzee and human genomes. *Nature*, 486, 527–31. DOI:10.1038/nature11128.

Ryu, H., Hill, D., and Furuichi, T. (2015). Prolonged maximal sexual swelling in wild bonobos facilitates affiliative interactions between females. *Behaviour*, 152, 285–311.

Serckx, A., Kuhl, H.S., Beudels-Jamar, R.C., Poncin, P., Jean-Franscois, B., and Huynen, M.C. (2015). Feeding ecology of bonobos living in forest-savannah mosaics: diet seasonal variation and importance of fallback foods. *American Journal of Primatology*, 77, 948–62.

Stevens, J.R., Vervaecke, H., de Vries, H., and van Elsacker, L. (2007). Sex differences in the steepness of dominance hierarchies in captive bonobo groups. *International Journal of Primatology*, 28, 1417–30.

Surbeck, M., Mundry, R., and Hohmann, G. (2011). Mothers matter! Maternal support, dominance status and mating success in male bonobos (*Pan paniscus*). *Proceedings of the Royal Society B*, 278, 590–8.

Takemoto, H., Kawamoto, Y., and Furuichi, T. (2015). How did bonobos come to range south of the Congo River? Reconsideration of the divergence of *Pan paniscus* from other *Pan* populations. *Evolutionary Anthropology*, 24, 170–84.

Tan, J. and Hare, B. (2013). Bonobos share with strangers. *PLoS One*, 8(1), e51922. DOI:10.1371/journal.pone. 0051922.

Tokuyama, N. and Furuichi, T. (2016). Do friends help each other? Patterns of female coalition formation in wild bonobos at Wamba. *Animal Behaviour*, 119, 27–35.

Warneken, F. (2016). Insights into the biological foundation of human altruistic sentiments. *Current Opinion in Psychology*, 7, 51–6.

Warneken, F., Hare, B., Melis, A.P., Hanus, D., and Tomasello, M. (2007). Spontaneous altruism by chimpanzees and young children. *PLoS Biology*, 5(7), e184. DOI:10.1371/journal.pbio. 0050184.

White, F.J. (1994). Food sharing in wild pygmy chimpanzees, *Pan paniscus*. In: Roeder, J.J., Thierry, B., Anderson, J.R., and Herrenschmidt, N. (eds). *Current Primatology*, Vol. II. Strasbourg: Université Louis Pasteur, pp. 1–10.

White, F.J. and Wood, K.D. (2007). Female feeding priority in bonobos, *Pan paniscus*, and the question of female dominance. *American Journal of Primatology*, 69, 837–50.

Wrangham, R.W. (1975). The behavioural ecology of chimpanzees in Gombe National Park, Tanzania. PhD thesis, Cambridge University, UK.

Yamamoto, S. (2015). Non-reciprocal but peaceful fruit sharing in wild bonobos in Wamba. *Behaviour*, 152, 335–57.

Yamamoto, S. and Tanaka, M. (2009). How did altruism and reciprocity evolve in humans? Perspectives from experiments on chimpanzees (*Pan troglodytes*). *Interaction Studies*, 10, 150–82.

Yamamoto, S., Humle, T., and Tanaka, M. (2009). Chimpanzees help each other upon request. *PLoS One*, 4, e7416. DOI:10.1371/journal.pone.0007416.

Yamamoto, S., Humle, T., and Tanaka, M. (2012). Chimpanzees' flexible targeted helping based on an understanding of conspecifics' goals. *Proceedings of the National Academy of Sciences*, 109(9), 3588–92.

Yamamoto, S., Humle, T., and Tanaka, M. (2013). Basis for cumulative cultural evolution in chimpanzees: social learning of a more efficient tool-use technique. *PLoS One*, 8(1), e55768. DOI:10.1371/journal.pone. 0055768.

CHAPTER 10

Prosociality among non-kin in bonobos and chimpanzees compared

Jingzhi Tan and Brian Hare

Abstract Models of the origin of human prosociality towards non-kin have been primarily developed from chimpanzee studies. Substantially less effort has been made to consider the prosociality of bonobos. Like chimpanzees, bonobos cooperate with non-kin extensively but, unlike chimpanzees, immigrating members are central to bonobo cooperation. In experiments bonobos are tolerant during encounters with strangers and during co-feeding. They help strangers without immediate tangible reward, and forfeit monopolizable food to facilitate a physical interaction with them. Such prosociality seems proactive as it is not elicited by solicitation. Bonobos also seem to prefer sharing food over non-food objects, while chimpanzees reliably transfer non-food objects rather than food. These findings highlight the possibility that human sharing with strangers might have also evolved as a mutualistic endeavour to initiate a long-term partnership. Future models of human prosociality will need to incorporate findings from both *Pan* species.

Les modèles de l'origine de la prosocialité humaine entre non-parents ont été développées en majorité à partir d'études de chimpanzé. Beaucoup moins d'efforts ont été faits pour considérer la prosocialité des bonobos. Comme les chimpanzés, les bonobos coopèrent extensivement avec non-parents mais, contrairement au chimpanzés, les membres immigrants sont au centre de la coopération bonobo. Les bonobos sont tolérants, en expérimentation, durant les rencontres avec des étrangers et durant la co-alimentation. Ils aident les étrangers sans récompense immédiate, et abandonnent la nourriture monopolisable pour faciliter une interaction physique avec eux. Une prosocialité pareille paraît proactive vu qu'elle n'est pas sollicitée. Les bonobos, il paraît, préfèrent partager la nourriture que d'autres objets, alors que les chimpanzés préfèrent partager les objets non-aliments que la nourriture. Ces résultats soulignent la possibilité que le partage humain avec les étrangers a pu évoluer comme une enquête mutuelle pour initier un partenariat à longue durée. Les modèles futurs de la prosocialité humaine doivent inclure les résultats des deux espèces Pan.

Introduction

Prosocal behavior refers to voluntary behaviors that benefit others (Cronin, 2012). Humans are exceptionally prosocial. From donating blood to sharing data, prosocial behaviors can be seen regularly in our daily life (Cronin, 2012). The most intriguing phenomenon of human prosociality is that humans are willing to benefit unrelated strangers without immediate, tangible rewards (Camerer, 2003). Theoretically, such kind of prosociality is high risk given the lack of past experience or genetic relatedness to prevent cheating. However, this kind of prosociality has an early onset in development, is universally observed across cultures and is heritable, which suggests a biological basis (Cesarini et al., 2009; Henrich et al., 2005; Warneken and Tomasello, 2009). The evolution of this ultra-prosociality has spurred tremendous interests in comparative research focused on exploring whether this prosociality is derived in *hominins* or shared to some degree with other apes.

The first efforts to test for the phylogenetic origins of human prosociality examined chimpanzees. In the wild, chimpanzees are known for their diverse set of cooperative behaviors among non-kin (Langergraber et al., 2007; Mitani, 2009). This is most commonly observed among males but occasionally does occur among females too (Pusey et al., 2013). Chimpanzees are, to some degree, capable of understanding other's psychological states such as perception, intention and desires (Hare, 2011). These social cognitive skills suggested the possibility that

Tan, J. and Hare, B., *Prosociality among non-kin in bonobos and chimpanzees compared*. In: *Bonobos: Unique in Mind, Brain, and Behavior*. Edited by Brian Hare and Shinya Yamamoto: Oxford University Press (2017).
© Oxford University Press. DOI: 10.1093/oso/9780198728511.003.0010

chimpanzees also are capable of prosociality towards non-kin. The original prosocial choice task is a simplified version of the Dictator Game where subjects could, at no cost to self, choose between a prosocial option that delivers rewards to a recipient and an asocial option that does not (Jensen et al., 2006; Silk et al., 2005). This choice paradigm soon gained popularity for its straightforward set-up and quickly became the 'conventional' test of prosociality in nonhuman animals (Cronin, 2012; Tan et al., 2015). A second paradigm, known as the 'instrumental helping task', took a go/no-go approach to address the question and examine whether subjects would assist a recipient to retrieve out-of-reach items (Melis et al., 2011; Warneken et al., 2007).

Based on the first generation of chimpanzee prosociality studies, some concluded that failure in the first paradigm meant human prosociality was completely derived (e.g. Silk and House, 2011). This conclusion is problematic for three reasons: 1) the instrumental helping paradigm revealed strong positive results; 2) even if non-human apes lack altruism, or costly helping, it does not rule out mutualistic forms of prosociality—voluntary behaviors that benefit both self and others—that are more similar to most cases of human helping; and 3) chimpanzee are only one of our two species in the genus *Pan*. In support of the final reason, the first quantitative comparisons of the two *Pan* species find cognitive similarities but also suggest that bonobos differ significantly from chimpanzees in their social cognition and cooperative nature (Hare et al., 2007; Hare et al., 2012; Herrmann et al., 2010). Moreover, when examining other human traits (i.e. non-conceptive sexual behavior, juvenilized cognitive development, extended maternal dependence), bonobos are often more similar to humans than chimpanzees. Without knowledge of bonobos it would have been concluded that these shared bonobo–human traits were uniquely human (Hare and Yamamoto, 2015). This means to test the origins of human prosociality fully and understand what traits our last common ape ancestor might have possessed, innovative methods were needed that examined altruistic and mutualistic forms of prosociality in bonobos (Hare and Tan, 2011).

In this chapter we will first review how bonobos cooperate with non-kin and the cognitive, motivational and temperamental basis for bonobo cooperation (Table 10.1). We will then compare experimental evidence of prosociality between bonobos and chimpanzees (Table 10.2). We conclude by recommending the expansion of models of human evolution to incorporate data from bonobo cooperation.

Table 10.1 Major differences in cooperation and prosociality between bonobos and chimpanzees.
(Principales différences de coopération et de comportements prosociaux entre les bonobos et les chimpanzés.)

	Bonobos	Chimpanzees
Non-kin cooperation in natural settings	• Immigrating members cooperate extensively	• Non-dispersing members cooperate extensively
	• Cooperators are used to be strangers until they reach adulthood	• Cooperative relationships are primarily built upon kinship and familiarity
	• Smaller-sized members cooperate to dominate larger-sized members	• Larger-sized members cooperate to dominate smaller-sized members
Psychological foundations for prosociality	• Bonobos are more sensitive and attentive to other's gaze and intention	• Chimpanzees are less sensitive and attentive to other's gaze and intention
	• Bonobos show consolation and contagious yawning	• Chimpanzees show consolation and contagious yawning
	• Bonobos are more tolerant during co-feeding	• Chimpanzees are less tolerant during co-feeding
	• Bonobos show peaceful and affiliative responses to strangers during intergroup encounters and immigrations	• Chimpanzees show xenophobic violence during intergroup encounters and immigrations
Non-kin prosociality in experimental settings	• Bonobos voluntarily share monopolizable food with strangers and help strangers obtain out-of-reach food	• Chimpanzees cannot be paired with strangers in experiments due to ethical concerns over their xenophobia
	• Bonobo sharing and helping are proactive	• Chimpanzees help upon request or harassment
	• Bonobos actively transfer food but not non-food items	• Chimpanzees actively transfer non-food but not food items

Cooperation with and tolerance for non-kin in bonobos

Bonobos cooperate with non-kin extensively

Although there is no evidence that male bonobos engage in group hunting, territorial defence and cooperative intergroup aggression like chimpanzees do, bonobos do cooperate with non-kin extensively. Contrary to chimpanzees, whose social behavior is primarily exhibited by males, bonobo females are the primary cooperators (Furuichi, 2011; Hare et al., 2012). They are considered more gregarious than chimpanzees and they frequently associate with each other (Furuichi, 2011). These close bonds allow females to achieve dominance status and to gain feeding priority (e.g. Parish, 1996; Surbeck and Hohmann, 2013). Interestingly, hunting parties were composed primarily of females in the few hunting episodes that have been observed (Surbeck and Hohmann, 2008). Furthermore, females form coalitions and direct aggression towards males, possibly as a function to suppress male aggression towards themselves or towards immature members (Parish, 1996; Surbeck and Hohmann, 2013).

The pattern of cooperation observed in female bonobos is exceptional because, like chimpanzees, females are unrelated immigrants in a patrilocal social system (Gerloff et al., 1999). If the high level of cooperation observed in chimpanzee males can be explained by a combination of kinship and familiarity between non-kin, bonobo cooperation is remarkable given that it occurs among unrelated members who remain strangers until adulthood. Furthermore, unlike chimpanzee males who form coalitions to dominate smaller-bodied females, bonobo females form coalitions against larger-bodied males (Hare et al., 2012). Therefore, the initiation and maintenance of bonobo cooperation among non-kin is more challenging to explain theoretically. This pattern of cooperation also predicts that bonobos are a promising candidate species to examine prosocial behavior towards non-kin.

Cognitive, motivational and temperamental basis for prosociality towards non-kin

There is growing evidence that bonobos possess several cognitive and motivational traits fundamental to prosocial behaviors. First, social attention and an understanding of other's intention are both important for prosociality because a prosocial actor needs first to recognize there is a mismatch between the recipient's goal and the reality, and then direct actions to help the recipient meet the goal (Warneken, 2015). Current evidence suggests that relative to chimpanzees, bonobos are more inclined to pay attention to a conspecific's eyes, follow other's gaze and infer intentions from other's actions (Herrmann et al., 2010; Kano et al., 2015). Second, it has been suggested that bonobos might show empathic concerns towards others. After conflicts, bystanders provide voluntary consolation of victims (Clay and de Waal, 2013). Furthermore, contagious yawning has been proposed by some as a form of emotional contagion, and even a precursor of empathy. When bonobos were exposed to the yawns of other bonobos, they yawned contagiously (Demuru and Palagi, 2012; Tan et al., submitted).

Beside cognitive and motivational traits, social tolerance is necessary for any prosocial behavior to occur. It is certain temperamental profile that creates a relaxed social space for actors and recipients to perceive an encounter as safe, to come into close proximity and to initiate any subsequent interactions (Carter and Porges, 2011). Social tolerance is believed to be a prerequisite for the formation of social bonds (Carter and Porges, 2011), effective social learning (van Schaik and Pradhan, 2003) and flexible cooperation (Hare et al., 2007). In the next two sections, we will review evidence showing that, relative to chimpanzees, bonobos can show more tolerance during feeding and during encounters with unfamiliar conspecifics (Hare et al., 2012).

Social tolerance for food

In dyadic food-sharing tasks where pairs of subjects approached a feeding location simultaneously, bonobos were more likely to share access to food with each other than chimpanzees, and this difference was particularly pronounced when the food was placed in a single, monopolizable dish (Figure 10.1) (Hare et al., 2007). Bonobos became more successful in cooperating than chimpanzees because they were capable of sharing 32 per cent to 50 per cent of the highly monopolizable food reward, whereas

Figure 10.1 a. Isiro (right) shared a bottle of soy milk, a highly desirable and monopolizable food resource, with Sake (left), an unrelated groupmate; b. Two adult females, Kalina (left) and Maya (right), were struggling over a rock, a valuable tool for nut cracking. (Isiro (droite) partage une bouteille de lait de soja, une ressource alimentaire très désirée et monopolisable, avec Sake (gauche), une partenaire sans rapport familial.)

chimpanzee dyads only shared 0 to 7 per cent of the food (Hare et al., 2007; see also Hamann et al., 2011; Melis et al., 2011a). Furthermore, while chimpanzees showed a decline in sharing in this context as they aged, bonobos' sharing remained consistent during development (Wobber, Wrangham, et al., 2010). In these contexts, bonobos demonstrated more frequent social play and socio-sexual behaviors than chimpanzees, possibly as a tension reduction mechanism to maintain feeding tolerance (Hare et al., 2007; Wobber, Wrangham, et al., 2010). These results suggest that bonobos can tolerate others to feed on the same food source even when it can be easily monopolized (see also Koops et al., 2015). In fact, this might result from their aversion to social conflict. In anticipation of food contest, bonobo males showed an increase in cortisol, indicating a more passive coping style, whereas chimpanzee males experienced an increase in testosterone, consistent with status striving (Wobber et al., 2010).

Interestingly, several observational studies used group feeding in which a bag or a bundle of food was given to a single member among a group of chimpanzees or bonobos. After the individual was in possession of the food bag, subsequent social interactions were observed in the group setting. Bonobos were reported to be *less* likely than chimpanzees to transfer food to others once the food was in their possession (de Waal, 1992; Jaeggi et al., 2010, 2013; see also Cronin et al., 2014, 2015). The findings from this group-feeding paradigm might seem to be inconsistent with the dyadic experiments, but this is likely not to be the case. First, inter-observer reliability was often not reported in the above studies. Three out of four articles testing bonobos in the group feeding paradigm provided no reliability measure (de Waal, 1992; Cronin et al., 2015; Jaeggi et al., 2013), while in Jaeggi and colleagues (2010), kappa was only 0.63. Second, the dyadic sharing experiments measure the ability of subjects to co-feed not to proactively share food in one's physical possession. The group sharing tests require subjects to give food or allow for its theft. Co-feeding only requires tolerance as subjects feed in proximity to a stationary food tray. Third, the chimpanzee and bonobo groups used in the group feeding studies were not matched for group size, sex or kinship. For example, all bonobo groups had less than 10 individuals while the average size of chimpanzee groups was above 20 (from 13 to 42). The coalitionary sex was outnumbered in the chimpanzee groups (adult male to adult female ratio: de Waal, 1992, 1:8; Jaeggi et al., 2010 and 2013, 3:8; Cronin et al., 2014, 6:8, 3:17, 4:5 and 5:5), but not in the bonobo groups (adult female to adult male ratio: de Waal, 1992, 2:1; Jaeggi et al., 2010 and 2013, 3:3 and 3:3; Cronin et al., 2015, 2:2). Without controlling these confounding factors, the sharing pattern in bonobos could be an artefact of the small group size consisting of a majority of the coalitionary sex (e.g. see Koops et al., 2015 for evidence of feeding tolerance in two large, wild bonobo groups with 28 and 31 individuals respectively). Finally, there was little information in these studies about the initial possessor of the food bag. In a tolerant species, it is likely that low-ranking individuals might have a greater chance to monopolize food in the first place, or, as de Waal put it, monopolization might be prevalent because 'the probability of escalation of aggression in bonobos is lower' (de Waal, 1992, p. 48). For example, a recent group feeding study comparing bonobos and chimpanzees following de Waal (1992) found that juveniles could become the initial food possessors in bonobos but not in chimpanzees (Byrnit et al., 2015). Essentially, because bonobos are so tolerant they can get away with not proactively sharing/allowing tolerated theft even if they are not dominant.

Social tolerance for strangers

Another prominent difference in social tolerance between bonobos and chimpanzees becomes apparent when looking at their spontaneous responses towards unfamiliar conspecifics. Chimpanzees are highly xenophobic while bonobos are generally affiliative with strangers. This difference is suggested by two lines of evidence: territoriality and immigration.

First, chimpanzees are highly territorial. They patrol borders in groups, raid into neighbouring territories, attack and kill strangers (Mitani et al., 2010; Wilson et al., 2014). This strong territoriality is a major contributor to adult mortality and has potential fitness benefits as it leads to territory expansion (Mitani et al., 2010). Intergroup encounters in bonobos are relatively peaceful: in some cases adults from different groups can even engage in affiliative behaviors such as co-feeding in the same tree, genital rubbing, mating and grooming (Furuichi, 2011; Wilson et al., 2014). Tensions can rise during encounters but aggressive interactions remain at the level of displays and rarely escalate into physical aggression (Hohmann and Fruth, 2002). Lethal intergroup aggression has never been observed among wild bonobos even though the level of observation suggest it should have been seen by this point if it occurred with any regularity (Wilson et al., 2014).

The second line of evidence is immigration. For chimpanzees, a lone stranger appearing on the border is typically attacked to prevent their entrance into the territory unless they are a sexually receptive female (Williams et al., 2004). Even when a female immigrant manages to transfer into the new social group, she is likely to become a victim of cooperative (and occasionally lethal) aggression from residential females and thus seek protection from residential males (Pusey et al., 2013). Given that female chimpanzees are usually not gregarious due to intense food competition, the motivation to attack new immigrants is so strong that it even promotes female cooperation. In contrast, field reports on bonobo immigration suggest that new immigrants and residents (both males and females) quickly engage in affiliative behaviors such as grooming and socio-sexual behaviors (Furuichi, 2011; Hohmann and Fruth, 2002; Sakamaki et al., 2015). While chimpanzee immigrants seek protection from residential males to avoid aggression from residential females, bonobo immigrants seek to affiliate with residential females (Sakamaki et al., 2015).

This differential reaction to immigrants is also seen in captivity when new individuals are introduced into a pre-existing group (Figure 10.2, Plate 10). Bonobos, including infants, juveniles and adults, can be integrated swiftly and peacefully (1 day, Gold, 2001; 14 days, Pfalzer et al., 1995). In contrast, chimpanzee integration is much more challenging. Their initial response towards newcomers is primarily agonistic and successful integration requires extensive familiarization periods (e.g. 4 to 17 months, Schel et al., 2013; Thunström et al., 2013), stepwise, dyadic introduction (Baker and Aureli, 2000; Schel et al., 2013; Seres et al., 2001) and careful matching of the two parties' age, sex and reproductive status (Alford et al., 1995; Baker and Aureli, 2000; Schel et al., 2013; Seres et al., 2001).

Taken together, the two species seem to show marked difference in their responses towards unfamiliar conspecifics. An alternative explanation to bonobos' unusual tolerance for strangers is that bonobos possess a general preference for novelty. However, general neophilia is not supported. When bonobos are presented with novel objects or unfamiliar humans, they are less exploratory and more reluctant to approach these novel stimuli than chimpanzees (Herrmann et al., 2011). In non-social foraging contexts, bonobos are also more risk-averse than chimpanzees (Heilbronner et al., 2008; Rosati and Hare, 2013). Furthermore, it is important to point out that the species difference in tolerance for strangers is unlikely an artefact of captivity

Figure 10.2 On the first day of her transfer to a new social group, Kalina (bottom) was engaging in affiliative, socio-sexual behavior with Salonga (top), a resident female. (see Plate 10)
(Durant le premier jour de son transfert vers un nouveau groupe social, Kalina (dessous) a participé dans des comportements affiliatifs socio-sexuels avec Salonga (dessus), une femelle résidente.)

(Forss et al., 2015), since captivity does not alter the xenophobic response of chimpanzees (see evidence of captive introduction earlier).

Overall, bonobos and chimpanzees share a set of behavioral and psychological prerequisites for prosociality towards non-kin: they regularly engage in cooperation with non-kin, they are sensitive to the attention and intention of others, and they have shown evidence consistent with empathic concern for others. They also have several critical differences in their behavior and psychological bases for prosociality towards non-kin—bonobos demonstrate higher level of tolerance during feeding and during interactions with unfamiliar conspecifics.

Why are bonobos more tolerant than chimpanzees in these two contexts? We speculate that this might be due to the different origins of social bonds in the two species. The non-kin cooperation of chimpanzees is likely to be built upon long-term relationships among unrelated partners (Mitani, 2009). These partners might have belonged to the same age cohort and their partnerships are at least based on long-term familiarity (Langergraber et al., 2007). In contrast, in bonobos the resident sex is not the dominant sex. Instead, it is unrelated females that play the central role in forming strong bonds within a group. This means these partnerships are between unrelated individuals and could not have started until adulthood. This would predict that the initiation and maintenance of non-kin cooperation in bonobos should be supported by psychological mechanisms (e.g. tolerance for strangers and co-feeding) that can quickly build trust and that can tolerate more social uncertainty and higher social risk.

Next, we review bonobos' performances in various prosociality tests and compare them to those of chimpanzees (Table 10.2). We first review two popular paradigms that measure low-cost prosociality: the standard prosocial choice task and the instrumental helping task. Both paradigms require prosocial actors to pay minimal energetic costs or no cost at all to benefit the recipient. Both paradigms yield no selfish benefits for the actors either because the actors are rarely rewarded and the pairings usually do not allow for reciprocity. We then review experiments where subjects will need to give up desirable resources in their possession in order to benefit others.

Low- and high-cost prosociality in bonobos

Low-cost prosociality in the standard prosocial choice task

As reviewed earlier, the standard prosocial choice task was first developed and repeatedly implemented in chimpanzees. In the experimental condition of this task subjects would choose between a prosocial option that benefits self and other, and an asocial option that is only self-rewarding. The amounts of benefits to self in both options are equal, making the choice of the prosocial option cost-free

Table 10.2 Summary of experiments on bonobo prosociality towards non-kin.
(Résumé des expériences sur les comportements prosociaux des bonobos envers les individus sans lien de parenté.)

Paradigm	R's Identity	Costs to S	Potential benefits to S	Results of bonobo prosociality towards non-kin
Will S let R in to eat together?	Groupmate or stranger	Food loss	Physical interaction with R	• S shared only when R was a stranger • Solicitation or harassment did not affect sharing • Ref: Hare & Kwetuenda, 2010; Tan & Hare, 2013; Bullinger et al., 2013
Will S let R in the middle room?	Groupmate or stranger	Energy and playing time (and food loss)	None	• S helped both strangers and groupmates obtain out-of-reach food, when S could not obtain the food • S helped even when R could not make any gesture • S did not help any R, when S could obtain the food • Ref: Tan & Hare, 2013; Tan et al., submitted
Will S/R transfer the nuts/rock?	Groupmate	Food loss	None	• S actively transferred edible nuts to R • R's gestures had not effect on subject's sharing • Transfer of non-food objects occurred rarely • Ref: Krupenye et al., submitted; Liebal et al., 2014
Will S pick the tray that delivers food to R?	Groupmate or stranger	None	None	• Bonobos developed a strong bias towards a specific location during the pretests, which masked any prosociality during the test (like chimpanzees) • This paradigm is considered to be lacking validity as a measure of prosociality in nonhuman primates • Ref: Tan et al., 2015; Amici et al., 2014

to subjects. In the control condition, the subjects would be presented with the same two options but the recipient is absent, which theoretically gauges subjects' baseline tendency to choose the prosocial option. The underlying logic of this paradigm is that given prosociality is at no cost to the subjects, they should show a preference for the prosocial option in the experimental condition but not in the control, if they are motivated to benefit the recipient.

Chimpanzees from four different populations consistently showed no evidence of prosocial preference in this task (see references in Tan et al., 2015). Some hypothesized that this absence of prosociality was due to subjects not even perceiving the task as social (Hare and Tan, 2011). Chimpanzees were preoccupied with obtaining food for themselves and thus did not even perceive how their choice impacted the recipient (e.g. Warneken and Tomasello, 2009). Horner and colleagues (2011) modified the prosocial choice task to minimize the food preoccupation effect. The experimenters trained subjects to exchange tokens with food and then used tokens as indirect rewards in the prosocial choice task. When subjects exchanged tokens for food with the experimenters, food was delivered in paper wrappings to further reduce visibility. In partial support of the food preoccupation hypothesis, chimpanzees in this version of prosocial choice task preferred to benefit others. The food preoccupation even prevents chimpanzees from benefiting *themselves*. When chimpanzees had to cooperate to obtain food reward, they often attempted to monopolize the food with no regard to their partner's contribution (Hamann et al., 2011; Melis et al., 2011).

Plate 1a Dr Yoshitaka Kano, the pioneer of the first wild bonobo study, and Dr Brian Hare, the co-editor of this book. Photo taken in Kyoto by Tetsuro Matsuzawa.
(Dr Takayoshi Kano, pionnier de la première étude sur les bonobos sauvages, et le Dr Brian Hare, coéditeur de ce livre.)

Plate 1b International collaboration, particularly with Congolese, is essential for bonobo studies. On the left is Dr Shinya Yamamoto, the co-editor of this book. Photo taken in Wamba by Ryu Heungjin.
(La collaboration internationale, notamment avec les Congolais, est essentielle pour les études sur les bonobos. À gauche, le Dr Shinya Yamamoto, coéditeur de ce livre.)

Plate 2 An ex-alpha female, Kame (left), is groomed by the ex-alpha male, Ibo (centre), and other offspring. Photo taken in Wamba by Takeshi Furuichi. (see Chapter 2, see pages 26 and 27)
(Une ex-femelle alpha, Kame (à gauche), se fait toiletter par l'ex-mâle alpha, Ibo (au centre), et d'autres petits. Photo prise à Wamba par Takeshi Furuichi. (voir Chapitre 2))

Plate 3 Apollo carrying his maternal brother on his back during travelling. Photo by LuiKotale Bonobo Project / Tim Lewis Bale. (see Chapter 3)
(Apollo voyageant avec son frère maternel sur le dos. (voir Chapitre 3))

Plate 4 Adult male Keza fear grimacing to mother-reared juvenile Poli. Photo taken in Lola ya Bonobo by Jingzhi Tan. (see Chapter 4, see pages 58 and 59)
(Le mâle adulte, Keza, faisant une grimace destinée à effrayer Poli, un jeune élevé par sa mère. (voir Chapitre 4))

Plate 5 Bonobos are playful even among adults. Photo taken in Lola ya Bonobo by Jingzhi Tan. (see Chapter 5)
(Même les bonobos adultes sont joueurs. (voir Chapitre 5))

Plate 6 We humans know what others see and know. Do bonobos have such social cognitive abilities? Photo of Yolo taken in Lola ya Bonobo by Alexandra Rosati. (see Chapter 6)
(Nous, les humains, savons ce que les autres voient et savent. Les bonobos ont-ils de telles capacités cognitives sociales ? (voir Chapitre 6))

Plate 7 Savage-Rumbaugh communicating with Kanzi using the lexigram keyboard. (see Chapter 7, see page 97)
(Savage-Rumbaugh communiquant avec Kanzi à l'aide du clavier à lexigrammes. (voir Chapitre 7)) Picture from Wikicommons: File:Kanzi,_conversing.jpg.

Plate 8 Mimi, the dominant female, calls. Bonobos communicate with each other with various gestures and vocalization. Photo taken in Lola ya Bonobo by Vanessa Woods. (see Chapter 8)
(Mimi, la femelle dominante, appelle. Les bonobos communiquent entre eux grâce à divers gestes et sons vocaux. (voir Chapitre 8))

Plate 9 Bonobos seem to use begging for social purposes, such as to make friends, during food sharing. Photo taken in Wamba by Takeshi Furuichi. (see Chapter 9, see page 128)
(Les bonobos semblent utiliser la mendicité à des fins sociales, notamment pour se faire des amis lors du partage de la nourriture. (voir Chapitre 9))

Plate 10 On the first day of her transfer to a new social group, Kalina (bottom) was engaging in affiliative, socio-sexual behaviour with Salonga (top), a resident female. Photo taken in Lola ya Bonobo by Jingzhi Tan. (see Chapter 10, see pages 144 and 145)
(Le premier jour de son transfert vers un nouveau groupe social, Kalina (en bas) s'engageait dans un comportement affiliatif socio-sexuel avec Salonga (en haut), une femelle résidente. (voir Chapitre 10))

Plate 11 Wild bonobos enjoy some delicious fruit. The relatively stable food availability bonobos experience may have affected their foraging strategies and cognition. Photo taken in Wamba by Shinya Yamamoto. (see Chapter 11)
(Des bonobos sauvages dégustent de délicieux fruits. L'accès à la nourriture relativement stable pour les bonobos a peut-être affecté leur cognition et leurs stratégies en matière de recherche de nourriture. (voir Chapitre 11))

Plate 12 Bonobos in captivity demonstrate diverse forms of sophisticated tool use, although this same form of extractive foraging is absent in wild populations. Photo of Lukaya taken at Lola ya Bonobo by Jingzhi Tan. (see Chapter 12)
(Les bonobos en captivité démontrent diverses formes d'utilisation sophistiquée des outils, bien que cette même forme d'extraction de la nourriture soit absente chez les populations sauvages (voir Chapitre X))

Plate 13 Personality dimensions in bonobos are correlated with sex, age and behaviours; some are similar to humans and chimpanzees but other are different. Photo taken at Wamba by Shinya Yamamoto. (see Chapter 13)
(Les dimensions de la personnalité chez les bonobos sont liées au sexe, à l'âge et aux comportements. Certaines sont similaires à celles des humains et des chimpanzés, mais d'autres sont différentes.)

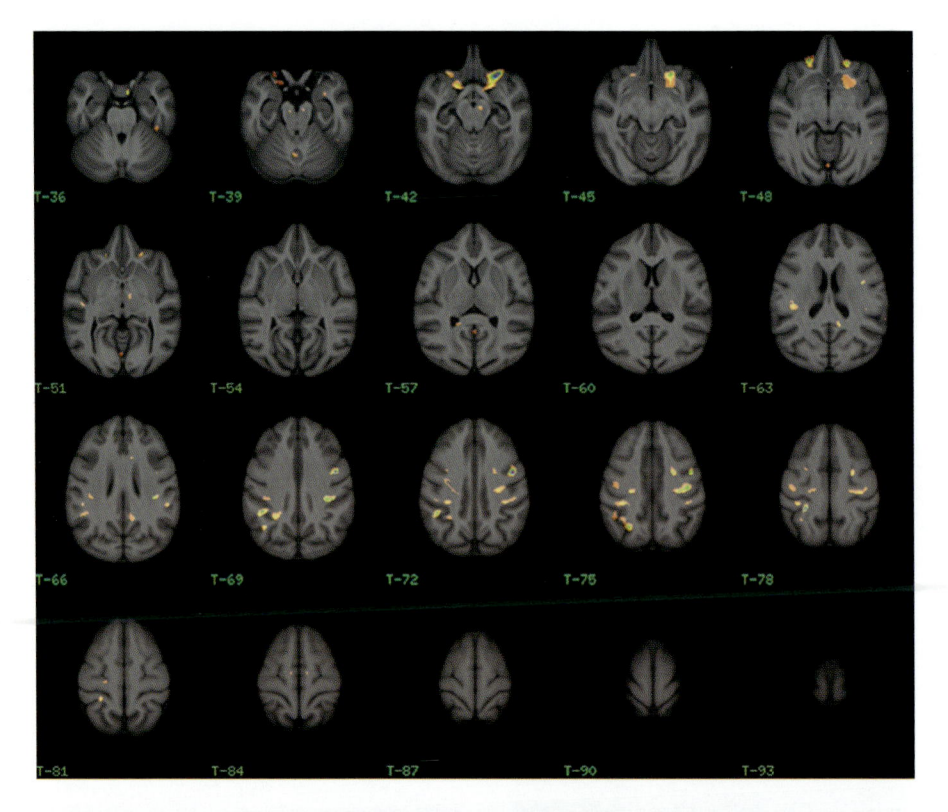

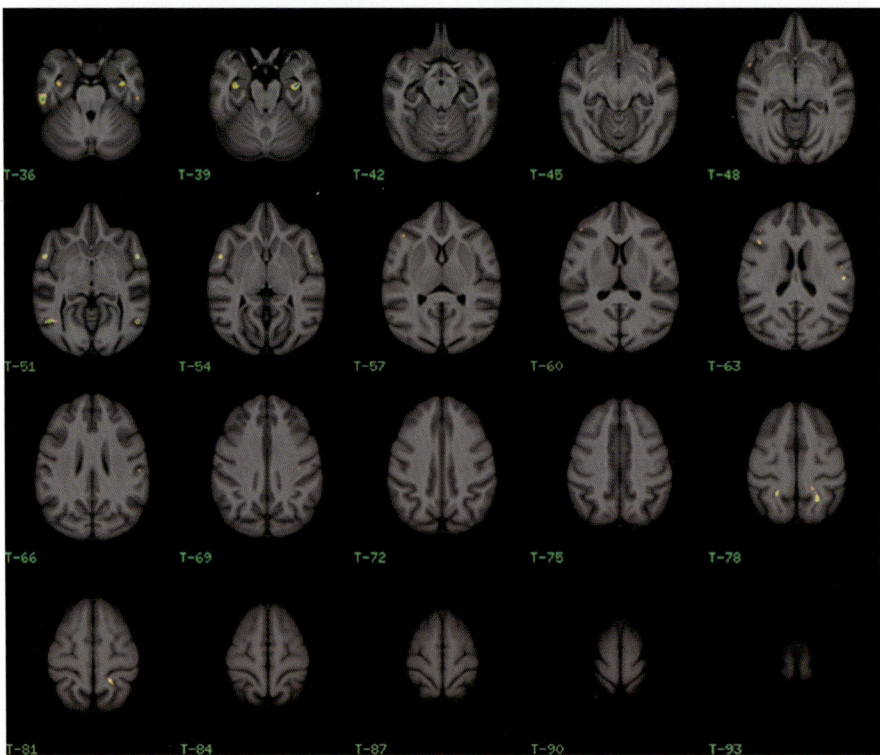

Plate 14 Clusters with higher grey matter values in bonobos compared to chimpanzees (a), and clusters with higher grey matter values in chimpanzees compared to bonobos (b). (see Chapter 14, see page 205)
(Amas contenant davantage de matière grise chez les bonobos que chez les chimpanzés (a) et amas contenant davantage de matière grise chez les chimpanzés que chez les bonobos (b).)

Plate 15 Etumbe, the bushmeat orphan, raised in a deprived lab environment in Kinshasa shows affection to Crispin Mahamba the vet who helped rescue her at Lola ya Bonobo. She since has been successfully released back to the wild. The highly cooperative and tolerant features that she and other bonobos show may have evolved through self-domestication, similar to Pleistocene *Homo sapiens*. Photo taken in Lola ya Bonobo by Vanessa Woods. (see Chapter 15)
(Etumbe, orpheline victime du braconnage élevée dans un environnement de laboratoire défavorisé à Kinshasa, montre de l'affection à Crispin Mahamba, le vétérinaire qui a aidé à la sauver à Lola ya Bonobo. Elle a depuis été relâchée avec succès dans la nature. Les caractéristiques hautement coopératives et tolérantes qu'elle et d'autres bonobos présentent ont peut-être évolué grâce à l'autodomestication, de la même façon que l'Homo sapiens du Pléistocène.)

Plate 16 Where do bonobos come from? What are bonobos? Where are bonobos going? Photo of Noiki taken at Lola ya Bonobo by Vanessa Woods. (see Chapter 16)
(D'où viennent les bonobos ? Qu'est-ce qu'un bonobo ? Où vont les bonobos ?)

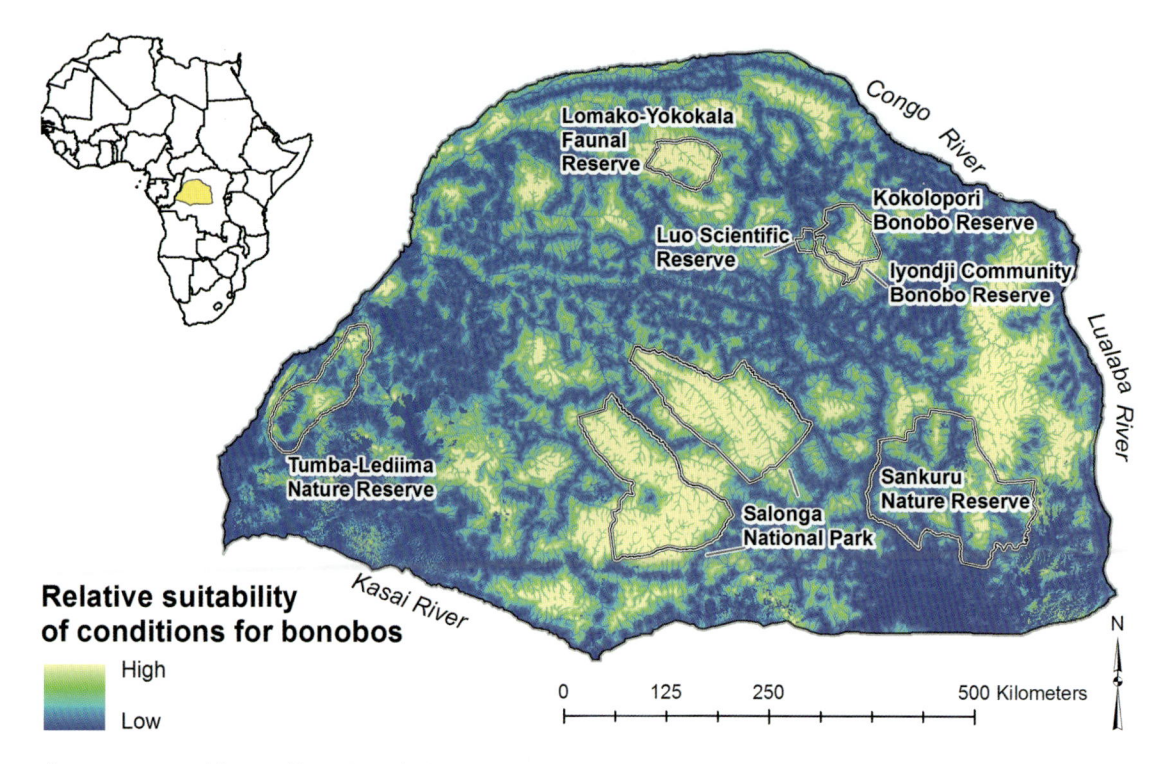

Plate 17 A map modelling suitable conditions (yellow to green) for bonobos within their natural range South of the Congo River in the Congo Basin. Existing protected areas indicated with lines. (see Chapter 17)
(Une carte de modélisation des conditions appropriées (en vert) pour les bonobos dans leur aire de répartition naturelle au sud du fleuve Congo dans le bassin du Congo. Zones protégées existantes indiquées avec des lignes.)

Plate 18 Bonobos are not pets or food, but they are gravely threatened by the bushmeat trade. It is our responsibility to carefully plan for the healthy growth of captive bonobo population as well as to conserve wild population. Photo of newly arrived bushmeat orphan at Lola ya Bonobo by Jingzhi Tan. (see Chapter 18)
(Les bonobos ne sont pas des animaux domestiques ou des aliments, mais ils sont gravement menacés par le commerce de la viande de brousse. Il est de notre responsabilité de planifier soigneusement la croissance saine de la population des bonobos en captivité ainsi que de préserver la population sauvage. Photo d'un orphelin victime du braconnage fraîchement arrivé à Lola ya Bonobo.)

Despite the fact they were highly skilful and experienced in this task, their food preoccupation led to the breakdown in cooperation that could have been mutually beneficial (Hare et al., 2007).

With the higher level of feeding tolerance seen in bonobos it seems they should demonstrate more prosocal behaviors in the food-related context. Tan and colleagues (2015) tested bonobos in the standard prosocial choice task and, like most chimpanzee studies, found that when aiding the recipient was cost-free, bonobos did not prefer the prosocial option more in the experimental than control. However, this is largely because they had a strong bias always to choose the prosocial option in both the experimental and control conditions (i.e. this is very different from *avoiding* the prosocial option in the experimental; also see experiment 1 in Jensen et al., 2006). Tan and colleagues (2015) concluded that the negative results and those of most chimpanzees studies were likely methodological. They highlighted three weaknesses of the standard prosocial choice task: few studies have conducted a pre-test to measure subjects' understanding of the consequence of their choices; for those studies with a proper pre-test, subjects developed a strong location bias towards one of the two options during the pre-test that persisted into the test phase, resulting in a false negative; finally, this paradigm is cognitively taxing even for young children because of the need to keep track simultaneously of two potential outcomes for oneself plus two more for the recipient. In corroboration, Amici and colleagues (2014) carried out a direct comparison between chimpanzees and bonobos (in addition to gorillas, orangutans, capuchin monkeys and spider monkeys) in two variants of the prosocial choice task. They found no consistent evidence of prosociality in any species tested. They also demonstrated that the paradigm was difficult for subjects to understand but easy for them to develop a baseline bias.

Taken together, we currently have yet to observe any difference between bonobos and chimpanzees in this prosocial choice task. However, because this paradigm is vulnerable to type II error, has a high cognitive demand and is difficult to implement properly (even with human children), it appears to be a less valid measure than other more intuitive paradigms. Any results from this paradigm should be treated with great caution (Amici et al., 2014; Tan et al., 2015).

Low-cost prosociality in the instrumental helping tasks

Another popular paradigm for measuring prosociality is the instrumental helping task. In this paradigm, subjects could usually assist a recipient to fulfil its desire/intention to acquire desirable items (food or non-food objects). The assistance includes unlocking a door that is blocking the recipient's access to desirable items or facilitating the delivery of the items to the recipient by dropping them or directly handing them to the recipient. For example, Melis and colleagues (2011) placed a bag of food or a valuable token up a slope that was beyond the reach of a chimpanzee recipient. The recipient could only acquire the item if the subject on the opposite side released the item so that it could slide down the slope to the recipient's room. This condition was compared to the control condition where no recipient was present. Chimpanzee subjects were found to release the item more often when there was a recipient to acquire the item. Importantly, subjects could not obtain the food themselves, they understood the consequences of their behaviors prior to the test and they were never rewarded for the prosocial behavior. Taken together, these findings suggest that chimpanzee subjects are motivated to assist the recipient to acquire the desirable item. Similar positive results are observed when chimpanzees brought out-of-reach food within the reach of their collaborative partner (Greenberg et al., 2010), released a locked recipient into a room baited with food (Warneken et al., 2007) or handed tools directly to the recipient in the adjacent room (Yamamoto et al., 2012; but see Liebal et al., 2014).

Bonobos are also prosocial towards others in some of these contexts. In one recent experiment bonobos could pull a rope from an adjacent room to release a recipient into a blocked tunnel baited with food (Tan and Hare, 2013, experiment 3). Like the chimpanzee experiments (e.g. Warneken et al., 2007), bonobos could not access the food themselves and would not obtain any reward for helping. Furthermore, they all passed pre-tests showing an understanding of the physical properties of the task and an ability to inhibit any intrinsic motivation to pull the rope as play. Subjects pulled the rope more often when the unrelated recipient was present (i.e.

when they could help) than when it was absent (i.e. they could not help).

This experiment reveals that bonobos will voluntarily help an unrelated recipient to access desirable items as chimpanzees do. It further points out an important distinction from chimpanzees: bonobos extend their prosociality towards strangers (i.e. bonobos are *xenophilic*). In this experiment, half of the unrelated recipients were strangers from another social group. As predicted based on their unusual tolerance for strangers, bonobos helped the unfamiliar recipient as often as they did when the recipient was a familiar group member (Tan and Hare, 2013). Importantly, the roles of the subject and the recipient were never reversed and the pairs would return to their own social groups after the test, precluding any possibility for reciprocity within or outside the experiment. As a result, this experiment demonstrates that bonobos are motivated to help a stranger acquire food even though they receive no benefits for themselves.

The phenomenon was replicated by a follow-up experiment (Tan et al., submitted). Bonobo subjects were each paired with a stranger recipient that had never entered in the same enclosure with the subjects. There was a piece of apple hanging above the ceiling of the recipient room, and the food was beyond the reach of the recipient. The subjects, however, could climb up, reach out from the ceiling of the subject room which is adjacent to the recipient room, and then release the apparatus, causing the food to drop within reach of the recipient. Prior to the test, all subjects understood the contingency of the apparatus and the fact that they could not obtain the food from inside the subject room. In the test, subjects released the apparatus more often when the stranger recipient was in the recipient room than when it was absent. Again, bonobos voluntarily helped a stranger from an out-group even when they received no selfish benefits.

In addition to xenophilia, another interesting feature of bonobo prosociality in these two experiments is its *proactive nature*. It has been proposed that human prosociality is not only reactive (i.e. prosociality is a response to extrinsic stimuli) but also proactive (i.e. prosociality occurs without extrinsic stimuli) (Warneken and Tomasello, 2008). The latter component presumably requires a more sophisticated understanding of the recipient's intention, because only contextual and indirect cues are available to infer the need for help. For instance, humans make donations to anonymous recipients in one-shot interactions (Camerer, 2003) and infants help others instrumentally in the absence of behavioral or communicative cues (Warneken, 2013). It has been argued by theorists that this proactive component is a derived feature from cooperative breeding and thus *Pan* prosociality should be primarily reactive (Burkart et al., 2009).

This prediction is generally consistent with the majority of chimpanzee experiments where ostensive signals of the recipient's desires or overt requests for help are crucial for helping to occur (Warneken, 2015). For instance, Yamamoto and colleagues (2009) found that chimpanzee subjects transferred out-of-reach tools to a recipient when it actively gestured for help approximately three times as often as than when the recipient remained passive (see also Warneken et al., 2007; Yamamoto et al., 2012). Melis and colleagues (2011) systematically manipulated the opportunity for solicitation from the recipient to test if overt signals were necessary in a food-/tool-delivery task. They found that chimpanzees *only* helped when the recipient was actively signalling its desire and attempting to capture subjects' attention. However, two chimpanzee studies have found evidence consistent with proactive helping (Greenberg et al., 2010; Horner et al., 2011). Greenberg and colleagues found that chimpanzees helped a partner to obtain out-of-reach food even when there was no request for help. Helping in this study occurred almost immediately (90 per cent of trials) and thus the recipient requested help at a very low rate. It remains unclear if a request has any effect on prosociality in this context. Horner and colleagues found that chimpanzees actually behave more selfishly when the recipient showed 'attention-getting' behaviors or 'directed requests and pressure' behaviors. However, their operational definitions differed from what others previously measured since they included aggressive behaviors that may have prohibited prosocial behavior (e.g. threatening displays) and non-communicative (e.g. self-scratching).

In the instrumental helping experiments where bonobos have shown prosocial behaviors, the emergence of prosociality is consistent with the definition of proactive prosociality (Burkart et al., 2009; Warneken and Tomasello, 2008). Bonobos unlocked the recipient into the tunnel baited with food even when the recipient did not make any signal showing its desire (Tan and Hare, 2013, experiment 3). An active recipient did not increase the chances of prosociality received from the subjects. In Tan and colleagues (submitted), we manipulated the recipient's signalling behaviors by following Melis and colleagues (2011b). In the reaching condition, the ceiling of the recipient room had wide mesh so that the recipient could reach out an entire arm if it attempted to acquire the hanging fruit. In the blocked condition, the ceiling was blocked by narrow mesh so it was impossible for this kind of signalling behaviors. Subjects released the food for the recipient equally often in the two conditions. Furthermore, the recipient could not direct any gestures to the subjects as a way of harassment or begging because they were always separated by narrow mesh. These findings are clearly in contrast with the observed reactive prosociality in chimpanzees (Melis et al., 2011; Warneken et al., 2007; Yamamoto et al., 2012). This contrast seems to indicate that motivationally bonobos have a stronger prosocal tendency, and/or cognitively they possess a stronger sensitivity to other's intention without ostensive cues (Herrmann et al., 2010). Further research should be conducted to examine these hypotheses.

Finally, there appears to be a third difference in the instrumental helping task between bonobos and chimpanzees: bonobos seem to be less motivated to help retrieve non-food objects. In an instrumental helping task, subjects usually need to release an apparatus to give the recipient access to desirable items (Melis et al., 2011b; Tan et al., submitted), or they need to directly pass the recipient a non-food item that is of no value to the subjects (Warneken and Tomasello, 2009). The bonobo helping experiments mentioned belong to the first version because subjects basically had to unlock an apparatus which would then give the recipient access to food (Tan and Hare, 2013, experiment 3; Tan et al., submitted). Following Warneken and Tomasello (2006), we also tested whether bonobos would pick up a wooden stick placed out of the reach of an experimenter and hand it back to him (Krupenye et al., submitted). Informally during everyday interactions with care staff, chimpanzees often retrieve objects for human caregivers when requested; however, bonobos do not. In fact bonobos usually take the objects requested and play with them for extended periods of time. Recent experimental results are consistent with this observation. Bonobos typically inspected an out-of-reach stick but rarely 'helped' by handing it back to the experimenter. We then designed an experiment to compare the intrinsic motivation bonobos have for sharing objects versus food (Krupenye et al., submitted). In the experimental condition, one bonobo was given a handful of nuts that could only be cracked with a rock. A second bonobo was in an adjacent room separated by open mesh. This bonobo was given two rocks but no nuts. In the control condition, each bonobo was given a complete nut-cracking kit: a handful of nuts and a rock. Prior to the test, both subjects understood that a rock was necessary for nut cracking and they could either pass a rock, nuts or both across the mesh in order to eat the nuts. We found that bonobos transferred nuts significantly more often than rocks. They also transferred nuts more often in the experimental condition than in the control, while there was no difference in rock transfer between conditions. Because the subject in possession of the rock had two rocks and it only needed one to crack nuts, rock transfer was theoretically less costly than nut transfer. This overwhelming preference for food transfer to object transfer is a striking deviation from the pattern observed in chimpanzees. We hypothesize this bonobo pattern is in part caused by a stronger feeding tolerance (Hare et al., 2012) as well as a different motivation to interact with objects—in several studies, bonobos appear less motivated than chimpanzees to look at, manipulate and transfer objects (Kano et al., 2015; Koops et al., 2015; Liebal et al., 2014).

Taken together, bonobos reveal three important differences in the instrumental helping paradigm: they voluntarily help strangers; the emergence of this prosociality does not require ostensive cues signalling the recipient's desire; their prosocial preference seems to be stronger in contexts involving food items than those with non-food items.

High-cost prosociality in food-sharing paradigms

In the standard prosocial choice task and the instrumental helping task, prosocial behaviors usually incur a relatively low cost (e.g. energetic cost to pull a rope) or no cost at all (e.g. choosing two options with identical payoff to oneself). Experiments that incur a high cost (e.g. food loss) to the actor have been rare. However, two high-cost tasks have been implemented in bonobos.

The first of these tasks used a sharing paradigm where one subject could unlock a side door and release a recipient into the same room to feed on a pile of desirable food (Bullinger et al., 2013; Hare and Kwetuenda, 2010; Tan and Hare, 2013). Subjects had complete control over the food because the recipient could not enter the room, reach for the food or harass the subjects. Prior to the test, subjects had to demonstrate the ability to inhibit unlocking the door for no obvious incentives. Chimpanzees in this situation preferred to feed alone and ignored the familiar recipient blocked in an adjacent room (Bullinger et al., 2013; Melis et al., 2006). Bonobos showed mixed results. When they were paired with a familiar recipient from the same social group, they, like chimpanzees, tended to prefer feeding alone (Tan and Hare, 2013, experiment 2; Bullinger et al., 2013, but see Hare and Kwetuenda, 2010, for evidence of a preference to share in one pairing between groupmates). However, when they were paired with a stranger from a neighbouring group, they reliably forfeited monopolizable food and release the stranger to feed together (Hare and Kwetuenda, 2010; Tan and Hare, 2013, experiments 1 and 2). Importantly, it was always possible for subjects to release the recipient into the same room after they finished the food but the subject chose to eat together with the stranger. As a result, this xenophilic food sharing suggests that being in the same room with a stranger is highly rewarding for bonobos, given they are willing to lose their own food in order to engage in this interaction. This social reward is so strong that the released bonobo would often release the subject's groupmate that was unknown to it, effectively splitting the food in three and leaving itself outnumbered by strangers (Tan and Hare, 2013, experiment 1). Although it may not be possible to test chimpanzees in this paradigm due to ethical concerns, field playback experiments suggest that they spontaneously avoid a stranger unless they outnumber the stranger by a factor of three (Wilson et al., 2001). When they outnumber the stranger, their response is aggressive and sometimes lethal (Wilson et al., 2014).

This food-sharing paradigm reveals that a physical interaction with strangers is highly desirable for bonobos. This social reward seems to be strongest for strangers, given that food sharing did not typically occur when bonobos were paired with group members. This further suggests that bonobo prosociality is not only driven by an unselfish motivation to benefit others, as seen in the low-cost helping task, but also driven by selfish motivation to pursue social rewards. In the presence of certain social rewards, bonobos are willing to give up resources that they would otherwise monopolize.

In corroboration of this prediction, we designed a high-cost instrumental helping task in which it was impossible to acquire the social reward for physically interacting with a stranger (Tan and Hare, 2013, experiment 4). In this experiment, subjects and the recipient (a group member or a stranger) were always separated in different room. Subjects could pull a rope to unlock a door and allow the recipient to enter a room baited with a pile of food. However, this prosocial behavior was costly since the subjects could instead simply monopolize the food pile. Releasing the recipient results in food loss but not physical interaction. In this low-reward, high-cost context, no prosocial behavior was observed even when the recipient was a stranger (Tan and Hare, 2013, experiment 4).

These two high-cost tasks shared similar costs of food loss but differed in the potential to gain social reward from physically interacting with a stranger. Their contrasting results show that bonobos can be prosocial for both selfish and unselfish motivations. The unselfish motivation is evident in the observed helping in the low-cost tasks, while the interplay between the chance to interact with a stranger physically and the potential for food loss in the high-cost paradigms suggests a more strategic side of bonobo prosociality.

Like chimpanzees, it seems that in may contexts bonobos may also be reluctant to transfer food in their possession to fellow group members (Bullinger et al., 2013; Jaeggi et al., 2010; but see Byrnit et al., 2015). However, bonobos will pass edible nuts to a familiar partner (Krupenye et al., submitted), suggesting that feeding tolerance allows bonobos to show active food transfer in certain contexts (i.e. perhaps with medium- to low-value food). Furthermore, tolerance for strangers allows bonobos to show prosociality in contexts where chimpanzees do not.

Conclusion

Bonobos are prosocial towards non-kin. In many ways, bonobo prosociality is similar to that of chimpanzees. When the recipient is a familiar group member, they struggle with the standard prosocial choice task (Amici et al., 2014; Tan et al., 2015), are willing to help a recipient gain access to desirable items (Melis et al., 2011b; Tan et al. submitted), but show a reluctance to transfer high-quality food in their possession to a group member with whom they cannot physically interact (Bullinger et al., 2013; Tan and Hare, 2013).

Important differences have also been uncovered between bonobo and chimpanzee prosociality. First and foremost, bonobos are prosocial towards strangers. They will help a stranger from a neighbouring group access out-of-reach food, although they receive no benefits to themselves (Tan and Hare, 2013; Tan et al., submitted). They are willing to forfeit their own food in exchange for a desirable physical interaction with a stranger (Hare and Kwetuenda, 2010; Tan and Hare, 2013). Bonobo xenophilia is therefore driven in part by an unselfish motivation, which is consistent with an 'other-regarding preference' or genuine concern for another's welfare (Silk and House, 2011). This provides strong evidence that non-human primates can show a human-like unselfishness towards strangers (Silk and House, 2011). The unselfishness of bonobo xenophilia challenges various models on the origin of human co-operation, which propose that any kind of prosocial sharing with non-group members is unique to humans due to social norms, language and/or cooperative breeding (Bowles and Gintis, 2011; Burkart et al., 2009). However, it is perhaps the selfish side of bonobo xenophilia that provides new insight into the origin of prosociality towards strangers. It has been (incorrectly) assumed by many that xenophilia

can only be unselfish because it is difficult to secure reciprocation from the strangers (see reviews by Hagen and Hammerstein, 2006). The unusual pattern of cooperation in bonobos might provide a novel context that allows selfish benefits for xenophilic actors—strangers might be valuable partners in the future (Tan et al., submitted). This new framework can have important implications for explaining the origin of human cooperation because it shifts the exclusive focus on altruism to emphasize mutualism instead (Hare and Tan, 2011).

The second feature that distinguishes bonobo prosociality from that of chimpanzees is its proactive nature. We found no evidence that behavioral signals from the recipient are necessary to trigger the emergence of bonobo prosociality. This pattern seems different from the reliance on behavioral cues in chimpanzee prosociality and is consistent with the findings that bonobos might show greater sensitivity to others' attention and intention than chimpanzees (Herrmann et al., 2010; Kano et al., 2015). Again, this challenges the models developed from a focus on chimpanzees that proposes proactivity is unique to humans or limited to cooperative breeders (Burkart et al., 2009). However, it remains unclear whether bonobos could demonstrate more sophisticated forms of proactivity. For example, can bonobos, like human children, infer the needs of the recipient solely based on situational cues (Warneken, 2013)?

The third difference between bonobo and chimpanzee prosociality is the context of prosociality. Bonobos are more tolerant with food but less motivated to interact with non-food objects (Hare et al., 2012; Kano et al., 2015; Koops et al., 2015). Correspondingly, although chimpanzees would directly hand back non-food items upon request (Warneken et al., 2007; Yamamoto et al., 2012), bonobos are more motivated to actively transfer food (un-cracked nuts) than objects (Krupenye et al., submitted). This is an interesting pattern given that human children are flexibly prosocial in both food- and object-related contexts (Warneken, 2015). More comparative research is needed to further determine the (in)flexibility of prosociality in bonobos and chimpanzees.

Taken together, it becomes clear that the form of prosociality among non-kin differs between bonobos and chimpanzees. While we have gained significant insights into the evolution of cooperation by studying chimpanzees in the past decade, the distinct features of bonobo prosociality raise new and exciting questions. For example, what are the extent and limits of bonobo xenophilia? How do bonobos solve collective action problems with non-kin? Do bonobos trust strangers? Do only some bonobos sex–age classes of bonobos show these preferences towards strangers and when does the identity of the stranger matter? However, the most puzzling question remains how humans evolved such a mosaic pattern of prosociality among non-kin which, like bonobos, extends to strangers and includes food-sharing while, like chimps, shows clear ingroup favouritism and also occurs in non-food contexts. Answers will not be uncovered without an equal understanding of both *Pan* species.

References

Alford, P.L., Bloomsmith, M.A., Keeling, M.E. and Beck, T.F. (1995). Wounding aggression during the formation and maintenance of captive, multimale chimpanzee groups. *Zoo Biology*, 14(4), 347–59.

Amici, F., Visalberghi, E. and Call, J. (2014). Lack of prosociality in great apes, capuchin monkeys and spider monkeys: convergent evidence from two different food distribution tasks. *Proceedings of the Royal Society of London B*, 281, 1–9.

Baker, K.C. and Aureli, F. (2000). Coping with conflict during initial encounters in chimpanzees. *Ethology*, 106(6), 527–41.

Bowles, S. and Gintis, H. (2011). *A Cooperative Species: Human Reciprocity and Its Evolution.* Princeton, NJ: Princeton University Press.

Bullinger, A. F., Burkart, J. M., Melis, A. P. and Tomasello, M. (2013). Bonobos, (*Pan paniscus*), chimpanzees, (*Pan troglodytes*), and marmosets, (*Callithrix jacchus*), prefer to feed alone. *Animal Behaviour*, 85(1), 51–60.

Burkart, J.M., Hardy, S.B., and Van Schaik, C.P. (2009). Cooperative breeding and human cognitive evolution. *Evolutionary Anthropology*, 18(5), pp.175–186.

Byrnit, J.T., Høgh-olesen, H., and Makransky, G. (2015). Share your sweets: chimpanzee (*Pan troglodytes*) and bonobo (*Pan paniscus*) willingness to share highly attractive, monopolizable food sources. *Journal of Comparative Psychology*, 129(3), 218–28.

Camerer, C. (2003). *Behavioral Game Theory: Experiments in Strategic Interaction.* Princeton, NJ: Princeton University Press.

Carter, C.S. and Porges, S.W. (2011). The neurobiology of social bonding and attachment. In: Decety, J. and Cacioppo, J. (eds). *The Oxford Handbook of Social Neuroscience.* Oxford: Oxford University Press, pp. 151–63.

Cesarini, D., Dawes, C.T., Johannesson, M., Lichtenstein, P. and Wallace, B. (2009). Experimental game theory and behavior genetics. *Annals of the New York Academy of Sciences*, 1167, 66–75.

Clay, Z. and de Waal, F.B.M. (2013). Bonobos respond to distress in others: consolation across the age spectrum. *PloS One*, 8(1), e55206.

Cronin, K.A. (2012). Prosocial behaviour in animals: the influence of social relationships, communication and rewards. *Animal Behaviour*, 84(5), 1085–93.

Cronin, K.A., De Groot, E., and Stevens, J.M.G. (2015). Bonobos show limited social tolerance in a group setting: a comparison with chimpanzees and a test of the relational model. *Folia Primatologica*, 86(3), pp. 164–77.

Demuru, E. and Palagi, E. (2012). In bonobos yawn contagion is higher among kin and friends. *PLoS One*, 7(11).

Forss, S.I.F., Schuppli, C., Haiden, D., Zweifel, N., and Van Schaik, C. P. (2015). Contrasting responses to novelty by wild and captive orangutans. *American Journal of Primatology*, (April).

Furuichi, T. (2011). Female contributions to the peaceful nature of bonobo society. *Evolutionary Anthropology*, 20(4), 131–42.

Gerloff, U., Hartung, B., Fruth, B., Hohmann, G., and Tautz D. (1999). Intracommunity relationships, dispersal pattern and paternity success in a wild living community of Bonobos (*Pan paniscus*) determined from DNA analysis of faecal samples. *Proceedings of the Royal Society of London B*, 266(1424), 1189–95.

Gold, K. (2001). Group formation in captive bonobos: sex as a bonding strategy. In: *The Apes: Challenges for 21st Century.* Conference Proceedings Brookfield: Brookfield Zoo, Chicago Zoological Society, pp. 90–3.

Greenberg, J.R., Hamann, K., Warneken, F. and Tomasello, M. (2010). Chimpanzee helping in collaborative and noncollaborative contexts. *Animal Behaviour*, 80(5), 873–80.

Hagen, E.H. and Hammerstein, P. (2006). Game theory and human evolution: a critique of some recent interpretations of experimental games. *Theoretical Population Biology*, 69(3), 339–48.

Hamann, K., Warneken, F., Greenberg, J.R. and Tomasello, M. (2011). Collaboration encourages equal sharing in children but not in chimpanzees. *Nature*, 476(7360), 328–31.

Hare, B. (2011). From hominoid to hominid mind: what changed and why? *Annual Review of Anthropology*, 40(1), 293–309.

Hare, B. and Kwetuenda, S. (2010). Bonobos voluntarily share their own food with others. *Current Biology*, 20(5), R230–R231.

Hare, B. and Tan, J. (2011). How much of our cooperative behavior is human? In: de Waal, F.B.M. and Ferrari, P.F. (eds). *The Primate Mind: Built to Connect with Other Minds*. Cambridge, MA: Harvard University Press, pp. 175–93.

Hare, B., Wobber, V., and Wrangham, R.W. (2012). The self-domestication hypothesis: evolution of bonobo psychology is due to selection against aggression. *Animal Behaviour*, 83(3), 573–85.

Hare, B. and Yamamoto, S. (2015). Moving bonobos off the scientifically endangered list. *Behaviour*, 152, 247–58.

Hare, B., Melis, A. P., Woods, V., Hastings, S. and Wrangham, R.W. (2007). Tolerance allows bonobos to outperform chimpanzees on a cooperative task. *Current Biology*, 17(7), 619–23.

Heilbronner, S.R., Rosati, A.G., Stevens, J.R., Hare, B., and Hauser, M. D. (2008). A fruit in the hand or two in the bush? Divergent risk preferences in chimpanzees and bonobos. *Biology Letters*, 4(3), 246–9.

Henrich, J., Boyd, R., Bowles, S., Camerer, C., Fehr, E., Gintis, H., McElreath, R., Alvard, M., Barr, A., Ensminger, J., Henrich, N.S., Hill, K., Gil-White, F., Gurven, M., Marlowe, F.W., Patton, J.Q. and Tracer, D. (2005). 'Economic man' in cross-cultural perspective: behavioral experiments in 15 small-scale societies. *Behavioral and Brain Sciences*, 28(6), 795–815; discussion 815–55.

Herrmann, E., Hare, B., Call, J. and Tomasello, M. (2010). Differences in the cognitive skills of bonobos and chimpanzees. *PloS One*, 5(8), e12438.

Herrmann, E., Hare, B., Cissewski, J. and Tomasello, M. (2011). A comparison of temperament in nonhuman apes and human infants. *Developmental Science*, 14(6), 1393–405.

Hohmann, G. and Fruth, B. (2002). Dynamics in social organization of bonobos (*Pan paniscus*). In: Boesch, C., Hohmann, G., and Marchant, L.F. (eds). *Behavioral Diversity in Chimpanzees and Bonobos*. Cambridge: Cambridge University Press, pp. 138–50.

Horner, V., Carter, J.D., Suchak, M. and de Waal, Frans B.M.M (2011). Spontaneous prosocial choice by chimpanzees. *Proceedings of the National Academy of Sciences*, 108(33), 13847–51.

Jaeggi, A.V., De Groot, E., Stevens, Jeroen M.G.G. and van Schaik, C.P. (2013). Mechanisms of reciprocity in primates: testing for short-term contingency of grooming and food sharing in bonobos and chimpanzees. *Evolution and Human Behavior*, 34(2), 69–77.

Jaeggi, A.V., Stevens, J.M.G., and Van Schaik, C.P. (2010). Tolerant food sharing and reciprocity is precluded by despotism among bonobos but not chimpanzees. *American Journal of Physical Anthropology*, 143(1), 41–51.

Jensen, K., Hare, B., Call, J. and Tomasello, M. (2006). What's in it for me? Self-regard precludes altruism and spite in chimpanzees. *Proceedings of the Royal Society B*, 273(1589), 1013–21.

Kano, F., Hirata, S., and Call, J. (2015). Social attention in the two species of *Pan*: bonobos make more eye contact than chimpanzees. *PloS One*, 10(6), e0129684.

Koops, K., Furuichi, T., and Hashimoto, C. (2015). Chimpanzees and bonobos differ in intrinsic motivation for tool use. *Scientific Reports*, 5, 11356.

Langergraber, K.E., Mitani, J.C., and Vigilant, L. (2007). The limited impact of kinship on cooperation in wild chimpanzees. *Proceedings of the National Academy of Sciences*, 104(19), 7786–90.

Liebal, K., Vaish, A., Haun, D. and Tomasello, M. (2014). Does sympathy motivate prosocial behaviour in great apes? *PloS One*, 9(1), e84299.

Melis, A.P., Warneken, F., Jensen, K., Schneider, Anna-Claire Call, J. and Tomasello, M. (2011). Chimpanzees help conspecifics obtain food and non-food items. *Proceedings of the Royal Society B*, 278(1710), 1405–13.

Melis, A.P., Hare, B., and Tomasello, M. (2006). Chimpanzees recruit the best collaborators. *Science*, 311(5765), 1297–300.

Melis, A.P., Hare, B., and Tomasello, M. (2006). Engineering cooperation in chimpanzees: tolerance constraints on cooperation. *Animal Behaviour*, 72(2), 275–86.

Melis, A.P., Schneider, A.-C.C., and Tomasello, M. (2011). Chimpanzees, *Pan troglodytes*, share food in the same way after collaborative and individual food acquisition. *Animal Behaviour*, 82(3), 1–9.

Mitani, J.C. (2009). Male chimpanzees form enduring and equitable social bonds. *Animal Behaviour*, 77(3), 633–40.

Mitani, J.C., Watts, D.P., and Amsler, S.J. (2010). Lethal intergroup aggression leads to territorial expansion in wild chimpanzees. *Current Biology*, 20(12), R507–R508.

Parish, A. (1996). Female relationships in bonobos (*Pan paniscus*): evidence of bonding, cooperation, and female dominance in a male-philopatric species. *Human Nature*, 7(1), 61–96.

Pfalzer, S., Ehret, G.G., and Ulm, U. (1995). Social integration of a bonobo mother and her dependent daughter into an unfamiliar group. *Primates*, 36(3), 349–60.

Pusey, A.E. and Schroepfer-Walker, K. (2013). Female competition in chimpanzees. *Philosophical Transactions of the Royal Society of London B, Biological Sciences*, 368(1631), 20130077.

Rosati, A.G. and Hare, B. (2013). Chimpanzees and Bonobos Exhibit Emotional Responses to Decision Outcomes *PLoS One*, 8(5), e63058.

Sakamaki, T., Behncke, I., Laporte, M., Mulavwa, M., Ryu, H., Takemoto, H., Tokuyama, N., Yamamoto, S. and Furuichi, T. (2015). Intergroup transfer of females and social relationships between immigrants and residents in bonobo (*Pan paniscus*) societies. In: *Dispersing Primate Females: Life History and Social Strategies in Male-Philopatric Species*. Springer, Springer Japan. pp. 127–64.

Schel, A.M., Rawlings, B., Claidière, N., Wilke, C., Wathan, J. and Richardson, J. (2013). Network analysis of social changes in a captive chimpanzee community following the successful integration of two adult groups. *American Journal of Primatology*, 75(3), 254–66.

Seres, M., Aureli, F., and de Waal, F.B.M.M. (2001). Successful formation of a large chimpanzee group out of two preexisting subgroups. *Zoo Biology*, 20(2001), 501–15.

Silk, J.B., Brosnan, S.F., Vonk, J., Henrich, J., Povinelli, D.J., Richardson, A.S., Lambeth, S.P., Mascaro, J. and Schapiro, S.J. (2005). Chimpanzees are indifferent to the welfare of unrelated group members. *Nature*, 437(7063), 1357–9.

Silk, J.B. and House, B.R. (2011). Evolutionary foundations of human prosocial sentiments. *Proceedings of the National Academy of Sciences*, 108 Suppl, 10910–17.

Surbeck, M. and Hohmann, G. (2008). Primate hunting by bonobos at LuiKotale, Salonga National Park. *Current Biology*, 18(19), R906–R907.

Surbeck, M. and Hohmann, G. (2013). Intersexual dominance relationships and the influence of leverage on the outcome of conflicts in wild bonobos (*Pan paniscus*). *Behavioral Ecology and Sociobiology*, 67(11), 1767–80.

Tan, J. and Hare, B. (2013). Bonobos share with strangers. *PLoS One*, 8(1), 1–11.

Tan, J., Kwetuenda, S., and Hare, B. (2015). Preference or paradigm? Bonobos show no evidence of other-regard in the standard prosocial choice task. *Behaviour*, 152(3–4), 521–44.

Thunström, M., Persson, T., and Björklund, M. (2013). Integration of a hand-reared chimpanzee (*Pan troglodytes*) infant into a social group of conspecifics. *Primates*, 54(1), 13–19.

van Schaik, C.P. and Pradhan, G.R. (2003). A model for tool-use traditions in primates: implications for the coevolution of culture and cognition. *Journal of Human Evolution*, 44(6), 645–64.

Warneken, F., Hare, B., Melis, A.P., Hanus, D. and Tomasello, M. (2007). Spontaneous altruism by chimpanzees and young children. *PLoS Biology*, 5(7), e184.

de Waal, F.B.M. (1992). Appeasement, celebration, and food sharing in the two *Pan* species. *Topics in Primatology*, 1, 37–50.

Warneken, F. (2013). Young children proactively remedy unnoticed accidents. *Cognition*, 126(1), 101–8.

Warneken, F. (2015). Insights into the biological foundation of human altruistic sentiments. *Current Opinion in Psychology*.

Warneken, F., and Tomasello, M. (2006). Altruistic helping in human infants and young chimpanzees. Science 311, 1301–3.

Warneken, F. and Tomasello, M. (2008). Roots of human altruism in chimpanzees. *Primate Eye*, 96, 16.

Warneken, F. and Tomasello, M. (2009). Varieties of altruism in children and chimpanzees. *Trends in Cognitive Sciences*, 13(9), 397–402.

Williams, J. M., Oehlert, G. W., Carlis, J. V. and Pusey, A. E. (2004). Why do male chimpanzees defend a group range? *Animal Behaviour*, 68(3), 523–32.

Wilson, M. L., Boesch, C., Fruth, B., Furuichi, T., Gilby, I.C., Hashimoto, C. and Hobaiter, C. L. (2014). Lethal aggression in *Pan* is better explained by adaptive strategies than human impacts. *Nature*, 513(7518), 414–17.

Wilson, M.L., Hauser, M.D., and Wrangham, R.W. (2001). Does participation in intergroup conflict depend on numerical assessment, range location, or rank for wild chimpanzees? *Animal Behaviour*, 61(6), 1203–16.

Wobber, V., Hare, B., Maboto, J., Lipson, S., Wrangham, R. W. and Ellison, P. T. (2010). Differential changes in steroid hormones before competition in bonobos and chimpanzees. *Proceedings of the National Academy of Sciences*, 107(28), 12457–62.

Wobber, V., Wrangham, R.W., and Hare, B. (2010). Bonobos exhibit delayed development of social behavior and cognition relative to chimpanzees. *Current Biology*, 20(3), 226–30.

Yamamoto, S., Humle, T., and Tanaka, M. (2009). Chimpanzees help each other upon request. *PLoS One*, 4(10), e7416.

Yamamoto, S., Humle, T., and Tanaka, M. (2012). Chimpanzees' flexible targeted helping based on an understanding of conspecifics' goals. *Proceedings of the National Academy of Sciences*, 109(9), 3588–92.

PART V

Foraging Strategies

CHAPTER 11

Ecological variation in cognition: Insights from bonobos and chimpanzees

Alexandra G. Rosati

Bonobos and chimpanzees are closely related, yet they exhibit important differences in their wild socio-ecology. Whereas bonobos live in environments with less seasonal variation and more access to fallback foods, chimpanzees face more competition over spatially distributed, variable resources. This chapter argues that bonobo and chimpanzee cognition show psychological signatures of their divergent wild ecology. Current evidence shows that despite strong commonalities in many cognitive domains, apes express targeted differences in specific cognitive skills critical for wild foraging behaviors. In particular, bonobos exhibit less accurate spatial memory, reduced levels of patience and greater risk aversion than do chimpanzees. These results have implications for understanding the evolution of human cognition, as studies of apes are a critical tool for modelling the last common ancestor of humans with nonhuman apes. Linking comparative cognition to species' natural foraging behavior can begin to address the ultimate reason for why differences in cognition emerge across species.

Les bonobos et les chimpanzés sont prochement liés, pourtant ils montrent d'importantes différences dans leur sociologie naturelle. Alors que les bonobos vivent dans des environnements avec peu de diversité de climat entre saisons et plus d'accès à des ressources de nourriture alternatives, les chimpanzés ménagent une compétition étalée spatialement et des ressources plus variées. Je soutiens que la cognition des chimpanzés et bonobos montre les signatures psychologiques de leur écologie naturelle divergente. Les témoignages courants montrent que, malgré les forts points communs dans en cognition, les grands singes expriment des différences au niveau de compétences cognitives importantes au butinage. En particulier, les bonobos démontrent une mémoire spatial moin précise, moin de patience, et plus d'aversion de risques que les chimpanzés. Ces résultats fournissent des signes dans l'étude de l'évolution de la cognition humaine. Les études des grands singe sont un outil d'importance majeure dans la modélisation du dernier ancêtre commun des humains et grands singes non-humains. Faire des liens cognitives comparatives entre le butinage des différentes espèces peut commencer à dévoiler les raisons pour les différences de cognition entre espèces.

Introduction

One of the most pressing problems for the comparative study of cognition is explaining the emergence of variation in psychological abilities across species. How and why do such differences in cognition arise? This is an especially important question with respect to understanding the evolutionary history of our own species: although humans show important cognitive continuities with animals, we also exhibit a suite of abilities that markedly differ from other species. These differences span domains of behavior ranging from complex reasoning and planning to thinking about other's minds and to learning cultural norms of behavior. Consequently, illuminating the evolutionary mechanisms that shape cognitive capacities across species in general can shed light into the evolution of human-unique cognition specifically.

Many theoretical accounts argue that cognitive traits are shaped by natural selection, much like morphological traits. In primates, social organization in particular has long been thought to be critical in shaping the evolution of intelligent behavior

Rosati, A. G., *Ecological variation in cognition: Insights from bonobos and chimpanzees*. In: *Bonobos: Unique in Mind, Brain, and Behavior*. Edited by Brian Hare and Shinya Yamamoto: Oxford University Press (2017). © Oxford University Press.
DOI: 10.1093/oso/9780198728511.003.0011

(Byrne and Whiten, 1988; de Waal, 1982; Dunbar, 1998; Humphrey, 1976; Jolly, 1966); that is, primates are thought to evolve especially complex cognitive skills in order to navigate their especially complex social world. However, there are also compelling arguments for the importance of ecological environments in shaping cognition across taxa (Byrne, 1997; Gibson, 1986; Milton, 1981). These proposals suggest that species facing especially difficult foraging problems may evolve psychological capacities in response to these environmental factors.

This chapter will focus on this second, less well-explored hypothesis by examining recent empirical research comparing cognition in bonobos (*Pan paniscus*) and chimpanzees (*Pan troglodytes*). Comparative studies of the traits of different species are one of the most powerful tools in evolutionary biology for illuminating the historical process of natural selection (Clutton-Brock and Harvey, 1979; Harvey and Purvis, 1991; MacLean et al., 2012; Mayr, 1982). This chapter will examine whether there is a psychological 'signature' of these species' natural history in the cognitive capacities that apes utilize in foraging contexts; that is, I will examine evidence that apes show differences in their cognition that parallel the sorts of differences in dentition or other morphological characters that map onto dietary ecology across many primate species (see also Rosati, in press).

Bonobos and chimpanzees are a particular useful model for examining the role of ecology in shaping cognition, as these species are closely related but they exhibit systematic differences in their wild feeding ecology. *Pan* can therefore serve as a model to test hypotheses about when and why different species differ in their cognition. Chimpanzees and bonobos are also our two closest living relatives, so they are the best model for the cognition and behavior of the last common ancestor of humans with other apes. In other words, these apes can provide special insights into the evolution of cognitive capacities in our own lineage.

The ecological hypothesis

Foraging problems are ubiquitous as all animals must identify appropriate food resources, locate them in their environment and make trade-offs between the benefits of pursuing a particular resource and the costs necessary to acquire that resource. However, not all dietary niches necessarily require the same cognitive abilities to be efficiently exploited. Resources whose distribution varies in time (seasonal variability) or space (patchiness), as well as resources that require extensive behavioral processing (such as through extractive foraging) may require specific cognitive and behavioral skills relevant to these problems. In contrast, diets focused on foods that involve more homogenous spatial or temporal distributions, or whose consumption depends more on morphological adaptations in dentition and the digestive tract, may require a different set of abilities.

Along these lines, several proposals have been put forward arguing that cognitive evolution may be shaped by the different cognitive demands posed by various features of the diet. Some flavours of the ecological intelligence hypothesis have focused on the idea that animals consuming patchy foods, like fruit, may need to keep track of more information than animals that eat less variably distributed resources, such as folivores who consume more leaves (Milton, 1981). Other theories focus on the cognitive demands of extractive foraging or other aspects of 'technical' intelligence—the idea that processing food in several steps may require especially complex cognition (Byrne, 1997; Gibson, 1986; Parker and Gibson, 1997). Finally, some authors have linked variability in ecological conditions to the need for behavioral flexibility and innovation in cognitive skills (Deaner et al., 2003; Sol et al., 2002). Studies of brain evolution have provided some support for this ecological hypothesis. For example, frugivorous species tend to have larger brains than folivorous species (Barton, 1996; 2006; Clutton-Brock and Harvey, 1980; DeCasien et al., 2017; MacLean et al., 2009).

However, many of these proposals focus on the idea that ecology may spur the emergence of complex cognition in a general sense rather than specifying the particular cognitive skills that may play critical roles in foraging. Along the same lines, much of the evidence from variation in overall brain size (or the size of very large, functionally diverse areas such as neocortex of even frontal cortex) is at best a rough index of more specific cognitive functions. Yet many modern views from psychology suggest that there are, in fact, different domains of cognition that may operate—and evolve—relatively independently (Barrett and Kurzban, 2006; Herrmann et al., 2007; Hirschfeld and Gelman, 1994; Shettleworth, 2012; Spelke and Kinzler, 2007; Tomasello and Call, 1997). Comparative studies of brain anatomy further show that neurobiological systems may evolve in a mosaic fashion, with certain functional areas changing in size or structure independent of others (Barton, 2006; Striedter, 2005). Thus, testing the ecological hypothesis requires more specific proposals about what cognitive abilities may be especially relevant for foraging.

Defining 'foraging cognition'

What psychological capacities are actually utilized in foraging contexts? Many mental functions could fall under this rubric, ranging from perceptual abilities to detect certain classes of food (Dominy and Lucas, 2001), to collaborative interactions that allow individuals to exploit resources they could not acquire alone (Tomasello et al., 2012). Here, I will focus on three sets of skills: spatial memory capacities allowing individuals to remember the location of resources; patience, or decisions about deferring immediate gratification to seek out more valuable but delayed resources; and risk preferences, or willingness to accept variability in rewards. This is obviously not an exhaustive list of foraging-relevant skills in chimpanzees and bonobos; for example, there are clear differences in chimpanzee and bonobo propensities to engage in tool use and extractive foraging behaviors (Furuichi et al., 2015; Herrmann et al., 2010). However, both decision-making and spatial memory are intimately related to core problems presented by foraging, are relatively well-understood sets of psychological capacities and there is evidence that these skills vary across species with known differences in socio-ecology. These abilities are therefore good targets for considering the role of ecology in chimpanzee and bonobo cognition.

First, spatial memory is a foundational cognitive skill for foraging behaviors. It is essential for solving foraging problems that involve locating widely distributed resources and navigating between patchy food sources efficiently, two problems that all primates face to a greater or lesser degree (Janson and Byrne, 2007). Memory systems allow organisms to use information acquired from past experiences to alter their currently behavior. It is clear from wild observations that animals can move through familiar spaces and successfully locate food (Normand et al., 2009; Normand and Boesch, 2009), but many potential underlying cognitive mechanisms could support such behaviors. Indeed, modern views from cognitive science and neurobiology highlight the fact that there are parallel but distinct memory systems in the brain that could support foraging behavior, not all of which actually involve thinking about spatial locations (Burgess, 2008; Sherry and Schacter, 1987). For example, animals could recall their own physical movements without tracking any relevant environmental features (Burgess, 2006; Packard, 2009). This review will focus specifically on spatial memory: how animals form cognitive maps of the locations of resources in space based on their previous experiences (Bird and Burgess, 2008; Menzel, 1973), in the absence of direct perceptual cues.

Studies of spatial memory comprise the strongest evidence that ecology can shape cognition across species. For example, variation in the spatial memory skills of different bird species (as well as the size of their hippocampus, a key brain region involved in encoding spatial locations) is related to the degree to which they depend on caching or storing food in the wild (Balda and Kamil, 1989; Basil et al., 1996; Bednekoff et al., 1997; Clayton, 1998; Clayton and Krebs, 1994; Healy et al., 2005; Kamil et al., 1994; Krebs et al., 1990; Pravosudov and Clayton, 2002; Shettleworth, 1990; Shettleworth et al., 1990). Although there has been less research in other taxa, there are some hints that similar relationships between ecology and spatial memory capacities hold in other species as well. For example, spatial memory varies by sex in vole species with sexually dimorphic spatial ranging patterns, but not in species that do not exhibit sex differences in ranging behaviors (Gaulin and FitzGerald, 1986; Gaulin and Fitzgerald, 1989; Jacobs et al., 1990). There is also accumulating evidence that ecology shapes spatial memory in some primates. Marmoset species are obligate gummivores that gouge holes in trees that are located in smaller home ranges, whereas closely related tamarin species exhibit much larger ranging patterns as they feed more on patchily distributed fruit and insects. Accordingly, tamarins show more accurate spatial memory over longer time intervals than do marmosets in several spatial and visual memory tasks (Platt et al., 1996). Similarly, more frugivorous lemur species, who must recall the locations of patchily distributed fruits in their territories, show more accurate and flexible performance on a variety of spatial memory problems compared to folivorous species that eat homogenously distributed leaves (Rosati et al., 2014).

Decision-making is another critical and universal cognitive component of foraging behaviors. At the heart of foraging is a series of decisions about value: given all the available alternatives, what is the best thing to eat? This sort of choice involves the integration of many disparate types of information, including the quality of the food, the likelihood that the food is are actually available at any given time and the temporal and energetic costs of pursuing foods found in more distant locations. Here I focus on two foundational components of value-based decision-making: decisions about time, which involve trade-offs between the value of the reward and the temporal costs necessary to acquire it; and decisions about risk, which involve trade-offs between options that provide more constant rewards and options that provide more variable or risky outcomes.

Although decision-making has not been studied as extensively from a comparative perspective as spatial memory, existing evidence suggests that feeding ecology may also have a profound influence on different species' economic preferences. For example, cotton-top tamarins (*Saguinus oedipus*) and common marmosets (*Callithrix jacchus*) show different patterns of choice in both temporal discounting (Rosati et al., 2006; Stevens et al., 2005a) and spatial discounting (Stevens et al., 2005b) contexts. Specifically, marmosets are more willing to wait in temporal tasks to acquire more food, whereas tamarins are more willing to travel longer distances to acquire more food. This maps onto the aforementioned differences in their feeding ecology: marmosets may be more temporally patient because they are obligate gummivores who must wait for sap to exude from trees, whereas tamarins may be more willing to pay effort costs because they typically travel longer distances to forage on insects and fruit. Moreover, work from behavioral ecology highlights the critical role of environmental contexts on decision-making more generally (Rosati and Stevens, 2009; Santos and Rosati, 2015). For example, risk-sensitivity theory proposes that decisions about variability in pay-offs should depend on energetic state, such that individuals in a poor condition are more risk-prone than individuals in a better condition (Caraco, 1981). That is, the impact of a pay-off on fitness depends on context: if an animal's current nutritional requirements exceed the pay-off offered by the safe option, then they should be more risk-prone, as the alternative is potentially death (Stephens, 1981; Stephens and Krebs, 1986). Overall, this theoretical work highlights the fact that decision-making can be tailored to the environmental context that individuals face, suggesting that species with consistent differences in ecology might exhibit consistent differences in their psychological preferences.

Predictions from bonobo and chimpanzee natural history

Chimpanzees and bonobos are human's closest extant relatives, sharing approximately 98 per cent of their DNA with humans (Prüfer et al., 2012). These species are therefore our best model for cognitive profile of the last common ancestor of humans with non-human apes. Importantly, chimpanzees and bonobos also exhibit a suite of morphological and behavioral difference even though they diverged from one other less than 1 mya (Won and Hey, 2005). In particular, studies from both captivity and the wild show that chimpanzees exhibit more pronounced sexual dimorphism and increased rates of escalated aggression, whereas bonobos exhibit increased socio-sexual behaviors and a social organization with stronger bonds between females (Hare et al., 2012; Parish, 1996; Surbeck et al., 2011; Wrangham and Pilbeam, 2001).

An influential hypothesis specifically links these behavioral differences to apes' feeding ecology: chimpanzees and bonobos are thought to live in environments with important ecological differences that alter the character of the foraging problems faced by these species (Kano, 1992; Wrangham and Peterson, 1996). In particular, compared to bonobos chimpanzees are thought to utilize more patchy, seasonally variable fruit resources, feed on less-abundant food patches, and have less access to homogenously distributed terrestrial herbaceous vegetation (THV) as fallback food (Malenky and Wrangham, 1993; White, 1989, 1998; White and Wrangham, 1988; Wrangham, 2000). Furthermore, wild chimpanzees at multiple sites regularly use tools to engage in temporally costly extractive foraging techniques, including insect fishing and nut cracking, whereas wild bonobos have not been observed using tools for feeding (Furuichi et al., 2015). Finally, wild chimpanzees exhibit high rates of hunting monkeys whereas bonobos only occasionally do so (Stanford, 1999; Surbeck and Hohmann, 2008), and hunting requires the investment of time and energy in pursuing an uncertain outcome (Gilby and Wrangham, 2007).

These ecological differences mean that bonobos spend less time and effort to find food and face less feeding competition while doing so than chimpanzees. This may have resulted in relaxed selection on aggressive behaviors and contribute to differences in their social structure (Hare et al., 2012; Wrangham and Pilbeam, 2001). Although apes may show important diversity in behavior (and ecology) across populations (Boesch et al., 2002; Hohmann et al., 2010), overall these observations suggest that chimpanzees may face more 'difficult' foraging problems in the sense that their diets are broadly characterized by high levels of effortful, temporally costly food processing, variation in pay-offs, and more intense feeding competition.

What does this ecological hypothesis predict concerning foraging cognition in bonobos and chimpanzees? First, chimpanzees should exhibit more accurate spatial memory abilities to locate their patchy food resources, given that bonobos depend more on resources that are homogenously distributed in the environment like terrestrial herbs. Second, chimpanzees should exhibit higher levels of patience than bonobos, given that they must tolerate longer delays to access food both in terms of search times when locating patches, as well as increased temporal and work effort when engaging in extractive foraging. Third, chimpanzees should be more risk-seeking than bonobos, as chimpanzees feed on food that is more temporally, spatially and seasonally variable than that of bonobos, and also regularly engage in risky hunting behaviors. Overall, these ecological differences predict that several memory and decision-making capacities should systematically differ between chimpanzees and bonobos, specifically those cognitive skills that map onto these aspects of their natural environments.

Empirical evidence for divergence in *Pan* foraging cognition

Is there evidence from ape cognition supporting these predictions derived from wild ape ecology? Studies in captivity allow for controlled experiments that can rule out alternative psychological explanations for observed patterns of results (Tomasello and Call, 2008), which can be very challenging in wild populations (Zuberbühler, 2014). Moreover, studies of ape populations in zoos or sanctuaries are better at equating the environments that individuals of both species experience over their individual lifetime. This minimizes the possibility that individual chimpanzees and bonobos acquire different skills over ontogeny in response to foraging actively in different types of habitats, as in wild populations. For example, African sanctuaries care for apes following relatively standardized procedures across sites (Farmer, 2002; Wobber and Hare, 2011). While apes semi-free-range in large, naturalistic rainforest enclosures within complex social groups, they receive the majority of their food through provisioning. These environmental similarities allow species comparisons to hone in on the psychological 'signature' of their cognitive adaptations to typical wild environments. Unfortunately, many direct comparisons of bonobo and chimpanzee cognition involve small sample sizes of at least one species (typically bonobos, who are relatively rare in captivity), making it difficult to detect species differences. However, there is accumulating evidence from studies with larger samples that they differ in some specific capacities.

First, these species show divergent responses to some spatial memory problems. For example, when confronted with a problem where they must recall the location of food hidden under one of four containers over a period of one minute (while their view is occluded), chimpanzees are more accurate than are bonobos (Rosati and Hare, 2012a). Critically, this species difference in performance seems to emerge specifically when apes are faced with more complex problems where the location of food must be retained in memory for some time. In the same set-up, chimpanzees and bonobos had equivalent performance when they could make a response immediately; that is, when they did not have to remember the hiding location for as long. Along the same lines, chimpanzees and bonobos show similar performance when solving simple object permanence tasks where they must track the location of a hidden reward over short durations or across few possible locations (see Herrmann et al., 2010 for an example).

These differences in chimpanzee and bonobo spatial memory are further highlighted when apes must utilize their spatial memory skills in complex environments that emulate species-typical foraging problems. In one such test, apes saw a human hide ten pieces of food in a large, naturalistic outdoor enclosure; an additional ten pieces had also been hidden while the ape was out of view to control for the apes' ability to detect the food using other skills such as olfaction, rather than using spatial memory specifically (Rosati and Hare, 2012a). Apes were allowed to search for food after a 20-minute delay, requiring that they use long-term spatial memory abilities rather than short-term memory skills. Infant apes of both species exhibited similar (poor) performance when faced with the sort of problem. However, species differences emerged with age: older chimpanzees exhibited much more targeted searches—recalling the hiding locations of multiple pieces—while older bonobos showed search patterns similar to infants (see Figure 11.1). Importantly, when faced with a second version of the task where apes only had to recall four locations, and were allowed to search immediately after the experimenter hid the food, chimpanzees and bonobos were both quite successful. This shows that the bonobos' poor performance in the first version of the task was not due to a general lack of motivation or a basic unwillingness to search for food in the enclosure but rather due to the difficult memory-specific demands of the task. Overall, these results support the ecological hypothesis: although bonobos also clearly can utilize spatial memory to recall the location for resources in space, chimpanzees exhibit

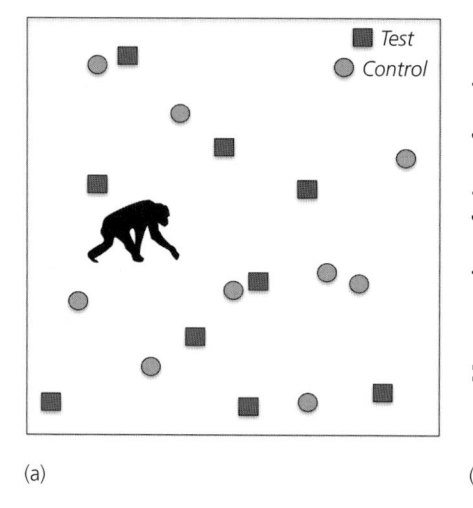

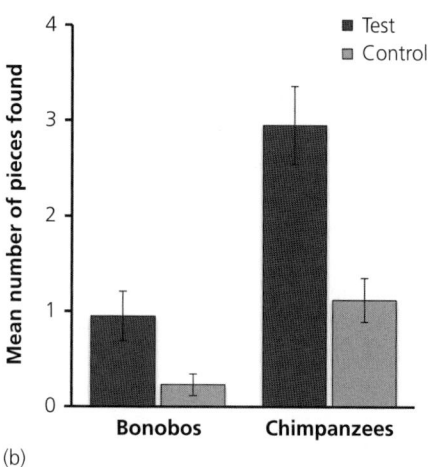

Figure 11.1 Divergence in *Pan* spatial memory. (a) Bonobos and chimpanzees observed food being hidden at ten test locations in a large outdoor arena; another ten pieces had previously been hidden at matched control locations. After a 20-minute delay, apes could search for food. (b) Chimpanzees were more successful than bonobos at retrieving the test pieces they had seen hidden. Adapted from Rosati and Hare (2012). (Divergence en mémoire spatiale. (a) Les bonobos et les chimpanzés ont observé la dissimulation de nourriture dans 10 emplacements dans une grande arène au plein air; 10 autres pièces ont été dissimulées à priori comme contrôle. Après une attente de 20 minutes, les singes pouvaient chercher la nourriture. (b) Les chimpanzés avaient plus de réussite que les bonobos à retrouver les pièces de contrôle. Adapté de Rosati and Hare (2012))

more accurate recall of such locations, especially in complex, naturalistic environments.

Second, patterns of decision-making in bonobos and chimpanzees also diverge in accordance with these ecological predictions. First, studies of intertemporal choice in *Pan* show that chimpanzees are more willing to pay temporal costs to acquire more valuable food rewards (see Rosati, 2017a). For example, in one study, apes were presented with a series of decisions between a smaller reward and a larger reward (Rosati et al., 2007). In this temporal discounting task, bonobos and chimpanzees made choices between a smaller reward that was always available immediately and a larger reward that was available after a delay. The time that apes had to wait to receive the larger reward was systematically adjusted across sessions to determine the delay at which each individual treated the larger and smaller rewards as equivalent. In fact, whereas chimpanzees were willing to wait 2 minutes on average before they switched to preferring the smaller, immediate option, bonobos were willing to wait only around 70 seconds (see Figure 11.2). This basic result—that bonobos are less willing to pay temporal costs than chimpanzees—has been replicated in different populations of apes using different experimental methods (Rosati and Hare, 2013), suggesting that it is robust to some extent across populations and paradigms.

Third, there is even stronger evidence that bonobos are more risk-averse than chimpanzees when faced with decisions about variability in pay-offs (see Rosati, 2017b). For example, in one comparison, apes chose between two options that differed in the amounts of rewards they provided: a 'safe' option reliably provided four pieces of food, whereas a 'risky' option offered one or seven pieces with equal probability (Heilbronner et al., 2008). Here, both options provided the same average pay-off (four pieces). They differed only in whether they provided a constant average pay-off or whether the particular outcome of any given decision was associated with variance: apes might receive a high pay-off from the risky option on some trials but a relatively low one on other trials. In fact, bonobos preferred the safe option, whereas chimpanzees preferred the risky option, a difference in preferences that widened as the apes gained more experience with the problem.

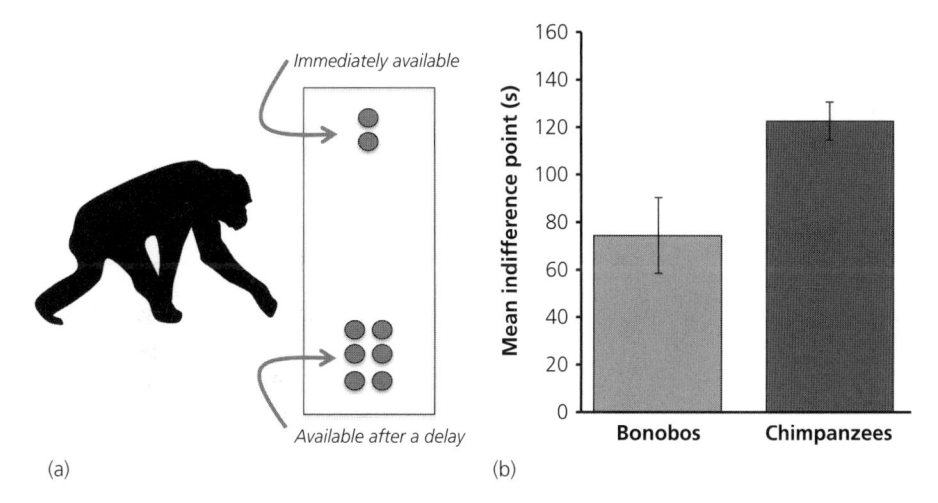

Figure 11.2 Divergence in *Pan* patience. (a) In a temporal discounting task, bonobos and chimpanzees made choices between a smaller reward that was always available immediately, and a larger reward that was available after a delay, in order to determine the how apes made tradeoffs between reward amount and reward delay. (b) Chimpanzees were willing to wait longer for larger reward than bonobos. Adapted from Rosati et al. (2007)
(Divergence en patience. (a) Durant une tâche d'actualisation temporelle, les bonobos et les chimpanzés ont pris des décisions entre une récompense plus petite qui est immédiatement disponible, et une récompense plus grande qui est faite disponible après une attente. L'objectif est comprendre comment les singes décident entre la somme de récompense et l'attente. (b) Les chimpanzés avaient plus de volonté à attendre plus de temps pour les grandes récompenses que les bonobos. Adapté de Rosati et al. (2007))

Other studies have examined apes' more spontaneous reactions to uncertainty using paradigms where animals can infer their chance of winning on a trial-by-trial basis. One example of such a risk paradigm tested how chimpanzees and bonobos make decisions about reward quality, where the risk involved concerned the *type* of food they received rather than the quantity (Rosati and Hare, 2012b; Rosati and Hare, 2013). Here, apes saw an intermediately preferred food type (such as peanuts) placed under one container (the safe option). A second container was baited with either a highly desirable (banana) or less-desirable (cucumber) food: apes initially saw both types but knew that only one had been placed in the container, so this option represented a risky choice (see Figure 11.3). In another inferential paradigm, apes could infer the hiding locations of rewards that were distributed under different containers (Haun et al., 2011). Whereas a smaller reward would be placed under one known container, a larger reward would be placed under one of a whole set of containers. Since the apes did not know which container in that set had the bigger piece of food, a selection of one of these containers similarly represented a risky choice as they might select an empty container. As in the choice situation involving risk in the amount of food received (e.g., Heilbronner et al., 2008), results from both of these inferential tasks showed that chimpanzees were more willing to gamble on the possibility of receiving the more valuable reward than were bonobos.

Evidence for targeted cognitive divergence in *Pan*

Importantly, this new formulation of the ecological intelligence hypothesis—which focuses on the relationship between ecology and specific domains of cognition rather than on general cognitive abilities—does not predict that chimpanzees and bonobos should differ across all possible cognitive skills. Rather, they should exhibit specific *targeted* differences in cognition related to specific differences in their typical foraging problems. Indeed, chimpanzees and bonobos share many commonalities in their social structure and diets, given that both are frugivores that live in male philopatric societies (Boesch et al., 2002; Kano, 1992; Stanford, 1998). Along these lines, there is increasing evidence that chimpanzees and bonobos show broad

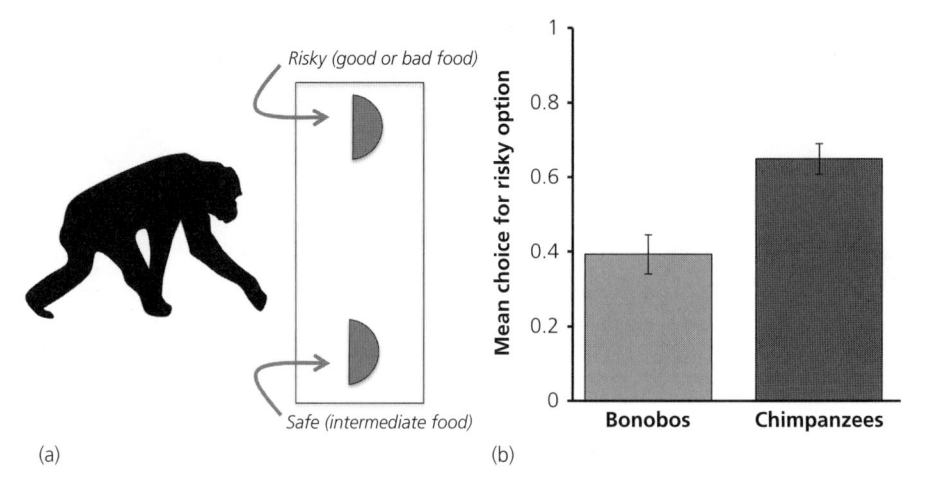

Figure 11.3 Divergence in *Pan* risk preferences. (a) In a risky choice task, bonobos and chimpanzees made choices between a safe bowl that they had seen baited with an intermediately preferred food option, and a risky bowl that they knew contained either a highly preferred food (good outcome) or an non-preferred food type (bad outcome). (b) Chimpanzees preferred to gamble on the possibility of receiving the good food outcome from the risky bowl, whereas bonobos preferred the safe bowl. Adapted from Rosati and Hare (2013).
(Divergence en préférences de risque. (a) Dans une tâche de choix risqués, les bonobos et les chimpanzés ont fait des choix entre un bol sûr qui contient de la nourriture à préférence intermédiaire et un bol risqué avec la nourriture plus préférée (bon résultat) ou la nourriture moins préférée (mauvais résultat). (b) Les chimpanzés ont préféré les bols risqués alors que les bonobos ont préféré le bol sûr. Adapté de Rosati and Hare (2013))

similarities across many cognitive domains unrelated to their socio-ecological differences. For example, when compared in a cognitive test battery probing a variety of cognitive skills—ranging from basic skills like numerical discrimination and simple spatial inferences to social learning and producing communicative gestures—both species show similar performance across many of these contexts (Herrmann et al., 2010; Wobber et al., 2014). This shows that bonobos and chimpanzees share many commonalities in the fundamental cognitive tools they use to think about the world.

Indeed, while chimpanzees and bonobos show robust variation in their decision preferences for risk and time, they do not necessarily show such differences in other decision-making contexts for which there are not strong predictions based on feeding ecology. One such example is preferences for ambiguity, or situations where decision-makers have incomplete or absent knowledge about reward contingencies. Although economists and psychologists tend to link ambiguity preferences with risky choice (Camerer and Weber, 1992), decisions about risk involve known variation, whereas decisions about ambiguity involve unknown variation where the set of potential outcomes is itself unclear. In contrast to the evidence that chimpanzees face greater seasonal variation and engage in risky hunting at higher rates than bonobos, there is no strong ecological prediction for ambiguity biases in these species. That is, there is no reason to suspect that one species has access to more or less knowledge about the resources available in their natural foraging environment, and therefore might be more tolerant of choices that involve ambiguity. In line with this analysis, chimpanzees and bonobos show similar responses to ambiguity in experimental paradigms when their basic risk preferences are first equated (Rosati and Hare, 2011). Similarly, chimpanzees and bonobos show similar responses to framing, where either the positive or negative attributes of potential options are highlighted (Krupenye et al., 2015)—another type of decision bias without a clear ecological prediction.

Other differences in chimpanzee and bonobo cognition further highlight the central importance of foraging behaviors in understanding these species' divergences. As mentioned previously, foraging almost always involves social interactions in gregarious species, including most primates. Consequently, social behavior and foraging behavior are almost always intertwined in species that must compete (or cooperate) with group mates to acquire food. In fact, chimpanzees and bonobos also differ in cognitive capacities that are relevant to these sorts of social foraging interactions. For example, wild bonobos face relaxed feeding competition compared to chimpanzees, a fact that is intimately related with the differences in patch size and resource distributions in their environments (White, 1989; White and Wrangham, 1988; Wrangham and Peterson, 1996). Accordingly, bonobos are more tolerant of conspecific partners when faced with experimental tests of social feeding (Hare et al., 2007; Wobber et al., 2010). Moreover, given these differences in feeding competition, chimpanzees are likely to require greater levels of inhibitory control in social contexts to control responses that are likely to be undesirable such as snatching a desirable food item out of a conspecific's grasp. In fact, chimpanzees seem to develop more robust social inhibition than bonobos (Wobber et al., 2010). Together, these results further support the hypothesis that these two species exhibit targeted differences for specific set of skills that relate to differences in their wild socio-ecology.

Implications for human evolution

As previously mentioned, the best current model for our last common ancestor's psychology is the living great apes. Although theoretical claims about behavioral and cognitive evolution in humans tend to use chimpanzees alone as a model for the last common ancestor (Wrangham and Pilbeam, 2001), recent research suggests that the last common ancestor may have had a mosaic of chimpanzee-like and bonobo-like traits (Hare and Yamamoto, 2015). This indicates that a true understanding of the origins of human cognition generally, as well as foraging cognition specifically, requires the integration of data on cognitive skills in both bonobos and chimpanzees. Comparative studies of these species can thus provide critical insights into the potential functions of uniquely human cognition (Rosati, in press).

In fact, human feeding ecology differs from that of other great apes in several important ways that parallel the socioecological differences seen in chimpanzees and bonobos. For example, human hunter–gatherers have larger home ranges than other apes and exhibit a unique pattern of central place foraging where individuals return to a centralized location with food (Hill et al., 2009; Marlowe, 2005). Consequently, humans are more reliant on distant food sources that are more temporally expensive to exploit. This shift in feeding ecology probably poses new problems for humans concerning locating food, navigating between resources, as well as anticipating and planning for future food consumption. These observations suggest that humans may also exhibit targeted differences in evolutionarily relevant skills supporting this human-specific pattern of foraging, such as the memory skills and patience needed to locate widely distributed resources and plan for their exploitation. In fact, humans are thought to develop an increasingly flexible representation of space integrating multiple cues (Hermer-Vazquez et al., 2001, 1999; Newcombe and Huttenlocher, 2006). Moreover, several theorists have argued that humans are uniquely well endowed in our ability to think about and plan for the future (Stevens and Stephens, 2008; Suddendorf and Corballis, 2007).

Similar inferences may be possible for other human decision-making traits as well. I have argued that the differences in risk preferences seen in chimpanzees and bonobos may be accounted for by the relative variability these species experience in the natural environments, with chimpanzees experiencing more variability in pay-offs when foraging than bonobos (both in terms of seasonal variability and hunting). This account is also applicable to humans, as several aspects of human hunter–gatherer foraging behavior have been hypothesized as mechanisms for risk reduction. In particular, strategies such as hunting have variable pay-offs: the majority of hunter–gatherer groups are successful at most only half the time, or at even lower rates (Hawkes et al., 2001; Hill and Hawkes, 1983). In other words, foragers face a food supply that is superabundant on some days but absent on others. Thus, one way for humans to reduce the variability in their resources is to share food with group-mates (Hill and Kaplan, 1993, Kaplan et al., 2012). This suggests that our species' willingness to accept risk may also be shaped by ecology, much like in apes. Moreover, the function of human risk preferences may actually be closely linked with human social decision-making strategies like reciprocity; sharing or other cooperative behaviors that are thought to depend on some type of reciprocal exchange can be construed as 'risky' in the sense that social partners may not repay those favours. Along these lines, although people are generally risk-averse for monetary rewards, humans show a fairly risk-prone pattern of decision-making when faced with choices about food emulating foraging decisions (Hayden and Platt, 2009), quite similar to chimpanzees when tested on a matched problem (Rosati and Hare, 2016).

Conclusions

I have argued that linking comparative cognitive studies with information on species' natural behavior can begin to address why differences in cognitive traits may evolve. Using bonobos and chimpanzees as models, I showed that even closely related species can exhibit variation in cognitive skills depending on their respective socio-ecological characteristics. *Pan* shows important differences in their wild feeding ecology: whereas chimpanzees depend more on seasonally and temporally variable food resources that are patchily distributed in space, bonobos can exploit more constant, homogenously distributed foods. Moreover, chimpanzees regularly pay temporal and risk costs to engage in extractive foraging or hunting, whereas bonobos much more rarely (if ever) exhibit these behaviors in the wild. In line with these ecological differences, bonobos and chimpanzees show a suite of targeted divergences in their foraging cognition: bonobos exhibit less accurate spatial memory, reduced patience and increased risk aversion compared to chimpanzees. These differences in cognitive capacities seem to be targeted, as *Pan* does not exhibit general divergences in cognition across the board but rather shows changes in skills specifically related to their respective socioecological niche. Overall, these findings suggest that studying bonobo cognition in tandem with that of chimpanzees can provide important insights both into the mechanisms of cognitive evolution as well as our understanding of human evolutionary history.

Acknowledgements

I thank Felix Warneken for comments on an earlier version of the manuscript.

References

Balda, R.P. and Kamil, A.C. (1989). A comparative study of cache recovery by three corvid species *Animal Behavior*, 38, 486–95.

Barrett H.C. and Kurzban, R. (2006). Modularity in cognition: framing the debate. *Psychological Review*, 113, 628–47.

Barton, R.A. (1996). Neocortex size and behavioral ecology in primates. *Proceedings of the Royal Society B*, 263, 173–7.

Barton, R.A. (2006). Primate brain evolution: integrating comparative, neurophysiological, and ethological data. *Evolutionary Anthropology*, 15, 224–36.

Basil, J.A., Kamil, A.C., Balda, R.P., and Fite, K.V. (1996). Differences in hippocampal volume among food storing corvids. *Brain, Behavior and Evolution*, 47, 156–64.

Bednekoff, P.A., Balda, R.P., Kamil, A.C., & Hile, A.G. (1997). Long-term memory in four seed-caching corvid species. *Animal Behavior*, 53, 335–341.

Bird, C.M. and Burgess, N. (2008). The hippocampus and memory: insights from spatial processing. *Nature Reviews Neuroscience*, 9, 182.

Boesch, C., Hohmann, G., and Marchant, L.F. (eds). (2002). *Behavioral Diversity in Chimpanzees and Bonobos*. Cambridge: Cambridge University Press.

Burgess, N. (2006). Spatial memory: how egocentric and allocentric combine. *Trends in Cognitive Science*, 10, 551–7.

Burgess, N. (2008). Spatial cognition and the brain. *Annals of the New York Academy of Sciences*, 1124, 77–97.

Byrne, R.W. (1997). The technical intelligence hypothesis: An additional evolutionary stimulus to intelligence? In: Whiten, A. and Byrne, R.W. (eds). *Machiavellian Intelligence II: Extensions and Evaluations*. Cambridge: Cambridge University Press, pp. 289–311.

Byrne, R.W. and Whiten, A.W. (1988). *Machiavellian Intelligence: Social Expertise and the Evolution of Intellect in Monkeys, Apes, and Humans*. Oxford: Clarendon Press.

Camerer, C. and Weber, M. (1992). Recent developments in modeling preferences. *Journal of Risk and Uncertainty*, 5, 325–70.

Caraco, T. (1981). Risk sensitivity and foraging. *Ecology*, 62, 527–31.

Clayton, N.S. (1998). Memory and the hippocampus in food-storing birds: a comparative approach. *Neuropharmacology*, 37, 441–52.

Clayton, N.S. and Krebs, J.R. (1994). One trial associative memory: Comparison of food-storing and nonstoring species of birds. *Animal learning and Behavior*, 22, 366–72.

Clutton-Brock, T.H. and Harvey, P.H. (1979). Comparison and adaptation. *Proceedings of the Royal Society of London*, 205, 547–65.

Clutton-Brock, T.H. and Harvey, P.H. (1980). Primates, brains, and ecology. *Journal of Zoology*, 190, 309–23.

de Waal, F.B.M. (1982). *Chimpanzee Politics: Power and Sex Among Apes*. New York, NY: Harper & Row.

Deaner, R.O., Barton, R.A., and van Schaik, C.P. (2003). Primate brains and life histories: renewing the connection. In: Kappeler, P.M. and Pereira, M.E. (eds). *Primate Life Histories and Socioecology*. Chicago, IL: University of Chicago Press, pp. 233–65.

DeCasien, A.R., Williams, S.A., and Higham, J.P. (2017) Primate brain size is predicted by diet but not sociality. *Nature Ecology and Evolution*, 1, 0112.

Dominy, N.J. and Lucas, P.W. (2001). Ecological importance of trichromatic vision to primates. *Nature*, 410, 363–6.

Dunbar, R.I. (1998). The social brain hypothesis. *Evolutionary Anthropology*, 6, 178–90.

Farmer, K.H. (2002). Pan-African Sanctuary Alliance: status and range of activities for great ape conservation. *American Journal of Primatology*, 58, 117–32.

Furuichi, T., Sanz, C., Koops, K., Sakamaki, T., Ryu, H., Tokuyama, N. & Morgan, D. (2015). Why do wild bonobos not use tools like chimpanzees do? *Behaviour*, 152, 425–460.

Gaulin, S.J.C. and FitzGerald, R.W. (1986). Sex differences in spatial ability: an evolutionary hypothesis and test. *American Naturalist*, 127, 74–88.

Gaulin, S. J. C., and FitzGerald, R. W., 1989, Sexual selection for spatial-learning ability, *Animal Behaviour*, 37, 322–331.

Gibson, K.R. (1986) Cognition, brain size, and the extraction of embedded food resources. In Primate Ontogeny, Cognition, and Social Behaviour (Else, J.G. and Lee, P.C., eds), Cambridge University Press, pp. 93–105.

Gilby, I.C. and Wrangham, R.W. (2007). Risk-prone hunting by chimpanzees (*Pan troglodytes schweinfurthii*) increases during periods of high diet quality. *Behavioral Ecology and Sociobiology*, 61, 1771–9.

Hare, B., Melis, A.P., Woods, V., Hastings, S., and Wrangham, R. (2007). Tolerance allows bonobos to outperform chimpanzees on a cooperative task. *Current Biology*, 17, 619–23.

Hare, B., Wobber, V., and Wrangham, R. (2012). The self-domestication hypothesis: evolution of bonobo psychology is due to selection against aggression. *Animal Behaviour*, 83, 573–85.

Hare, B. and Yamamoto, S. (2015). Moving bonobos off the scientifically endangered list. *Behaviour*, 152, 247–58.

Harvey, P.H. and Purvis, A. (1991). Comparative methods for explaining adaptations. *Nature*, 351, 619–23.

Haun, D.B.M., Nawroth, C., and Call, J. (2011). Great apes' risk-taking strategies in a decision making task. *PLoS One*, 6, e28801.

Hawkes, K., O'Connell, J.F., and Blurton Jones, N.G. (2001). Hadza meat sharing. *Evolution and Human Behavior*, 22, 113–42.

Hayden, B.Y. and Platt, M.L. (2009). Gambling for Gatorade: risk-sensitive decision making for fluid rewards in humans. *Animal Cognition*, 12, 201–7.

Healy, S.D., de Kort, S.R., and Clayton, N.S. (2005). The hippocampus, spatial memory, and food hoarding: a puzzle revisited. *Trends in Ecology and Evolution*, 20, 17–22.

Heilbronner, S.H., Rosati, A.G., Hare, B., and Hauser, M.D. (2008). A fruit in the hand or two in the bush? Divergent risk preferences in chimpanzees and bonobos. *Biology Letters*, 4, 246–9.

Hermer-Vazquez, L., Moffet, A., and Munkholm, P. (2001). Language, space, and the development of cognitive flexibility in humans: the case of two spatial memory tasks. *Cognition*, 79, 263–99.

Hermer-Vazquez, L., Spelke, E., and Katnelson, A.S. (1999). Sources of flexibility in human cognition: dual-task studies of space and language. *Cognitive Psychology*, 39, 3–36.

Herrmann, E., Call, J., Hernandez-Lloreda, M.V., Hare, B., and Tomasello, M. (2007). Humans have evolved specialized skills of social cognition: the cultural intelligence hypothesis. *Science*, 317, 1360–6.

Herrmann, E., Hare, B., Call, J., and Tomasello, M. (2010). Differences in the cognitive skills of bonobos and chimpanzees. *PLoS One*, 5, e12438.

Hill, K., Barton, M., and Hurtado, M. (2009). The emergence of human uniqueness: characters underlying behavioral modernity. *Evolutionary Anthropology*, 18, 187–200.

Hill, K. and Hawkes, K. (1983). Neotropical hunting among the Ache of Eastern Paraguay. In: Hames, R. and Vickers, W. (eds). *Adaptive Responses of Native Amazonians*. New York, NY: Academic Press, pp. 139–88.

Hill, K. and Kaplan, H. (1993). On why male foragers hunt and share food. *Current Anthropology*, 34, 701–6.

Hirschfeld, L.A. and Gelman, S.A. (eds). (1994). *Mapping the Mind: Domain Specificity in Cognition and Culture*. Cambridge: Cambridge University Press.

Hohmann, G., Potts, K., N'Guessan, A., Fowler, A., Mundry, R., Ganzhorn, J.U. & Ortmann, S. (2010). Plant foods consumed by *Pan*: exploring the variation of nutritional ecology across Africa. *American Journal of Physical Anthropology*, 141, 476–485.

Humphrey, N.K. (1976). The social function of intellect. In: Bateson, P.P.G. and Hinde, R.A. (eds). *Growing Points in Ethology*. Cambridge: Cambridge University Press, pp. 303–17.

Jacobs, L.F., Gaulin, S.J.C., Sherry, D.F., and Hoffman, G.E. (1990). Evolution of spatial cognition: Sex-specific patterns of spatial behavior predict hippocampal size. *Proceedings of the National Academy of Sciences*, 87, 6349–52.

Janson, C.H. and Byrne, R. (2007). What wild primates know about resources: opening up the black box. *Animal Cognition*, 10, 357–67.

Jolly, A. (1966). Lemur social behavior and primate intelligence. *Science*, 153, 501–6.

Kamil, A.C., Balda, R.P., and Olson, D.J. (1994). Performance of four seed-caching corvid species in the radial-arm maze analog. *Journal of Comparative Psychology*, 108, 385–93.

Kano, T. (1992). *The Last Ape: Pygmy Chimpanzee Behavior and Ecology*. Stanford, CA: Stanford University Press.

Kaplan, H., Schiniter, E., Smith, V.L., and Wilson, B.J. (2012). Risk and the evolution of human exchange. *Proceeding of the Royal Society of London B*, 279, 2930–5.

Krebs, J.R., Healy, S.D., and Shettleworth, S.J. (1990). Spatial memory of Paridae: comparison of a storing and non-storing species, the coal tit *Parus ater*, and the great tit *P. major*. *Animal Behavior*, 39, 1127–37.

Krupenye, C., Rosati, A.G., and Hare, B. (2015). Bonobos and chimpanzees exhibit human-like framing effects. *Biology Letters*, 11, 20140527.

MacLean, E.L., Barrickman, N.L., Johnson, E.M., and Wall, C.E. (2009). Sociality, ecology, and relative brain size in lemurs. *Journal of Human Evolution*, 56, 471–8.

MacLean, E., Matthews, L.J., Hare, B., Nunn, C.L., Anderson, R.C., Aureli, F., Brannon, E.M., Call, J., Drea, C.M., Emery, N.J., Haun, D.B.M., Herrmann, E., Jacobs, L.J., Platt, M.L., Rosati, A.G., Sandel, A.R., Schroepfer, K.K., Seed, A.M., Tan, J., van Schaik, C.P., & Wobber, V. (2012). How does cognition evolve? Phylogenetic comparative psychology. *Animal Cognition*, 15, 223–238.

Malenky, R.K. and Wrangham, R.W. (1993). A quantitative comparison of terrestrial herbaceous food consumption by *Pan paniscus* in the Lomako Forest, Zaire, and *Pan troglodytes* in the Kibale Forest, Uganda. *American Journal of Primatology*, 32, 1–12.

Marlowe, F.W. (2005). Hunter–gatherers and human evolution. *Evolutionary Anthropology*, 14, 54–67.

Mayr, E. (1982). *The Growth of Biological Thought*. Cambridge, MA: Harvard University Press.

Menzel, E.W. (1973). Chimpanzee spatial memory organization. *Science*, 182, 943–5.

Milton, K. (1981). Distribution patterns of tropical plant foods as an evolutionary stimulus to primate mental development. *American Anthropology*, 83, 534–48.

Newcombe, N.S., & Huttenlocher, J. (2006). Development of spatial cognition. In D. Kuhn & R.S. Siegler (Eds.), *Handbook of child psychology* (6th edn., pp. 734–776). New York: John Wiley and Sons.

Normand, E., Ban, S.D., and Boesch, C. (2009). Forest chimpanzees (*Pan troglodytes verus*) remember the location of numerous fruit trees. *Animal Cognition*, 12, 797–807.

Normand, E. and Boesch, C. (2009). Sophisticated Euclidean maps in forest chimpanzees. *Animal Behaviour*, 77, 1195–1201.

Packard, M.G. (2009). Exhumed from thought: basal ganglia and response learning in the plus-maze. *Behavioral Brain Research*, 199, 24–31.

Parish, A.R. (1996). Female relationships in bonobos (*Pan paniscus*): evidence for bonding, cooperation, and female dominance in a male-philopatric species. *Human Nature*, 7, 61–96.

Parker, S.T. and Gibson, K.R. (1997). Object manipulation, tool use and sensorimotor intelligence as feeding adaptations in cebus monkeys and great apes. *Journal of Human Evolution*, 6, 623–41.

Platt, M.L., Brannon, E.M., Briese, T.L., and French, J.A. (1996). Differences in feeding ecology predict differences in performance between golden lion tamarins (*Leontopithecus rosalia*) and Wied's marmosets (*Callithrix kuhli*) on spatial and visual memory tasks. *Animal Learning and Behavior*, 24, 384–93.

Pravosudov, V.V. and Clayton, N.S. (2002). A test of the adaptive specialization hypothesis: population differences in caching, memory, and the hippocampus in black-capped chickadees (Poecile atricapilla). *Behavioural Neuroscience*, 116, 515–22.

Prüfer, K., Munch, K., Hellmann, I., Akagi, K., Miller, J.R., Walenz, B., Koren, S., Sutton, G.C., Kodira, C., Winer, R., Knight, J.R., Mullikin, J.C., Meader, S.J., Ponting, C.P., Lunter, G., Higashino, S., Hobolth, A., Dutheil, J., Karakoc, E., Alkan, C., Sajjadian, S., Catacchio, C.R., Ventura, M., Marques-Bonet, T., Eichler, E.E., André, C., Atencia, R., Mugisha, L., Junhold, J., Patterson, N., Siebauer, M., Good, J.M., Fischer, A., Ptak, S.E., Lachmann, M., Symer, D.E., Mailund, T., Schierup, M.H., Andrés, A.M., Kelso, J. and Pääbo, S. 2012. The bonobo genome compared with the chimpanzee and human genomes. Nature, 486, 527–531.

Rosati, A.G. (2017). Foraging cognition: Reviving the ecological intelligence hypothesis. Trends in Cognitive Sciences. DOI: http://dx.doi.org/10.1016/j.tics.2017.05.011

Rosati, A.G. (2017a). 'The evolution of primate executive function: from response control to strategic decision-making.' In: Kaas, J. and Krubitzer, L. (eds). Evolution of Nervous Systems, 2nd edn, Vol. 3, Amsterdam: Elsevier, pp. 423–37.

Rosati, A.G. (2017b). 'Decisions under uncertainty: preferences, biases, and choice.' In: Call, J. (ed.). APA Handbook of Comparative Psychology.

Rosati, A.G. and Hare, B. (2011). Chimpanzees and bonobos distinguish between risk and ambiguity. *Biology Letters*, 7, 15–18.

Rosati, A.G. and Hare, B. (2012a). Chimpanzees and bonobos exhibit divergent spatial memory development. *Developmental Science*, 15, 840–53.

Rosati, A.G. and Hare, B. (2012b). Decision-making across social contexts: competition increases preferences for risk in chimpanzees and bonobos. *Animal Behaviour*, 84, 869–79.

Rosati, A.G. and Hare, B. (2013). Chimpanzees and bonobos exhibit emotional responses to decision outcomes. *PLoS One*, 8, e63058.

Rosati, A.G. and Hare, B. (2016). Reward type modulated human risk preferences. *Evolution and Human Behavior*, 37, 159–68.

Rosati, A.G., Rodriguez, K., and Hare, B. (2014). The ecology of spatial memory in four lemur species. *Animal Cognition*, 17, 947–61.

Rosati, A.G. and Stevens, J.R. (2009). Rational decisions: The adaptive nature of context-dependent choice. In: Watanabe, A.P., Blaisdell, L., Huber, L., and Young, A. (eds). Rational Animals, Irrational Humans. Tokyo: Keio University Press, pp. 101–17.

Rosati, A.G., Stevens, J.R., Hare, B., and Hauser, M. (2007). The evolutionary origins of human patience: temporal preferences in chimpanzees, bonobos, and human adults. *Current Biology*, 17, 1663–8.

Rosati, A.G., Stevens, J.R., and Hauser, M.D. (2006). The effect of handling time on temporal discounting in two New World primates. *Animal Behaviour*, 71, 1379–87.

Santos, L.R. and Rosati, A.G. (2015). The evolutionary roots of human decision making. *Annual Review of Psychology*, 66, 3221–347.

Sherry, D.F. and Schacter, D.L. (1987). The evolution of multiple memory systems. *Psychological Review*, 94, 439–54.

Shettleworth, S.J. (1990). Spatial memory in food-storing birds. *Philosophical Transactions of the Royal Society B*, 329, 143–51.

Shettleworth, S.J. (2012). Modularity, comparative cognition and human uniqueness. *Philosophical Transactions of the Royal Society, B*, 367, 2794–802.

Shettleworth, S.J. Krebs, J.R., Healy, S.D., and Thomas, C.M. (1990). Spatial memory of food-storing tits (*Parus ater* and *P. atricapillus*): Comparison of storing and non-storing tasks. *Journal of Comparative Psychology*, 104, 71–81.

Sol, D., Timmermans, S., and Lefebvre, L. (2002). Behavioral flexibility and invasion success in birds. *Animal Behaviour*, 63, 495–502.

Spelke, E. and Kinzler, K.D. (2007). Core knowledge. *Developmental Science*, 10, 89–96.

Stanford, C.B. (1998). The social behavior of chimpanzees and bonobos—empirical evidence and shifting assumptions. *Current Anthropology*, 39, 399–420.

Stanford, C.B. (1999). *The Hunting Apes: Meat Eating and the Origins of Human Behavior*. Princeton, NJ: Princeton University Press.

Stephens, D.W. (1981). The logic of risk-sensitive foraging preferences. *Animal Behaviour*, 29, 628–9.

Stephens, D.W. and Krebs, J.R. (1986). *Foraging Theory*. Princeton, NJ: Princeton University Press.

Stevens, J.R., Hallinan, E.V., and Hauser, M.D. (2005a). The ecology and evolution of patience in two New World monkeys. *Biology Letters*, 1, 223–6.

Stevens, J.R., Rosati, A.G., Ross, K., and Hauser, M.D. (2005b). Will travel for food: spatial discounting in two New World monkeys. *Current Biology*, 15, 1855–60.

Stevens, J.R. and Stephens, D.W. (2008). Patience. *Current Biology*, 18, R11–R12.

Striedter, G.F. (2005). Principles of brain evolution. Sunderland, MA: Sinauer Associates.

Suddendorf, T. and Corballis, M.C. (2007). The evolution of foresight: what is mental time travel, and is it unique to humans? *Behavioral and Brain Sciences*, 30, 299–351.

Surbeck, M. and Hohmann, G. (2008). Primate hunting by bonobos at LuiKotale, Salonga National Park. *Current Biology*, 18, R906–R907.

Surbeck, M., Mundry, R., and Hohmann, G. (2011). Mothers matter! Maternal support, dominance status and mating success in male bonobos (*Pan paniscus*). *Proceedings of the Royal Society of London*, 278, 590–8.

Tomasello, M. and Call, J. (1997). *Primate Cognition*. Oxford: Oxford University Press.

Tomasello, M. and Call, J. (2008). Assessing the validity of ape-human comparisons: a reply to Boesch (2007). *Journal of Comparative Psychology*, 122, 449–52.

Tomasello, M., Melis, A.P., Tennie, C., Wyman, E., and Herrmann, E. (2012). Two key steps in the evolution of human cooperation: the interdependence hypothesis. *Current Anthropology*, 53, 673–92.

White, F.J. (1989). Ecological correlates of pygmy chimpanzee social structure. In Standen, V., and Foley, R.A. (eds.), *Comparative Socioecology. The Behavioural Ecology of Humans and Other Mammals*. Blackwell Scientific, Oxford, pp. 151–164.

White, F.J. (1998). Seasonality and socioecology: the importance of variation in fruit abundance to bonobo sociality. *International Journal of Primatology*, 19, 1013–27.

White, F. J., and Wrangham, R. W. (1988). Feeding competition and patch size in the chimpanzee species *Pan paniscus* and *Pan troglodytes*. *Behaviour*, 105, 148–163.

Wobber, V. and Hare, B. (2011). Psychological health of orphan bonobos and chimpanzees in African sanctuaries. *PLoS One*, 6, e17147.

Wobber, V., Herrmann, E., Hare, B., Wrangham, R., and Tomasello, M. (2014). Differences in the early cognitive development of children and great apes. *Developmental Psychobiology*, 56, 547–73.

Wobber, V., Wrangham, R., and Hare, B. (2010). Bonobos exhibit delayed development of social behavior and cognition relative to chimpanzees. *Current Biology*, 20, 226–30.

Won, Y.J. and Hey, J. (2005). Divergence population genetics of chimpanzees. *Molecular Biology and Evolution*, 22, 297–307.

Wrangham, R. (2000). Why are male chimpanzees more gregarious than mothers? A scramble competition hypothesis. In: Kappeler, P. (ed.). *Primate Males. Causes and Consequences of Variation in Group Composition*. Cambridge: Cambridge University Press, pp. 248–58.

Wrangham, R. & Peterson, D. (1996). Demonic males: apes and the origins of human violence. Houghton Mifflin, Boston, MA.

Wrangham, R. and Pilbeam, D. (2001) African apes as time machines. In: Galdikas, B., Briggs, N., Sheeran, L., Shapiro, G., Goodall, J. (eds) All apes great and small. Kluwer/Plenum Publishers, New York, pp 5–18.

Zuberbühler, K. 2014. Experimental field studies with non-human primates. *Current Opinion in Neurobiology* 28: 150–156.

CHAPTER 12

Bonobos, chimpanzees and tools: Integrating species-specific psychological biases and socio-ecology

Josep Call

Abstract Over the years there has been some controversy regarding the comparison between chimpanzees and bonobos. Whereas some authors have stressed their differences, others have stressed their similarities. One striking difference between wild chimpanzees and bonobos is tool use, especially in foraging contexts. While several chimpanzee populations possess tool kits formed by multiple tools (and their associated techniques) to exploit embedded resources, bonobos display no such tool specialization. However, studies in the laboratory have shown that bonobos are perfectly capable of using tools. In fact, several studies devoted to investigate the cognitive abilities underlying tool use have failed to detect any substantial differences between the two species. This chapter explores three aspects that could explain the difference between chimpanzees and bonobos in their propensity to use tools in the wild: socio-ecological factors, social versus technical cognition, and personality profiles.

Au cours du temps, il y a eu beaucoup de controverse en relation aux comparaisons entres les chimpanzés et les bonobos. Alors que certains auteurs ont stressé les différences entre eux, d'autres ont stressé les similarités. Une grande différence entre les chipmanzés et les bonobos sauvages est l'utilisation des outils, spécialement en butinage. Tandis que plusieurs populations de chimpanzés possèdent des boîtes à outils diverses (et leur techniques respectives) pour exploiter les ressources, les bonobos ne montrent pas une spécialisation pareille. Cependant, les études en laboratoir ont montré que les bonobos sont capables d'utiliser des outils. En faite, plusieurs études des facultés cognitives dans l'utilisation des outils n'ont pas pu détecter de différences substantielles entre les deux espèces. Je vais explorer trois aspects qui pourraient expliquer les différences entre les chimpanzés et les bonobos en ce qui concerne leur tendance naturelle à utiliser les outils: facteurs socio-écologiques, cognition social vs. technique, et profils de personnalité.

My first attempt to study bonobos was a complete disaster. It took place in the autumn of 1991 at the Yerkes Primate Research Center Field station in Lawrenceville, GA. We confronted bonobos with a task in which they had to use a rake to obtain an out-of-reach piece of food placed on a table. When chimpanzees and orangutans had faced the same task, they quite readily, especially the latter, grabbed the rake and began to use it vigorously to bring the food within their reach. Bonobos took a quite different and unexpected approach. They did not touch the tool. Instead, they extended their hand in the direction of the food while alternated their gaze between us and the food. Despite our best efforts, which included increasing the amount of food and reducing its distance from the rake, bonobos refused to use the rake to get the food and instead persisted in attempting to enlist our collaboration in retrieving it.

I have to admit that I received such display of 'originality' with mixed feelings. On the one hand, I was unhappy because by refusing to play our game by our own rules, we were unable to study the bonobos. On the other hand, this experience taught me that despite their similarities, chimpanzees and bonobos could take quite different approaches to solve the same problem. It left me with the lasting impression that bonobos compared to chimpanzees were certainly socially sophisticated but perhaps a bit technologically challenged. Reports from field

Call, J., *Bonobos, chimpanzees and tools: integrating species-specific psychological biases and socio-ecology*. In: *Bonobos: Unique in Mind, Brain, and Behavior*. Edited by Brian Hare and Shinya Yamamoto: Oxford University Press (2017). © Oxford University Press.
DOI: 10.1093/oso/9780198728511.003.0012

studies showing that chimpanzees, unlike bonobos, possessed an impressive repertoire of tool-using and tool-making behaviors and, paired with bonobos' socio-sexual behavior that had recently emerged from the laboratory, this contributed to consolidate my initial impression.

The goal of this chapter is to explore the reasons for the remarkable difference that exists in tool use between chimpanzees and bonobos in the wild. This is a topic that has received considerable research attention and has generated some debate (Furuichi et al., 2014; Stanford, 1998). For instance, Furuichi and colleagues (2014) examined some of the ecological determinants that may explain the difference in tool use between chimpanzees and bonobos. Their thorough examination, however, did not reveal a compelling link between ecology and tool use. My analysis will be focused not on socio-ecological aspects but on the cognitive, motivational and temperamental aspects that may contribute to explain the differences between chimpanzee and bonobo tool use in the wild. This discrepancy between wild and captive data had been described many years ago (McGrew, 1989) but a detailed analysis of the cognitive aspects that may underlie this difference was not possible until now when enough data have accumulated. My analysis will proceed as follows.

First, I will examine the social versus physical intelligence hypothesis by focusing on the evidence of the direct comparisons between chimpanzees and bonobos on various social and physical cognition tasks that have accumulated in recent years, paying special attention to tool-using tasks. Second, I will examine another more recent hypothesis about the temperamental differences between the two species. Next, I will propose an additional aspect with regard to bonobo emotional responsiveness that may also contribute to explain the observed differences. I will finish with some thoughts about the developmental and evolutionary scenarios that may have contributed to the differences that we observe, because ultimately the phenotypes arise through the interaction of the organism with its social and physical environment. Although my main focus will be on the psychological aspects, I will also consider some contributing socio-ecological factors in this section.

Bonobos and chimpanzees in the wild

One of the aspects of chimpanzee behavior that has had the greatest impact on various disciplines including anthropology, psychology and human evolution is their tool-using and tool-making behavior. This is not just because tool use is flexible but because the variation across groups is hard to miss and it has been considered cultural. Thus, chimpanzees use tools in various contexts including foraging, aggression, sex and play. They use tools to extract embedded food; for instance, using stones to crack open nuts or sticks to extract insects or bone marrow. In some cases, they use multiple tools in sequence (each with a different function) to obtain those resources (see McGrew, 2004; 2010 for reviews). For instance, they use a stick to puncture a beehive and another tool to extract the honey. Modification of tools is also prominent in many cases (McGrew, 2004; Sanz and Morgan, 2007). Chimpanzees fashion tools to fit a certain shape or size and they do so in anticipation of use. For instance, chimpanzees in the Gouologo triangle fashion brush-tipped tools to extract termites prior to their use, a feature that increases the effectiveness of the tools fivefold (Sanz et al., 2009). This suggests that they possess a mental template of the features that they want to reproduce. In contrast, bonobos, despite their close phylogenetic relation to chimpanzees and the fact that some populations inhabit very similar habitats, lack the rich repertoire that characterizes chimpanzee material culture, especially with regard to tools used in extractive foraging. Tool use, however, has been sporadically documented in other contexts and the use of leaves as rain cover appears to be customary (Hohmann and Fruth, 2003; Kano, 1982).

Two qualifications are necessary before we examine the empirical support for the two psychological hypotheses that have been proposed to explain the differences between chimpanzees and bonobos. First, there is high variability between chimpanzee populations in their propensity for using and making tools. While chimpanzees in Goulougo and Tai possess a repertoire of 22 and 12 distinct types of tool use, respectively, chimpanzees in Budongo and Kibale only possess 8 and 10 distinct types, respectively (Sanz and Morgan, 2007). Thus, when making comparisons with bonobos, who have a rather small repertoire of foraging tools, one has to keep in mind that from a quantitative point of view, some chimpanzee populations are more similar to bonobos than to other chimpanzee populations.

Second, although it is true that, taken together, chimpanzees display a massive predominance for tools used in extractive foraging compared to bonobos (only leaf sponging has been described in bonobos), the same is not necessarily true for other categories. Furuichi and colleagues (2014) reported that bonobos have been observed to use tools in six different ways in social contexts and in five different ways for comfort or protection. However, the complexity of bonobo tools is lower than that of chimpanzees'. Most bonobo tools do not require any elaborate manufacture other than detaching leaves or branches from the substrate. Moreover, handling a stick to navigate the tunnels in termite mounds requires a level of manual precision and skill that is not required when using leaves as rain covers or as a fly swatter. In fact, recent studies show that tool use puts a considerable strain on apes' cognitive resources, possibly related to executive function. More specifically, apes required to use tools in a task that demands that they avoid multiple obstacles show clear disruption in causal knowledge and planning compared to those apes that are confronted with the same task but not required to use tools (Seed et al., 2009; Voelter and Call, 2014a,b). Although this has not been tested directly, it is possible that using a branch as a rain cover or to repel insects is easier from a cognitive standpoint than using a stick to extract insects from their nests. Thus, whereas wild bonobos and chimpanzees may display basic tool use, only wild chimpanzees display the most complex forms of tool use, which include making tools and using tools in sequence.

Thus, bonobos use tools in the wild just like chimpanzees do, but they differ in the context in which they use them, the complexity of the tools and the task and probably the frequency of use. Next we examine whether the different cognitive and motivational dispositions of each species can, at least in part, explain those results.

Social versus technical intelligence

Research on captive and semi-captive populations has allowed us to compare chimpanzees and bonobos on the same tasks under nearly identical testing conditions, thus bringing under close scrutiny the cognitive mechanisms underlying behavior in those populations. Herrmann and colleagues (2010) compared chimpanzees and bonobos by directly measuring their responses to a battery of tasks covering both physical and social cognition. The battery was designed in a way that allowed researchers to obtain a quick assessment on multiple tasks even for subjects who had never been tested before and after minimal familiarization with the task materials. Herrmann and colleagues (2010) found similarities between the two species in most tasks including things like quantity discrimination, spatial memory and social learning. However, there were two exceptions. Bonobos outperformed chimpanzees in social tasks whereas chimpanzees outperformed bonobos in tool-using tasks.

Kano, Hirata and Call (2015) also found differences between chimpanzees and bonobos in their allocation of attention to pictures of conspecifics that were either resting or engaged in foraging tasks (e.g. using a tool to extract some food). Measuring attention is important because it may provide important clues about the information that is acquired and the effect that it may have on determining individuals' responses. Although both chimpanzees and bonobos focused on the elements of the stimuli (e.g. eyes, mouth, faces, genitals, tools), they found that the two species distributed their attention differently. Bonobos were more likely to focus on eyes compared to chimpanzees whereas chimpanzees were more likely to focus on tools compared to bonobos. This suggests that the two species had different attentional biases that may be crucial determinants of their subsequent overt responses. In fact, Koops, Furuichi and Hashimoto (2015) reported that chimpanzee youngsters played with objects more often than bonobos, even though both species spend the same amount of time playing.

The results of these studies fit well with other studies on other populations using more time-intensive methods. For instance, bonobos are more sensitive to gaze following than chimpanzees (Braeuer et al., 2005) and bonobos are more tolerant and can cooperate more easily than chimpanzees (Hare et al., 2007). There is also some indication that bonobos might be more prosocially oriented than chimpanzees

Figure 12.1 Bonobos and chimpanzees readily engage in a variety of extractive foraging tasks in the laboratory (Photo courtesy of the MPI-EVA). (Les bonobos et les chimpanzés participent dans des tâches de butinage variées dans le laboratoire.)

but these data need to be confirmed with a larger sample (Amici et al., 2014; Braeuer et al., 2009; but see Jaeggi et al., 2010). However, other studies have found no inter-specific differences in economic games (Kaiser et al., 2012), ape language learning (Savage-Rumbaugh, Brakke and Hutchins, 1992) or perspective-taking tasks (Kaminski et al., 2004; Liebal et al., 2004; but see MacLean and Hare, 2012).

Focusing on tool use, both chimpanzees and bonobos readily use tools for foraging purposes in the laboratory (see Figure 12.1). More importantly, studies focusing on various aspects of tool use have been unable to find consistent differences between chimpanzees and bonobos. More specifically, we have found no evidence of differences in sequential tool use (Martin-Ordas et al., 2012b), tool innovation (Marin-Manrique and Call, 2011), avoiding obstacles and planning (Martin-Ordas et al., 2012a,b; Voelter and Call, 2014a,b) and saving and making tools for future use (Braeuer and Call, 2015; Dufour and Sterck, 2008; Mulcahy and Call, 2006; Osvath and Osvath, 2008). Some of these aspects (e.g. avoiding obstacles) have been investigated in multiple studies and the results have been quite reliable. In general, one finds large inter-individual differences but no consistent inter-species differences. Such an outcome cannot solely be attributed to our sample sizes or testing methods because we have found differences between orangutans and chimpanzees and bonobos in tool innovation (Hanus et al., 2011; Marin-Manrique and Call, 2011; Mendes et al., 2007) and inhibitory control (Vlamings et al., 2010) with comparable sample sizes.

Furthermore, the lack of consistent interspecific differences between chimpanzees and bonobos cannot be attributed to floor effects because at least some individuals perform well in many of these tasks.

In sum, although the idea of a social bonobo versus technical chimpanzee is appealing and reasonably supported by the evidence, it is not totally convincing with regard to tool use because some bonobos have performed at the same level as chimpanzees. More specifically, multiple studies have failed to reveal any cognitive differences that could explain the differences observed between these two species in the wild. Next we turn our attention to motivational and temperamental aspects as a potential explanation for the differences observed in the wild.

Temperament and boldness

Although the technical versus social intelligence hypothesis discussed in the previous section could very well explain bonobos' social problem-solving approach to the rake task mentioned at the beginning of this chapter, recently another hypothesis has been proposed that could also account for bonobos' response. It is a hypothesis borne out the current interest that exists in individual differences, personality and temperament in animals. Researchers have been able to measure consistent individual differences that are stable over time, affect various behaviors and can be measured reliably by multiple methods including questionnaires, behavioral observations and experiments.

Individual differences in personality have been postulated as an explanation for certain behavioral differences between individuals including exploration (Verbeek et al., 1994), learning ability (Brust et al., 2013; Günther et al., 2014) and problem-solving (Cole and Quinn, 2011) which have been traditionally been the sole purview of cognition. Is it possible that personality traits may also explain to some extent the tendency of individuals to use tools? One of the most intensely investigated traits in personality research from a comparative perspective is boldness. This is typically measured by an individual's willingness to approach and interact with novel objects, food or social partners. Compared to chimpanzees and orangutans, bonobos score lower in tests measuring boldness (Herrmann et al., 2011). More specifically, bonobos show longer latencies to approach both human strangers and unknown objects compared to chimpanzees and orangutans. Two-year-old children were even slower than bonobos tested under similar conditions. Note, however, that bonobos seem less wary and less aggressive towards unknown conspecifics compared to chimpanzees (Furuichi, 2011; Hare et al., 2012). Nevertheless, it is conceivable that bonobos' reluctance to grab and use the rake may have had an important boldness component and not simply been a reflection of their preference for social problem-solving. Furthermore, this explanation could also make bonobos more reluctant to explore their environment (and therefore discover new resources) in the wild, especially if this tendency is accompanied by a strong social influence that may further curtail individuals' exploration.

However, this explanation does not offer a fully satisfactory account for the *Panin* tool-use puzzle. Bonobos' lesser predisposition to manipulate the new object may account for their initial reluctance to use it but countless studies have demonstrated that bonobos end up approaching new objects and using them as tools. Note that what has been measured is latency to approach, but virtually all individuals tested approached and made contact with the objects. This means that although they may be slower in coming off their marks, they eventually do and explore new objects and agents. They are not irreversibly limited by their putative shyer predisposition.

Thus, just as we saw for the social versus technical hypothesis, the temperament hypothesis only offers a partial resolution to our question.

Emotional responsiveness and emotional reactivity

Emotional responsiveness is another important aspect that could be contributing to the differences between chimpanzees and bonobos. This concept is different from boldness. If boldness determined how likely individuals were to approach and manipulate or interact with an object or subject, respectively, emotional responsiveness to certain stimuli determines whether and how individuals continue to respond to these stimuli after their first experience. This is a particularly important consideration given that much of the chimpanzee and orangutan foraging tool use is aimed at exploiting resources such as insects or spiny fruits which can deliver painful sensations if not handled properly. If bonobos were less tolerant to pain and discomfort than chimpanzees and orangutans, they may simply avoid those resources and seek more benign alternatives, even though they might have been as eager as the other species to obtain those foodstuffs.

The potential role of aversive stimulation following the exploration of new resources has been discussed in the literature in relation to the issue of cumulative culture. Some authors have argued that aversive stimulation may curtail innovation and promote the extreme conservatism displayed by some wild chimpanzee and orangutan populations (Forss et al., 2015; Gruber et al., 2009; 2011). Compared to their captive counterparts, wild individuals of these species typically display high neophobia and little innovation and they even fail to adopt more efficient methods to exploit known resources. However, unlike wild individuals, captive ones are exposed to numerous human-made objects and artefacts whose exploration often produces rewarding consequences. For instance, the introduction of new enrichment devices invariably produces tasty and nutritious food. Exploring unknown stimuli in the wild is much riskier than in captivity because noxious stimuli are more likely to occur there than in the captive setting. The effect of negative experiences with unknown stimuli can have long-lasting consequences and generalize to other new stimuli, especially if one considers that individuals may be exposed to them at an early age and the social environment is capable of exerting a strong (albeit passive) influence in dissuading individuals from manipulating new stimuli or areas of their home range. For instance, offspring may be more likely to exploit the same resources as their mothers (e.g. Jaeggi et al., 2010), at least initially, and only travel to those locations visited by their mother and other group members.

Although, the painful stimulation hypothesis remains untested, there is some evidence suggesting that bonobos seem to be less risk seeking than chimpanzees and orangutans in several food-gambling games (e.g. Haun et al., 2011; Heilbronner et al., 2008; Rosati and Hare, 2013). Although both bonobos and chimpanzees showed negative responses to receiving a bad outcome, only bonobos, especially those individuals who were the less risk seeking, tried to change their choice (Rosati and Hare, 2013). Another study reported that bonobos were more tolerant than chimpanzees when co-feeding from clumped resources (Hare et al., 2007). Hare and Tomasello (2005) have argued that the differential emotional reactivity measured mainly in terms of social tolerance and risk-proneness, may be an important determinant of the differences between chimpanzees and bonobos. It is unclear, however, whether emotional reactivity focuses on the same emotional dimension or even the same neural substrate as physical pain originating from an insect bite. Emotional reactivity seems to be a concept that is closer to boldness, whereas emotional responsiveness seems more related to tolerance for aversive stimulation. Nevertheless, additional research is needed to determine the precise relation between emotional reactivity and emotional responsiveness. For now, we can entertain the possibility that a reduced tolerance to aversive (or mildly aversive) stimuli in bonobos compared to chimpanzees and orangutans could be another contributing factor to the absence of extractive foraging tools in wild bonobos.

Panins, tools and cognitive development

None of the three hypotheses that we explored could fully explain the observed differences in tool use between chimpanzees and bonobos in the wild. However, each explanation seems to possess a grain of truth and interestingly, they are not mutually exclusive but they complement each other. In the final part of this chapter, I try to articulate an account that includes cognition, temperament and emotional responsiveness and integrate it within the socio-ecological framework in which chimpanzees and bonobos develop and evolve.

Let me start this synthesis by returning to the differences that we found between the two species in terms of attentional predispositions. Recall that bonobos were more oriented towards social stimuli than chimpanzees, who distributed their attention equally between social and non-social stimuli. This difference was consistent across laboratories and reliable when the same individuals were studied at various points in time. The fact that these differences between chimpanzees and bonobos exist even though the study groups are housed in the same location under very similar conditions with opportunities for social interactions and environmental enrichment (including tool use) is a strong indication that this may constitute a species-specific attentional bias. This is not to say that some bonobos may not show the chimpanzee pattern and vice versa, but as a group these seem to be consistent differences. Recall also that the difference in attentional focus was in turn reminiscent of their predisposition for manipulating objects and ultimately, the strengths and weaknesses of the two species tested in various social and technical tasks, although such differences (when they existed) appeared to be of degree not kind.

Boldness is another predisposition, not of a cognitive nature like attentional focus but a dispositional one that determines how individuals respond to challenges in general, and how they confront new objects and individuals more specifically. I refer to boldness as a predisposition because, being a personality trait, it is stable over time. Chimpanzees, by virtue of being bolder than bonobos, again something that has been documented by several studies, are placed in a privileged position to contact, explore and learn about objects and their interactions with other entities.

However, predispositions are not the sole determinant of behavior. The interaction with the socio-ecological environment is an essential component that contributes to the development of the individual and the evolution of the species. If predispositions put individuals in a position to learn, the environment plays a major role in channelling that learning. One environment that provides multiple opportunities for interaction with tools (and their potential benefits) may be more conducive to the adoption of certain behaviors and the development of the cognitive substrate supporting them compared to one that does not. This could explain bonobos' widespread use of tools in the laboratory and their level of proficiency in using them, comparable in every way to chimpanzees'. Recall that detailed comparisons in the laboratory have revealed no differences in tool-use proficiency between bonobos and chimpanzees. It is true that some differences have been reported between the two species (Herrmann et al., 2010), but this was always before individuals had had ample opportunity to use tools prior to the tests. Once the members of the two species are already using tools their performance is virtually indistinguishable, even when challenged with new tasks. One possibility is that the differences do exist but our tests are not sensitive enough to detect them. However, the extensive range of aspects that we have investigated casts some doubt on this interpretation, although it is conceivable that other more refined tests might reveal those differences.

Thus, it is clear that both bonobos and chimpanzees use tools (and can do so proficiently) both in laboratories and sanctuaries. Moreover, both species also use tools in the wild for non-foraging purposes (Furuichi et al., 2014). These similarities between species and across settings make the difference in the use of foraging tools in the wild even more conspicuous, more so given that as far as we know, ecological differences between species do not seem to explain their different propensity to use tools in the wild.

Overcoming psychological biases and theories of innovation

Three key psychological biases that distinguish chimpanzees from bonobos have been identified whose combination could channel cognitive development in certain ways. I refer to psychological biases and not cognitive biases because although some are cognitive in nature (e.g. attentional focus, problem-solving), others are not. Motivation, temperament and emotion are interrelated but distinct from cognition. Together they are four of the pillars of an individual's psychology.

Despite the strong influence that these biases may exert on individuals, they should be understood as general species-specific predispositions rather than factors that completely predetermine the outcome of development and behavior. In fact, biases can be overcome by certain influences that may alter the behavior of individuals in significant ways. Animal invention and innovation is an area where researchers have explored the various factors that may lead individuals to change their current behavior substantially. In general, individuals may increase the probability of exploiting new resources by either an increased exposure to those items or a decreased exposure to the alternatives available. An increased exposure to certain items can occur either by an overall increase in availability of that item (e.g. an unusually large food crop of an otherwise rare species) or by the exposure to other individuals who are exploiting those items. The influence of the social environment, initially and primarily exerted by the mother and later extended to the rest of the group in the case of social species, may be especially important for those species with higher levels of neophobia. A reduction of the alternative food items commonly available may have the same effect of increasing the probability that individuals may turn to items that were not commonly exploited. The first two are consistent with the so-called opportunity hypothesis while the last is consistent with the so-called necessity hypothesis.

Some authors argue that invention and innovation are products of necessity whereas other authors suggest that opportunity rather than necessity is the main driver behind change (Parker et al., 1999). However, opportunity and necessity are not mutually exclusive (Koops et al., 2014). In fact, not only could both be operating simultaneously but also one may in fact enable the other. For instance, when arboreal primates are forced to come down to the ground because of forest fragmentation, this may bring them into contact with resources that they did not normally exploit, thus creating an opportunity that did not exist previously. However, note that the opportunity arose because it was caused by the necessity to descend to the ground in the first place. Conversely, individuals could travel in the opposite direction. For instance, increasing the opportunity to exploit a rich but scarce food source by installing artificial feeding stations would directly reduce the necessity to rely on other food sources of a lower quality.

Bonobo nut-cracking in the Lola Ya Bonobo sanctuary is an excellent illustration of how environmental and social conditions can foster the appearance of a form of extractive foraging that has to-date not been observed in wild bonobos. Bonobos living at Lola Ya sanctuary routinely engage in nut-cracking. They gather stones and nuts and they take them to places suitable to crack them open. In the sanctuary setting three factors coalesce that make the appearance of nut-cracking possible. First, bonobos have multiple opportunities for nut-cracking, since both nuts and stones are readily available. Second, other individuals are already proficient nut-crackers, thus further facilitating the acquisition of this behavior. Third, some of the items in their typical diet are not present in a semi-captive setting, which means that bonobos must seek other alternatives.

The appearance of nut-cracking, which is perhaps the form of chimpanzee tool use that has received the most attention for its implications for human evolution, in a sanctuary for a species that is otherwise not known for nut-cracking in the wild is remarkable. It is remarkable but not entirely unexpected since McGrew (1989) already pointed out the multiple 'unexpected' appearances of tool-use behavior in the laboratory by gorillas, bonobos and orangutans even though there was no indication of such practices in their natural habitat. We know that some populations of orangutans use foraging tools regularly but many others still don't. In the case of bonobos at Lola Ya Bonobo and other captive settings, both opportunity and necessity could be playing a role, something that may also be true of other populations of other species and living in various settings.

A pure probabilistic account predicts that the more opportunities exist to interact with an item, the higher the likelihood of discovery will be, and this also includes the effect that population (or group) size per se can have. However, the same will happen when resources become scarce because individuals will begin trying things that they did not attempt before. Those who do not will be less likely to survive, which means that in the next generation, the balance between innovators and non-innovators will have shifted. Furthermore, population size will exacerbate this effect, given that more individuals present will translate into fewer resources available. Only those willing or able to invent new solutions or capable of adopting those solutions invented by others are likely to thrive.

Conclusion

Both chimpanzees and bonobos readily use tools for extractive foraging in the laboratory but only chimpanzees customarily use tools for extractive foraging in the wild. There is no evidence that socio-ecological factors such as rainfall, fruit availability or party size explain this difference. Furthermore, detailed tests in the laboratory revealed virtually no differences in the cognitive processes underlying tool use between chimpanzees and bonobos. Both species use multiple tools, solve multiple tasks, encode the same tool and obstacle properties and show the same depth of planning.

However, the combination of three species-specific psychological biases in cognition, temperament and emotion could contribute to explain the differences between chimpanzees and bonobos observed in the wild. First, bonobos seemed more socially oriented in their problem-solving than chimpanzees, who did not seem to show a clear preference for technological versus social solutions. This is something that could make bonobos less tool-oriented than chimpanzees. Second, bonobos are less bold compared to chimpanzees, especially in situations not involving conspecifics, something that may make them less prone to exploit new material resources, thus reducing their repertoire at the individual, group and species level. Third, bonobos might be less tolerant to painful stimulation than chimpanzees, something that may make them less likely than chimpanzees to exploit those resources protected by chemical or physical defences. The joint action of these three biases could determine how likely the two species would be to adapt to new social and environmental challenges. One can speculate that bonobos would be especially prepared to cope with environments requiring extensive coordination with conspecifics, including strangers, whereas chimpanzees would cope well with changing physical environments requiring technological innovation. One could say that bonobos would make good travelling salesmen and chimpanzees superb explorers of uncharted territories.

These biases, however, should be best understood as general species-specific predispositions rather than factors that predetermine the outcome, but combined with the physical and social environment that the individual experiences, they may create different phenotypes. External influences such as the relative availability of certain resources and the social environment create opportunities and necessities that paired with psychological biases could contribute to explain the inter-specific differences in the use of tools for extractive foraging between chimpanzees and bonobos in the wild.

References

Amici, F., Visalberghi, E., and Call, J. (2014). Lack of prosociality in great apes, capuchin monkeys and spider monkeys: convergent evidence from two different food distribution tasks. *Proceedings of the Royal Society B*, 281(1793). pii: 20141699. DOI: 10.1098/rspb.2014.1699.

Braeuer, J. and Call, J. (2015). Apes produce tools for future use. *American Journal of Primatology*, 77, 254–63.

Brust, V., Wuerz, Y., and Krueger, O. (2013). Behavioural flexibility and personality in zebra finches. *Ethology*, 119, 559–69.

Bräuer, J., Call, J., and Tomasello, M. (2005). All great ape species follow gaze to distant locations and around barriers. *Journal of Comparative Psychology*, 119, 145–54.

Bräuer, J., Call, J., and Tomasello, M. (2009). Are apes inequity averse? New data on the token-exchange paradigm. *American Journal of Primatology*, 71, 175–81.

Cole, E.F. and Quinn, J.L. (2012). Personality and problem-solving performance explain competitive ability in the wild. *Proceedings of the Royal Society B*, 279(1731):1168–75. DOI: 10.1098/rspb.2011.1539.

Dufour, V. and Sterck, E.H.M. (2008). Chimpanzees fail to plan in an exchange task but succeed in a tool-using procedure. *Behavioural Processes*, 79, 19–27.

Dugatkin, L.A. and Alfieri, M.S. (2003). Boldness, behavioral inhibition and learning. *Ethology, Ecology & Evolution*, 15, 43–9.

Forss, S., Schuppli, C., Haiden, D., Zweifel, N., and van Schaik, C.P. (2015). Contrasting responses to novelty by wild and captive orangutans. *American Journal of Primatology*, 77, 1109–21.

Furuichi, T. (2011). Female contributions to the peaceful nature of bonobo society. *Evolutionary Anthropology*, 20, 131–42.

Furuichi, T., Sanz, C., Koops, K., Sakamaki, T., Ryu HJin, Tokuyama, N., and Morgan, D. (2014). Why do wild bonobos not use tools like chimpanzees do? *Behaviour*. DOI: 10.1163/1568539X-00003226.

Gruber, T., Muller, M.N., Reynolds, V., Wrangham, R.W., and Zuberbühler, K. (2011). Community-specific evaluation of tool affordances in wild chimpanzees. *Scientific Reports*, 1. DOI:10.1038/srep00128.

Gruber, T., Muller, M.N., Strimling, P., Wrangham, R.W., and Zuberbühler, K. (2009). Wild chimpanzees rely on cultural knowledge to solve an experimental honey acquisition task. *Current Biology*, 19, 1806–10.

Günther, A., Brust, V., Dersen, M., and Trillmich, F. (2014). Learning and personality types are related in cavies (*Cavia aperea*). *Journal of Comparative Psychology*, 128, 74–81.

Hanus, D.; Mendes, N., Tennie, C., and Call, J. (2011). Comparing the performances of apes (*Gorilla gorilla, Pan troglodytes, Pongo pygmaeus*) and human children (*Homo sapiens*) in the floating peanut task. *PLoS One*, 6, e19555.

Hare, B., Melis, A., Woods, V., Hastings, S., and Wrangham, R. (2007). Tolerance allows bonobos to outperform chimpanzees in a cooperative task. *Current Biology*, 17, 619–23.

Hare, B. and Tomasello, M. 2005. The emotional reactivity hypothesis and cognitive evolution. *Trends in Cognitive Sciences*, 10, 464–5.

Hare, B., Wobber, V., and Wrangham, R.W. (2012). The self-domestication hypothesis: evolution of bonobo psychology is due to selection against aggression. *Animal Behaviour*, 83, 573–85.

Haun, D.B.M., Nawroth, C., and Call, J. (2011). Great apes' risk-taking strategies in a decision making task. *PLoS One*, 6, 118–28.

Heilbronner, S.R., Rosati, A.G., Stevens, J.R., Hare, B., and Hauser, M.D. (2008). A fruit in the hand or two in the bush? Divergent risk preferences in chimpanzees and bonobos. *Biology Letters*, 4, 246–9.

Herrmann, E., Call, J., Hare, B., and Tomasello, M. (2010). Differences in the cognitive skills of bonobos and chimpanzees. *PLoS One*, 5, e12438.

Herrmann, E., Hare, B., Cissewski, J., and Tomasello, M. (2011). A comparison of temperament in nonhuman apes and human infants. *Developmental Science*, 14, 1393–405.

Hohmann, G. and Fruth, B. (2003). Culture in bonobos? Between-species and within-species variation in behavior. *Current Anthropology*, 44, 563–71.

Jaeggi, A.V., Dunkel, L.P., Van Noordwijk, M.A., Wich, S.A., Sura, A.A.L., and van Schaik, C.P. (2010). Social learning of diet and foraging skills by wild immature Bornean orangutans: implications for culture. *American Journal of Primatology*, 72, 62–71.

Jaeggi, A.V., Stevens, J.M.G., and van Schaik, C.P. (2010). Tolerant food sharing and reciprocity is precluded by despotism among bonobos but not chimpanzees. *American Journal of Physical Anthropology*, 143, 41–51.

Kaiser, I., Jensen, K., Call, J., and Tomasello, M. (2012). Theft in an ultimatum game: Chimpanzees and bonobos are insensitive to unfairness. *Biology Letters*, 8, 942–5.

Kaminski, J., Call, J., and Tomasello, M. (2004). Body orientation and face orientation: two factors controlling apes' begging behavior from humans. *Animal Cognition*, 7, 216–23.

Kano, T. (1982). The use of leafy twigs for rain cover by the pygmy chimpanzees of Wamba. *Primates*, 19, 187–93.

Kano, F., Hirata, S., and Call, J. (2015). Social attention in the two species of Pan: Bonobos make more eye contact than chimpanzees. *PLoS One*, 10, e0129684.

Koops, K., Furuichi T., and Hashimoto, C. (2015). Chimpanzees and bonobos differ in intrinsic motivation for tool use. *Scientific Reports*, 5, 11356.

Koops, K., Visalberghi, E., and van Schaik, C.P. (2014). The ecology of primate material culture. *Biology Letters*, 10. DOI: 10.1098/rsbl.2014.0508.

Liebal, K., Pika, S., Call, J., and Tomasello, M. (2004). To move or not to move: how apes alter the attentional states of humans when begging for food. *Interaction Studies*, 5, 199–219.

MacLean, E.L. and Hare, B. (2012). Bonobos and chimpanzees infer the target of another's attention. *Animal Behaviour*, 83, 345–53.

Manrique, H.M. and Call, J. (2011). Spontaneous use of tools as straws in great apes. *Animal Cognition*, 14, 213–26.

Martin-Ordas, G., Jaeck, F., and Call, J. (2012a). Barriers and traps: Great apes' performance in two functionally equivalent tasks. *Animal Cognition*, 15, 1007–13.

Martin-Ordas, G., Schumacher, L., and Call, J. (2012b). Sequential tool use in great apes. *PLoS One*, 7, e52074. DOI:10.1371/journal.pone. 0052074.

McGrew, W.C. (1989). Why is ape tool-use so confusing? In: Standen, V. and Foley, R.A. (eds). *Comparative Socioecology. The Behavioural Ecology of Humans and Other Mammals.* Oxford: Blackwell, pp. 457–72.

McGrew, W.C. (2004). *The Cultured Chimpanzee.* Cambridge: Cambridge University Press.

McGrew, W.C. (2010). Chimpanzee technology. *Science*, 328, 579–580.

Mendes, N., Hanus, D., and Call, J. (2007). Raising the level: Orangutans use water as a tool. *Biology Letters*, 3, 453–5.

Mulcahy, N.J. and Call, J. (2006). Apes save tools for future use. *Science*, 312, 1038–40.

Osvath M. and Osvath H. (2008). Chimpanzee (*Pan troglodytes*) and orangutan (*Pongo abelii*) forethought: self-control and pre-experience in the face of future tool-use. *Animal Cognition*, 11, 661–74.

Parker, S.T., Mitchell, R.W., Miles, H.L., Fox, E.A., Sitompul, A.F., and van Schaik, C.P. (1999). Intelligent tool use in Sumatran orangutans. In: Parker, S.T., Mitchell, R.W., and Miles, H.L. (eds). *The Mentalities of Gorillas and Orangutans.* Cambridge: Cambridge University Press, pp. 99–116.

Rosati, A.G. and Hare, B. (2013). Chimpanzees and bonobos exhibit emotional responses to decision outcomes. *PLoS One*, 8, e63058. DOI: 10.1371/journal.pone.0063058.

Sanz, C.M. and Morgan, D.B. (2007). Chimpanzee tool technology in the Goualougo Triangle, Republic of Congo. *Journal of Human Evolution*, 52, 420–33.

Sanz, C., Call, J. & Morgan, D. (2009). Design complexity in the tool use of chimpanzees (*Pan troglodytes*) in the Congo basin. *Biology Letters*, 5, 293–296.

Savage-Rumbaugh, E.S., Brakke, K.E., and Hutchins, S.S. (1992). Linguistic development: Contrasts between co-reared *Pan troglodytes* and *Pan paniscus*. In: Nishida, T., McGrew, W.C., Marler, P., Pickford, M., and de Waal, F.B.M. (eds). *Topics in Primatology.* Tokyo: University of Tokyo Press, pp. 51–66.

Seed, A.M., Call, J., Emery, N.J., and Clayton, N.S. (2009). Chimpanzees solve the trap problem when the confound of tool-use is removed. *Journal of Experimental Psychology: Animal Behavior Processes*, 35, 23–34.

Stanford, C.B. (1998). The social behavior of chimpanzees and bonobos. Empirical evidence and shifting assumptions. *Current Anthropology*, 39, 399–407.

Verbeek, M., Drent, P.J., and Wiepkema, P. (1994). Consistent individual differences in early exploratory behaviour of male great tits. *Animal Behaviour*, 48, 1243–50.

Vlamings, P., Hare, B., and Call, J. (2010). Reaching around barriers: The performance of the great apes and 3- to 5-year-old children. *Animal Cognition*, 13, 273–85.

Voelter, C. and Call, J. (2014a). Younger apes and human children plan their moves in a maze task. *Cognition*, 130, 186–203.

Voelter, C.J. and Call, J. (2014b). The cognitive underpinnings of flexible tool use in great apes. *Journal of Experimental Psychology: Animal Behavior Processes*, 40, 287–302.

van Schaik, C.P. and Whiten, A. (2007). The evolution of animal 'cultures' and social intelligence. In: Emery, N., Clayton, N., and Frith, C. (eds). *Social Intelligence: From Brain to Culture.* Oxford: Oxford University Press.

PART VI

Mind and Brains Compared

CHAPTER 13

Bonobo personality: Age and sex effects and links with behavior and dominance

Nicky Staes, Marcel Eens, Alexander Weiss and Jeroen M.G. Stevens

Abstract The study described in this chapter examines whether individual differences in six rating-based bonobo personality dimensions—Assertiveness, Conscientiousness, Openness, Attentiveness, Agreeableness and Extraversion—are related to sex, age, behaviors and dominance. To these ends, the study tested predictions based on previous studies of human and chimpanzee personality, and bonobo behavior and socio-ecology. Sex and age differences in Assertiveness, Openness and Extraversion, and correlations between these personality dimensions and behavior were consistent with predictions. Conscientiousness showed associations with observed behaviors but requires further investigation as sex and age effects differed from those reported in humans and chimpanzees. Agreeableness and Attentiveness showed few associations with age, sex and behaviors, indicating the need to further investigate validity of these factors. This chapter shows that personality dimensions in bonobos are correlated with sex, age and behaviors in ways that are consistent with what is known for bonobos and their socio-ecology.

L'étude décrite dans ce chapitre examine si les différences individuelles dans six dimensions de personnalité bonobos basées sur évaluation—Affirmation de soi, Conscience, Ouverture, Attention, Agréabilité, et l'Extraversion—sont liées au sexe et l'âge et les comportements et la dominance. L'étude a testé les prédictions basées sur des études précédentes de la personnalité humaine et chimpanzé, et le comportement bonobo et la socioécologie. Les différences de sexe et d'âge dans l'Affirmation de soi, l'Ouverture et l'Extraversion et les corrélations entre ces dimensions de personnalité et de comportement étaient cohérents avec nos prédictions. La Conscience montre des associations avec les comportements observés mais a besoin plus de recherche vu que les effets du sexe et de l'âge diffèrent des effets rapportés chez les humains et les chimpanzés. L'Agréabilité et l'Attention n'avaient pas autant d'associations avec l'âge, le sexe et les comportements. Cela montre qu'il faut plus rechercher la validité de ces facteurs. Cette étude montre que les dimensions de personnalité chez les bonobos sont corrélé à l'âge, au sexe et aux comportements de manières qui sont cohérentes avec notre connaissance des bonobos et de leur socioécologie.

Introduction

Like other higher primates, pygmy chimpanzees are rich in individuality, and the personality of individuals probably exerts a strong influence on the character of social relationships between group members . . . In the ongoing research then, there is a great need to focus our attention on personality. Kano, 1992, p. IX

Bonobos, like all primates, vary in their behavior across individuals (Kano, 1992). While much research focuses on explaining the evolution of species-specific characteristics, only more recently have within-species individual differences become the focus of analyses. Personality is one construct that has been used to describe and explain individual differences within populations. While personality has initially and predominantly been used to help characterize human behavioral diversity, it is now being used to help explain individual differences in animals, including great apes (Gosling, 2001; Mehta and Gosling, 2008). Personality in non-human primates and other animals is defined in

---

Staes, N., Eens, M., Weiss, A., and Stevens, J. M. G., *Bonobo personality: age and sex effects and links with behavior and dominance*. In: *Bonobos: Unique in Mind, Brain, and Behavior*. Edited by Brian Hare and Shinya Yamamoto: Oxford University Press (2017).
© Oxford University Press. DOI: 10.1093/oso/9780198728511.003.0013

much the same way as is personality in humans, namely as inter-individual differences in behavior, affect and cognition that are stable across time and situations (Goldberg, 1990; McCrae and Costa, 1999; Réale et al., 2007). Animal personality has been documented in species ranging from insects to non-human primates (Carere and Eens, 2005; Freeman and Gosling, 2010; Gosling, 2001; Gosling and John, 1999), a fact that underlines its importance and early evolution in the animal kingdom.

Most often animal personality is measured using behavioral observations and behavioral tests (Freeman and Gosling, 2010; Staes, 2016). However, animal personality is also measured using trait ratings, which involves asking humans familiar with individual animals to rate these animals on a set of predefined traits (Freeman and Gosling, 2010). This approach was at first perceived as 'subjective' and anthropomorphic by behavioral biologists, but it is very likely that it reflects real personality constructs as ratings meet the same criteria as behavioral observations and behavioral tests (Baker et al., 2015; Gosling and Vazire, 2002; Vazire et al., 2007; Weiss et al., 2012). The first criterion is that ratings have to be consistent across observers, different times of measurement and different contexts (Capitanio, 1999; Dutton, 2008; King et al., 2008; Weiss et al., 2011). The second criterion is that the link between personality dimensions and life history parameters, such as sex and age, as well as between personality dimensions and other measures (typically behaviors) has to be established (Baker et al., 2015; Capitanio, 1999; Murray, 2011; Pederson et al., 2005; Uher and Asendorpf, 2008). If males and females face different selection pressures, sex differences in personality may emerge. For example, in many species, males vary more in reproductive success than females, and therefore they may rely more on fitness benefits associated with bolder and more exploratory behavior. In three-spined sticklebacks (*Gasterosteus aculeatus*) and great tits (*Parus major*), males are bolder than females (King et al., 2013; van Oers, 2005) and risky male behaviors have been shown to increase mating success in guppies (Godin and Dugatkin, 1996).

Factor analytic studies, which seek to uncover latent constructs that make up variation in human personality traits, typically yield five broad dimensions—Openness, Conscientiousness, Extraversion, Agreeableness and Neuroticism—collectively known as the Big Five or Five-Factor Model (Goldberg, 1990; McCrae and Costa, 1999). Variations of these dimensions and novel dimensions are found in non-human primates, including great apes (Adams et al., 2015; Freeman and Gosling, 2010). Bonobos are particularly interesting because although they are, together with chimpanzees, the closest living relatives of humans (Prado-Martinez et al., 2013), yet their social organization and behavior differs from chimpanzees and humans (Boesch et al., 2002; Stumpf, 2007). We therefore expect similar differences to be reflected in their personality. Unfortunately, despite the large number of chimpanzee personality studies, few studies have focused on bonobos, and those that did involved small sample sizes (Garai et al., 2016; Murray, 2011; Uher and Asendorpf, 2008) or focused on measuring single traits, such as boldness (Herrmann et al., 2011). To complement this existing literature we obtained personality ratings on 154 bonobos. All of the bonobos in our study were rated on the Hominoid Personality Questionnaire (HPQ; Weiss et al., 2015). We found six dimensions (Table 13.1): bonobo Assertiveness, Openness and Agreeableness were similar to chimpanzee Dominance, Openness and Agreeableness, respectively; Conscientiousness appeared to split into the dimensions Attentiveness and Conscientiousness in bonobos; bonobo Extraversion showed little overlap with its chimpanzee counterpart and we therefore considered it a bonobo-specific factor (Weiss et al., 2015).

Based on what is known about human and chimpanzee personality and about the ecology and social structure of bonobos, we made several predictions about sex and age differences in bonobo personality. We then tested these predictions; that is, we examined the 'comparative validity' of the dimensions found in our previous study (Weiss et al., 2015). In terms of sex differences, among chimpanzees, males tend to score higher than females on Dominance, which has high loadings of traits, such as *dominant, independent* and *bullying* (King et al., 2008; Weiss et al., 2007; Weiss and King, 2015). Also, compared to females, male chimpanzees score lower on Conscientiousness, which has high negative loadings on traits such as *aggressive* and *impulsive*

Table 13.1 Comparison of adjectival contents of rated personality dimensions for bonobos and chimpanzees (bonobo data from Weiss et al. 2015, chimpanzee data from Weiss et al. 2009).
(Comparaison des contenus adjectifs des dimensions de personnalité évalués chez les bonobos et les chimpanzés (données bonobo de Weiss et al. 2015, données chimpanzé de Weiss et al. 2009)).

Bonobo		Chimpanzee	
Dimension	Adjectives	Dimension	Adjectives
Assertiveness	+ **Independent** + **Dominant** + **Decisive** + **Persistent** + Cool + Stable − Excitable − **Dependent** − **Submissive** − **Timid** − **Vulnerable** − **Anxious** − Fearful	Dominance	+ **Independent** + **Dominant** + **Decisive** + **Persistent** + Stingy + Manipulative − **Dependent** − **Submissive** − **Timid** − **Vulnerable** − **Anxious**
Conscientious-ness	+ Gentle + Predictable − **Impulsive** − **Defiant** − **Reckless** − **Irritable** − **Aggressive** − **Jealous** − **Bullying** − Manipulative − Erratic − Stingy	Conscientious-ness	− **Impulsive** − **Defiant** − **Reckless** − **Irritable** − **Aggressive** − **Jealous** − **Bullying** − Disorganized − Excitable − Clumsy − Unperceptive − Distractible − Thoughtless
Openness	+ **Inquisitive** + **Inventive** + **Innovative** + **Curious** + Active + Playful + Imitative − Lazy − Conventional	Openness	+ **Inventive** + **Inquisitive** + **Innovative** + **Curious** + Cautious
Attentiveness	+ **Intelligent** − Clumsy − Unperceptive − Distractible − Thoughtless − Disorganized	Neuroticism	+ **Intelligent** + Erratic + Depressed + Fearful + Autistic − Stable −Unemotional − Cool
Agreeableness	+ **Friendly** + **Affectionate** + **Protective** + **Sympathetic** + **Helpful** + **Sensitive** + Sociable	Agreeableness	+ **Friendly** + **Affectionate** + **Protective** + **Sympathetic** + **Helpful** + **Sensitive** + Gentle + Conventional
Extraversion	− **Individualistic** − **Solitary** − Autistic − Depressed	Extraversion	+ Active + Playful + Sociable − **Individualistic** − **Solitary** − Lazy

Bold font indicates that variables are similar in these dimensions for both species

(King et al. 2008; Weiss et al. 2007; Weiss and King 2015). Compared to the male-dominated societies in chimpanzees (Boesch and Boesch-Acherman, 2000; Goodall, 1986), in bonobo societies females dominate males, but not all females are higher ranking than all males, a condition known as 'partial female dominance' (Furuichi, 2011; Stevens et al., 2007; Surbeck and Hohmann, 2013; Vervaecke et al., 2000a). We therefore expect sex differences in bonobos to be the opposite of those found in chimpanzees for these two factors. In other words, we expected that bonobo females will score higher on Assertiveness and lower on Conscientiousness than male bonobos.

Personality is also known to show moderate developmental changes. For example, cross-sectional and longitudinal studies in humans show that Agreeableness and Conscientiousness increase with age, while Extraversion and Neuroticism decrease, and Openness increases and then later decreases (Costa et al., 2001; McCrae et al., 1999; McCrae et al., 2000). Age differences in chimpanzee personality describe similar trends. Older chimpanzees score higher in Dominance, Conscientiousness and Agreeableness, and lower in Extraversion and Openness than younger individuals (Dutton, 2008; King et al., 2008; Weiss and King, 2015).

These patterns, therefore, appear to be conserved across chimpanzees and humans. We thus expect that, compared to younger bonobos, older bonobos will score higher on Assertiveness and Agreeableness and lower on Openness. On the other hand, it has been suggested that bonobos retain juvenile characteristics into adulthood, including high levels of social tolerance, play and non-conceptive sexual behavior (Hare et al., 2012; Wobber et al., 2010). If this is the case, age effects for dimensions that are associated with these behaviors may be less pronounced in bonobos. As the remaining three bonobo dimensions, Attentiveness, Conscientiousness and Extraversion, differ structurally from the human and chimpanzee dimensions, it is difficult to predict age effects for these dimensions based on our knowledge from human and chimpanzee personality. We will therefore explore differences in these dimensions as a basis for future studies.

In addition to testing comparative validity by examining sex and age differences in personality, it is important to test for construct validity of the dimensions by looking at how personality correlates with other measures in ways that are consistent with the meaning (and likely function) of these dimensions (Baker et al., 2015; Capitanio, 1999; Murray, 2011; Paunonen, 2003; Pederson et al., 2005; Uher and Asendorpf, 2008). One way to assess the construct validity of these ratings-based measures is to examine their correlations with naturally occurring behaviors (Konečná et al., 2008; Pederson et al., 2005; Uher and Asendorpf, 2008; Vazire et al., 2007) and/or behaviors in response to experimental tests (Baker et al., 2015; Carter et al., 2012). In behavioral tests, stimuli are introduced to provoke behaviors that occur infrequently or that are difficult to observe in a naturalistic context. For example, novel food/object tests or predator experiments have been used to assess novelty seeking and boldness in several species (Andersson et al., 2014; Baker et al., 2015; Carere and Eens, 2005; Carter et al., 2012; Massen et al., 2013).

Previous studies of the construct validity of animal personality have shown that the time it took for common squirrel monkeys (*Saimiri sciureus*) to approach novel objects was negatively associated with their scores on Sociability and that the time it took black-crested macaques (*Macaca nigra*) to touch novel objects was negatively associated with Dominance (Baker et al., 2015). Studies looking at the associations between personality traits and behaviors have been conducted in bonobos, but understandably these studies only involved five and four individuals, respectively (Murray, 2011; Uher and Asendorpf, 2008). The present study includes a larger number of individual bonobos, and thus benefits from greater statistical power. In chimpanzees, associations have been found for several factors and variables (Murray, 2011; Pederson et al., 2005; Vazire et al., 2007). For example, Dominance is positively correlated with aggression given and negatively correlated with submissive behaviors, Extraversion is positively correlated with social and affinitive behaviors, and Sociability is positively correlated with play (Buirski et al., 1978; Murray, 2011; Pederson et al., 2005; Vazire et al., 2007). Based on the findings in chimpanzees, we expect to find similar associations in bonobos. For example, bonobo Assertiveness should correlate positively with aggression and dominance rank and negatively with submissive behaviors. We expect Extraversion to correlate positively with social and affinitive behaviors like grooming, play and sexual interactions. We expect Openness to correlate positively with exploratory behaviors, like approaching new stimuli. Finally, to rule out the fact that researchers rate the individuals based on their own stereotypical beliefs about age, sex and social position of human individuals, we also tested for sex and age differences in the observed behavioral variables. If age and sex effects are found for rated dimensions, they should show similar effects in the behavioral variables that are correlated with the rated personality dimensions.

In summary, this study will: 1) assess the comparative validity of bonobo personality dimensions by testing whether sex and age differences in personality are consistent with what we would expect based on findings related to these differences in chimpanzees and humans and with differences between bonobos and chimpanzees in social structure, behavior and life history; and 2) test the construct validity of the dimensions by examining their associations with relevant behavioral variables derived from naturalistic observations and behavioral tests.

Methods

Questionnaire ratings

Personality ratings were independently collected by different raters for 154 bonobos (71 males, 83 females) with ages ranging from 2 to 62 years using the Hominoid Personality Questionnaire (Weiss et al., 2009) in two waves (first wave 2006–08, second wave 2012) (see Weiss et al. 2015 for details). The interrater reliabilities and repeatabilities of the six dimensions revealed by these ratings were high enough to justify their being used in further studies.

Naturalistic observations

For 44 individuals (18 males and 26 females) that were rated, behavioral data were collected by researchers (NS and 8 students she supervised) who did not rate the bonobos on the HPQ. Each student received at least three weeks of training. After each student was trained, their inter-observer reliability was tested by scoring two ten-minute bonobo focal video recordings. High Spearman rank correlations (mean r_s = 0.86) were found across all observers coding the behaviors (Martin and Bateson, 1993). Observations involved alternating continuous ten-minute focal observations with instantaneous group scan sampling. During focal observations, the main activity of the focal individual was recorded. All self-directed behaviors and social interactions were also recorded. Scans were used to determine spatial proximity measures for all individuals in the group in between each focal observation. During the entire observation period, bouts of agonistic behavior were recorded using all occurrence sampling. In total, 1666.15 hours of focal observations (mean 32.04 hours per individual), 10 472 group scans (mean 616 per individual) and 2132 hours of all occurrence sampling of aggression (mean 39.5 hours per individual) were collected. Behavioral observations were coded using the Observer (Noldus version XT 10, The Netherlands).

Variables based on durations of behavioral states (activity, grooming given, grooming received, individual play, social play, auto-scratching and auto-grooming) were calculated as the proportion of individual focal time performing the behavior. Behavioral variables based on occurrences of events (submission, aggression given, aggression received, socio-sexual behaviors) were calculated as frequencies per hour, corrected for individual total observation times (both focal observation times and all occurrence observation times). Sit alone was calculated as the proportion of all scans where the subject was recorded as being alone with no other individuals sitting within a 2 m radius. Number of neighbours was the average number of neighbours present in scans where the subject was recorded as being in proximity (within 2 m) of at least one other group member. Grooming density given and received was calculated as the proportion of available grooming partners that were groomed by the subject or who groomed the subject. Grooming diversity was calculated with the Shannon–Wiener index (Di Bitetti, 2000), corrected for group size effects as follows:

$$\text{Grooming diversity} = H/H_{max}$$

$$H = -\Sigma (p_i \times \ln(p_i))$$

in which p_i is the proportion of the individual's grooming effort given to the ith individual

$$H_{max} = \ln(N-1)$$

in which N is the number of individuals in the group. Grooming diversity results can take a value between 0 and 1, with 0 meaning perfect skew with all grooming directed to one individual, and 1 meaning perfect equality; that is, all grooming partners receive an equal amount of grooming.

Dominance rank was measured by assigning normalized David's scores to each individual, based on the occurrence of fleeing upon aggression (de Vries et al., 2006; Stevens et al., 2007). David's scores use dyadic dominance proportions to calculate cardinal ranks for each individual based on the proportions of wins and losses in agonistic encounters. An individual loses an agonistic interaction when it flees as a response to the aggression. We then standardized the David's scores by dividing them by group size to test overall associations of rank with personality traits.

Behavioral experiments

We conducted eight group experiments (Figure 13.1), adapted from previous work on chimpanzees (Massen et al., 2013). In the predator experiments, two model predators were used: a taxidermied leopard in crouching position with bared teeth and a four-metre-long snake made out of a fire hose and clay, and painted to resemble a python. As leopards are natural predators of bonobos in the wild (D'Amour et al., 2006) and pythons are assumed to be natural predators of bonobos in the wild (Kano, 1992), these model predators should be representative of threats that wild bonobos would face. Both models were placed in sight but out of reach of the bonobos and the bonobos' behavior towards the models was scored. These behaviors included number of approaches, time spent in proximity and the number of displays against the mesh, including both poking with sticks and banging the mesh, behind which the predator was placed. The snake experiment was dropped from further analysis as the bonobos did not respond to the model python and consequently, there was little variation in response.

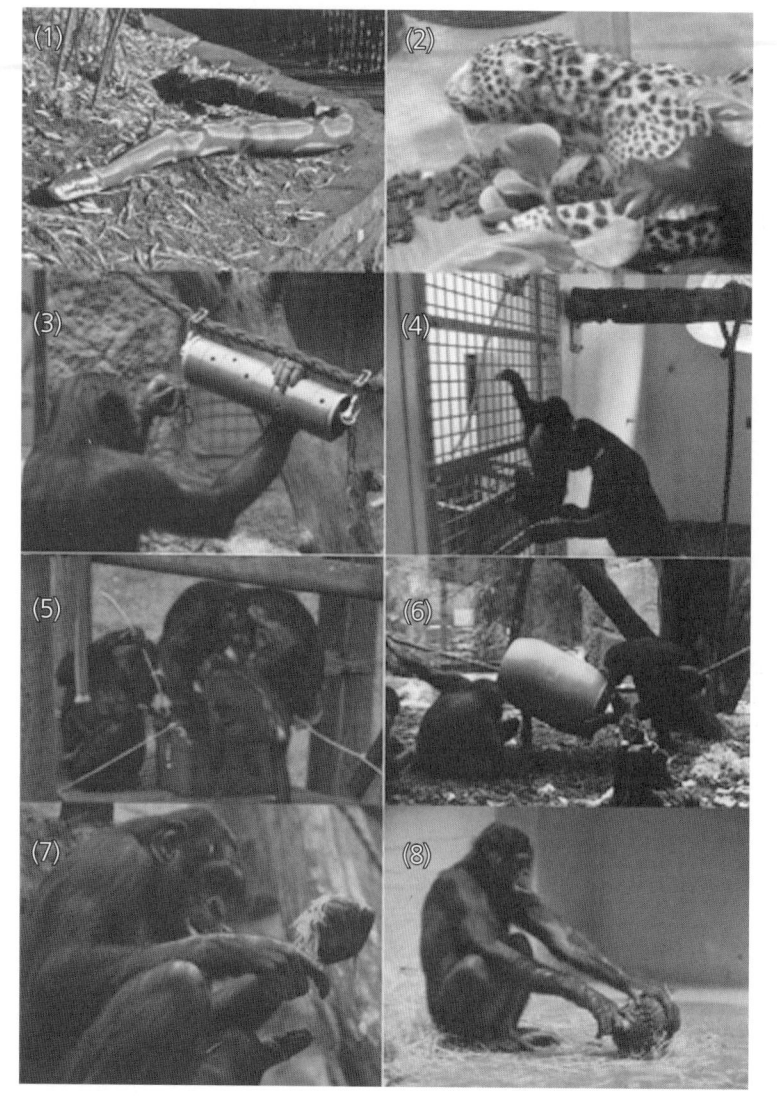

Figure 13.1 Behavioral variables from eight experiments were collected to assess their association with rated personality traits. Bonobos were tested in three different experimental contexts: Predator experiments included a fake python (1) and taxidermied leopard (2). Puzzle feeder experiments included turning tubes puzzle (3), reel and feed puzzle (4), barrel with mesh puzzle (5) and hanging barrel puzzle (6). Novel food experiments included blue-dyed pasta (7) and durian fruit (8).
(Des variables comportementales de huit expériences avaient été collectées pour évaluer leur association avec les caractères de personnalité évalués. Les bonobos étaient testés dans trois contextes expérimentaux: Expériences de prédateur comprenant un faux python (1) et un léopard taxidermié (2). Expériences d'alimentateur puzzle comprenant un puzzle de tubes tournants (3), puzzle de baril avec maillage (5) et le puzzle du baril suspendu (6). Expériences de nouvelle fourniture comprenant la pasta colorée bleue (7) et le fruit Durian (8).)

In two novel food experiments, bonobos were given durian fruit and pasta, the latter being dyed blue with an edible dye. During the durian fruit experiment, two fruits were placed in the enclosure a few metres apart to avoid their being monopolized by one individual. Two cameras were used to track both fruits at the same time. During the pasta experiment, 250 g of pasta was boiled, left to cool and then put in two piles in the enclosure, again, some metres apart to avoid their being monopolized by one individual. In these experiments we measured latency to approach the food closer than 2 m (in seconds) and whether the subject tasted the food item (0 = no, 1 = yes). When an individual did not approach the food item, it was given the maximum duration of 1800 seconds.

In each of four puzzle feeder experiments, bonobos were given a different foraging device. The first was a hanging barrel filled with seeds ('Hanging Barrel'). The second was a barrel filled with water and pieces of pear that sank to the bottom. This barrel was covered on top with a square mesh that was too small for an adult bonobo's hand to fit through ('Barrel with Mesh'). As sticks were needed to get to the pear in the 'Barrel with Mesh' condition, we measured the proportion of time during which subjects used sticks. The third puzzle feeder was a seed-filled hanging double-tube system that had to be rotated for it to release its contents ('Tubes'). The final puzzle feeder ('Reel and Feed') was a crate hanging from the outside of a mesh door next to the enclosure. To obtain the fruit and/or vegetables in this crate, bonobos had to tilt the crate using a rope that was hanging in the enclosure. For all puzzle feeders the following behavioral variables were measured: the time in seconds to approach the puzzle within 2 m; the proportion of time spent manipulating the puzzle; the proportion of time spent within 2 m of the puzzle without touching it; and the number of times the puzzle was approached.

All behavioral experiments were filmed (Canon Legria FS406, Japan) and recordings were coded using Observer Video-Pro (Noldus version XT 10, The Netherlands). Data recording started as soon as the group had access to the stimuli and was stopped after 30 minutes. All group members had access to the stimuli at the same time. In four out of six groups (FR, TW, ST, AP), group compositions differed between testing due to artificial fission–fusion of groups by zoo management. The order of the experiments was randomized and there were at least three days between experiments.

Analyses

Sex, age and rank differences in rated dimensions

To test whether personality dimensions were associated with sex, age or dominance rank, we used linear mixed models using the lme4 package (Bates et al., 2015) in R (http://www.r-project.org, version 3.1.0). In each model, sex, age and dominance rank were entered as fixed effects and group was entered as a random intercept to account for non-independence of observations within the same zoo. To test for the significance of the fixed effects we used F-tests with a Kenward–Roger correction for the number of degrees of freedom. The corrected F-tests were computed using the pbkrtest package (Halekoh and Hojsgaard, 2014). Inspection of residual plots did not reveal obvious deviations from homoscedasticity or normality.

Sex and age effects in behavioral variables

Independent-samples t-tests and Pearson correlations were used to test for sex and age differences, respectively, in the behavioral variables. We inspected qq-plots and plots of residuals against fitted values to check whether the assumptions of normally distributed and homogeneous error variance were fulfilled. These did not indicate severe violations of these two assumptions. As few personality studies apply a correction for multiple testing, and a Bonferroni correction in behavioral ecology may be overly conservative (Nakagawa, 2004), we reported results with and without Bonferroni correction.

Construct validity

These analyses were restricted to the 44 individuals (18 males and 26 females) for whom ratings and behavioral observations data were available. As the behavioral observations were made between 2011 and 2014, we only used ratings collected during the second wave in 2012. We used Spearman rank correlation coefficients to assess the association between the personality factors and the behavioral variables. Again, we reported results with and without Bonferroni correction.

Ethics

The study was approved by the Scientific Advisory Board of the Royal Zoological Society of Antwerp and the University of Antwerp (Belgium) and endorsed by the European Breeding Program for bonobos. All research complied with the ASAB (2012) guidelines.

Results

Relation of age, sex and rank with personality dimensions

Male bonobos scored significantly lower than females on Assertiveness, $b = -0.939$, s.e. = 0.185, $F(1,104) = 25.48$, $p < 0.001$, and Extraversion, $b = -0.612$, s.e. = 0.161, $F(1,103) = 14.28$, $p < 0.001$ (Figure 13.2). A significant association with age was found for Openness, $b = -0.052$, s.e. = 0.006, $F(1,106) = 70.54$, $p < 0.001$, and Extraversion, $b = -0.017$, s.e. = 0.007, $F(1,106) = 5.64$, $p = 0.019$, with older individuals scoring lower on both dimensions (Figure 13.3). The other dimensions did not differ significantly by sex, Conscientiousness: $b = 0.299$, s.e. = 0.191, $F(1,108) = 2.38$, $p = 0.126$; Openness: $b = -0.155$, s.e. = 0.140, $F(1,101) = 1.20$, $p = 0.274$; Agreeableness: $b = -0.241$, s.e. = 0.149, $F(1,98) = 2.60$, $p = 0.110$; Attentiveness: $b = -0.139$, s.e. = 0.149, $F(1,102) = 0.87$, $p = 0.355$, or by age, Assertiveness: $b = -0.006$, s.e. = 0.008, $F(1,106) = 0.64$, $p = 0.425$; Conscientiousness: $b = 0.008$, s.e. = 0.008, $F(1,109) = 0.97$, $p = 0.326$; Agreeableness: $b = 0.011$, s.e. = 0.007, $F(1,101) = 2.75$, $p = 0.100$; Attentiveness: $b = -0.006$, s.e. = 0.007, $F(1,105) = 0.74$, $p = 0.392$.

Sex and age differences in behavior

Tests for sex and age differences in the behavioral variables (Table 13.2) revealed that, after Bonferroni correction, females scored significantly

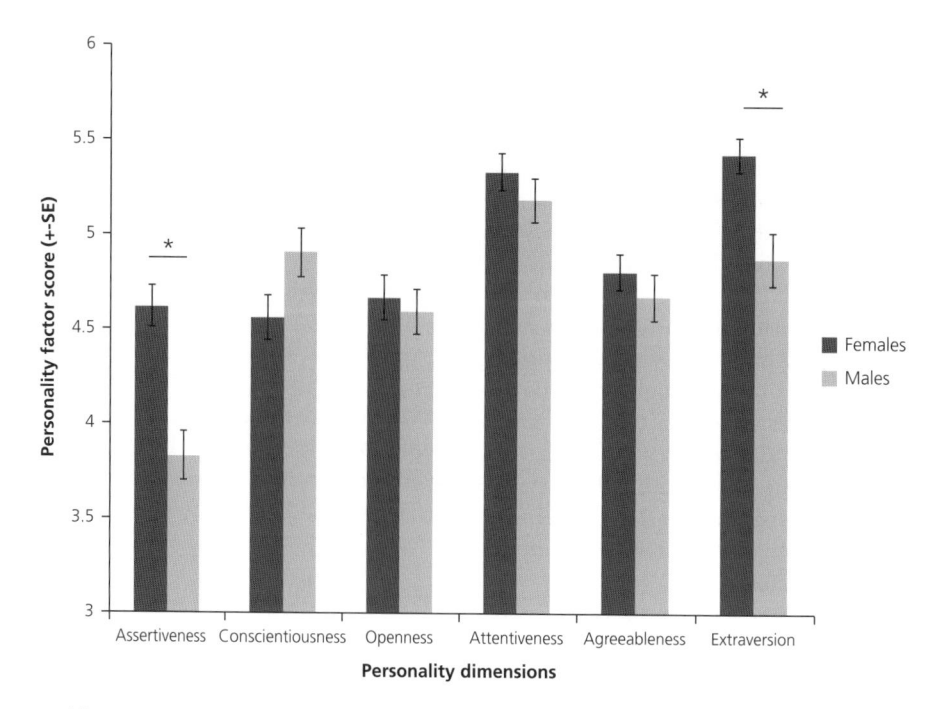

Figure 13.2 Sex difference in mean scores on personality dimensions (± standard error). Significant effects were found for Assertiveness and Extraversion with females scoring higher than males on both factors. The other dimensions showed no significant sex difference.
(Différences de sexe dans les marques moyennes sur les dimensions de personnalité (± erreur standard). Les effets importants sont dans l'Affirmation de soi et l'Extraversion où les femelles ont des marques plus élevées. Les autres dimensions n'ont pas montré de différences importantes entre les sexes.)

higher than males on number of neighbours sitting in proximity and time spent in proximity to the leopard and puzzles. Males scored significantly higher than females on levels of aggression received and frequency of fleeing upon aggression. Older individuals scored significantly lower on the number of times they approached the puzzle feeders.

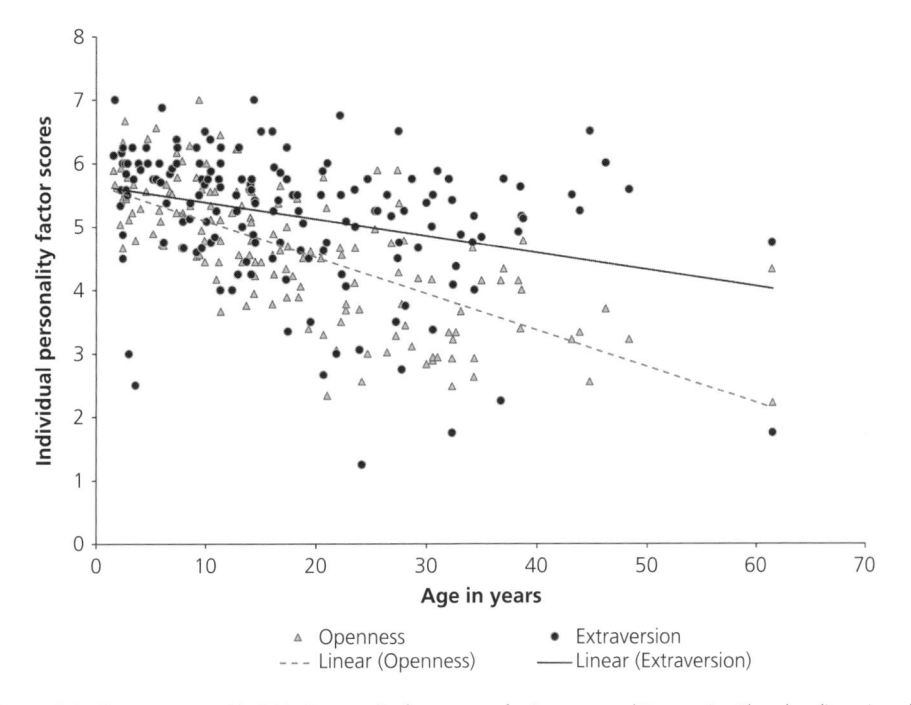

Figure 13.3 Correlation between age and individual personality factor scores for Openness and Extraversion. The other dimensions showed no significant sex difference.
(Corrélation entre l'âge et les marques du facteur de personnalité individu pour l'Ouverture et l'Extraversion. Les autres dimensions n'ont pas montré de différences importantes entre les sexes.)

Table 13.2 Sex and age differences for observed behavioral variables.
(Différences de sexe et d'âge dans les variables comportementales observées).

	Sex					95% CI		Age	
	b	se	t	df	p	Lower	Upper	r	p
Grooming Given	0.207	0.203	1.02	42	0.314	−0.203	0.617	−0.289	0.057
Grooming Received	**0.401**	**0.173**	**2.32**	**42**	**0.026**	**0.051**	**0.751**	−0.068	0.659
Grooming Density Given	0.025	0.075	0.33	42	0.746	−0.128	0.177	**−0.369**	**0.014**
Grooming Density Received	0.063	0.073	0.85	42	0.399	−0.086	0.211	0.006	0.970
Grooming Diversity index	0.020	0.072	0.28	42	0.784	−0.125	0.165	−0.168	0.276
N. of Neighbours	**0.193***	**0.051**	**3.78**	**42**	**< 0.001**	**0.090**	**0.296**	−0.098	0.526

(continued)

Table 13.2 (Continued)

	Sex							Age	
	b	se	t	df	p	95% CI		r	p
						Lower	Upper		
Be Approached	1.606	0.500	3.21	42	0.003	0.598	2.615	0.014	0.927
Approach	−0.373	0.684	−0.55	42	0.589	−1.754	1.008	**−0.444**	**0.003**
Play	−0.112	0.681	−0.16	42	0.870	−1.486	1.262	−0.259	0.090
Point Affinitive	−0.147	0.143	−1.03	20	0.317	−0.446	0.152	**−0.364**	**0.015**
Sexual interactions	0.300	0.199	1.51	42	0.139	−0.101	0.701	**−0.311**	**0.040**
Activity	−0.014	0.034	−0.41	42	0.686	−0.083	0.055	−0.295	0.052
Aggression Given	0.011	0.024	0.45	42	0.653	−0.038	0.060	−0.159	0.302
Aggression Received	**−0.277***	0.075	−3.68	42	**< 0.001**	−0.428	−0.125	**−0.327**	**0.031**
Flee upon aggression	**−0.670***	0.160	−4.19	29	**< 0.001**	−0.996	−0.343	−0.148	0.337
Autogroom	−0.126	0.397	−0.32	42	0.753	−0.926	0.675	−0.026	0.864
Nosewipe	−0.355	1.179	−0.30	42	0.765	−2.734	2.023	0.012	0.936
Scratch	1.666	0.964	1.73	42	0.091	−0.280	3.612	0.140	0.364
Leopard Latency	−61.681	75.260	−0.82	42	0.417	−213.562	90.201	**0.300**	**0.048**
Leopard Proximity	**18.413***	4.946	3.72	39	**< 0.001**	8.407	28.419	0.186	0.226
Leopard No. of Approaches	−0.001	0.001	−0.57	42	0.573	−0.002	0.001	**−0.375**	**0.012**
Leopard No. of Displays	0.001	0.001	1.25	42	0.219	0.000	0.001	−0.244	0.110
Novel Food Latency	−120.941	76.908	−1.57	26	0.128	−279.150	37.268	0.192	0.211
Novel Food Taste	0.038	0.074	0.52	42	0.607	−0.111	0.188	**−0.402**	**0.007**
Puzzle No. of Approaches	0.001	0.001	1.12	42	0.270	−0.001	0.002	**−0.553***	**0.001**
Puzzle Time Manipulated	**11.555**	3.821	3.02	30	**0.005**	3.750	19.361	−0.181	0.239
Puzzle Proximity	**8.956**	4.415	2.03	42	**0.049**	0.046	17.867	**−0.402**	**0.007**
Puzzle Latency	**−304.673**	115.380	−2.64	21	**0.015**	−544.905	−64.441	0.278	0.068
Puzzle Tool Use	4.894	8.090	0.60	42	0.548	−11.432	21.219	−0.078	0.613

Boldface indicates $p < 0.05$,
*Significant after Bonferroni correction ($p < 0.001$), $N = 44$

Personality dimensions and behavioral variables

Correlations between personality dimensions based on ratings and behaviors are shown in Table 13.3. The correlations that survived Bonferroni correction indicated that individuals that score higher on Assertiveness received more grooming, less aggression and showed fewer fleeing episodes upon aggression. Likewise, individuals scoring low on Conscientiousness showed higher frequencies of grooming interactions and individuals scoring high on Openness showed higher levels of approaching behavior towards puzzles and leopard, and were more inclined to taste novel food. Agreeableness, Extraversion and Attentiveness were not significantly correlated with the behavioral variables after correction for multiple tests.

Discussion

We tested whether bonobo personality dimensions derived from ratings demonstrated comparative validity by testing whether sex and age differences

Table 13.3 Spearman rank-order correlations between personality factors derived from HPQ and behavioral variables. (Les corrélations d'ordre-rang Spearman entre les facteurs de personnalité dérivés de HPQ et des variables comportementales).

		Assertiveness	Consientiousness	Openness	Agreeableness	Extraversion	Attentiveness
Observations	Grooming Given	0.21	−0.50*	0.26	−0.02	**0.31**	0.27
	Grooming Received	**0.43***	−0.50*	0.15	0.07	**0.31**	0.15
	Grooming Density Given	0.01	−0.33	**0.42**	0.01	0.19	0.17
	Grooming Density Received	0.13	−0.29	0.04	0.07	0.03	0.16
	Grooming Diversity index	0.01	−0.26	**0.34**	0.06	0.09	0.24
	No. of Neighbours	**0.31**	−0.30	0.12	0.19	0.26	0.13
	Be Approached	**0.32**	−0.04	−0.12	0.21	**0.31**	0.08
	Approach	−0.04	−0.23	**0.44**	−0.07	0.18	0.12
	Play	−0.04	−0.04	**0.32**	−0.14	0.26	−0.03
	Point Affinitive	−0.11	**−0.38**	**0.40**	−0.03	0.18	−0.01
	Sexual interactions	0.07	**−0.42**	**0.44**	0.02	**0.30**	−0.07
	Activity	0.02	**−0.39**	**0.51**	−0.13	0.19	0.12
	Aggression Given	−0.09	**−0.34**	0.16	−0.12	0.04	0.02
	Aggression Received	−0.39*	0.19	0.18	**−0.43**	−0.16	0.01
	Flee upon aggression	−0.57*	0.24	0.11	−0.20	−0.14	−0.06
	Dominance	0.29	**−0.34**	0.06	0.15	0.12	0.02
	Autogroom	−0.04	−0.13	−0.08	−0.27	−0.14	−0.17
	Nosewipe	−0.17	0.04	−0.04	0.14	−0.05	0.18
	Scratch	0.01	0.21	−0.12	0.09	−0.01	0.03
Predator test	Leopard Latency	−0.02	0.11	**−0.40**	0.06	**−0.30**	0.02
	Leopard Proximity	0.27	−0.14	**0.34**	0.26	0.13	0.12
	Leopard No. of Approaches	−0.10	−0.24	0.47*	−0.02	0.06	0.05
	Leopard No. of Displays	0.08	−0.29	0.29	−0.06	0.07	0.12
Novel food test	Novel Food Latency	−0.18	**0.37**	**−0.37**	−0.05	**−0.41**	−0.11
	Novel Food Taste	−0.16	**−0.38**	0.50*	0.03	0.15	−0.01
Puzzle feeder tests	Puzzle No. of Approaches	−0.05	**−0.37**	0.54*	−0.06	**0.40**	−0.03
	Puzzle Time Manipulated	**0.34**	−0.29	**0.30**	0.13	**0.38**	0.19
	Puzzle Proximity	−0.06	−0.27	**0.31**	−0.06	**0.31**	−0.06
	Puzzle Latency	−0.24	**0.45**	**−0.38**	−0.02	**−0.41**	−0.10
	Puzzle Tool Use	0.20	−0.15	0.27	0.19	**0.31**	0.15

Correlations in boldface are significant ($p < 0.05$).
*Significant after Bonferroni correction ($p < 0.001$), $N = 44$

were consistent with what we would expect based on findings related to these differences in chimpanzees and humans and with differences between bonobos and chimpanzees in social structure, behavior and life history. We also tested whether these dimensions demonstrated construct validity by examining their associations with behavioral variables derived from naturalistic observations and behavioral tests. The Assertiveness, Openness and Extraversion dimensions showed good comparative and construct validity. Conscientiousness showed evidence for construct validity but not comparative validity. Agreeableness and Attentiveness did not show evidence for comparative or construct validity in this study.

Previous studies that used different questionnaires and/or behavioral measures also reported modest to strong validity coefficients in apes (see Murray 2011). Many of the associations found in this study did not reach the $p < 0.001$ required by the Bonferroni correction, which is not surprising given the relatively small sample size in this study compared to similar studies of humans (Paunonen, 2003). Although a larger sample size would have increased our ability to detect associations between personality dimensions and behaviors, the sample size in this study was already relatively large for a study done on great apes and increasing it further was not possible due to time constraints. Moreover, it should be borne in mind that Bonferroni correction might be considered overly conservative, especially given the modest effect sizes common in the field of behavioral ecology research (Nakagawa, 2004). Despite the correlations not being significant after correction, many of the observed correlations were consistent with our predictions.

Assertiveness was associated with variables related to social attractiveness in bonobos. This is consistent with studies that show that high-ranking bonobos are groomed more often (Franz, 1999; Vervaecke et al., 2000b) and are less likely to be the targets of aggression (Paoli and Palagi, 2009). Studies in other non-human primates found similar results for similar personality dimensions. For example, in chimpanzees (Murray, 2011) and rhesus macaques (Capitanio, 1999), a personality dimension labelled Confidence was associated with receiving more grooming. Moreover, in gorillas (Kuhar et al., 2006) and chimpanzees (Pederson et al., 2005), personality dimensions labelled Dominance were negatively related to showing submissive behavior and positively related to receiving submissive behavior.

In line with our expectations, differences in personality between male and female bonobos were largely the opposite to what has been found in chimpanzees (King et al., 2008). For one thing, females scored higher on Assertiveness than males. This is consistent with the bonobo's social system where females occupy the higher ranks in the dominance hierarchy (Furuichi, 2011; Vervaecke et al., 2000a). It is also consistent with findings in hyena, another species in which the females are dominant, where females scored higher on Assertiveness than males (Gosling, 1998). This sex difference in Assertiveness was further supported by our finding of similar sex differences in behaviors: females were groomed more often, received less aggression and displayed submissive behaviors less often than did males.

Openness was associated with variables that reflect curiosity and play, two facets of Openness that have been identified across many species (Gosling and John, 1999). Further supporting the comparative validity of the dimension, as in humans and chimpanzees (Costa et al., 2001; King et al., 2008; McCrae et al., 2000; Weiss and King, 2015), Openness and behavioral variables related to Openness—for example, exploring puzzles, novel foods and leopards—were lower in older individuals. As Openness showed the strongest positive associations with behaviors related to play and affinitive and sexual interactions, the decline of Openness with age suggests that there is no support in this study for the hypothesis that the retention of juvenile levels of these traits in adult bonobos may be causing less pronounced age-related effects on personality. Similar age-related effects were found for Extraversion, a dimension that was positively associated with variables related to social integration, which was in line with our prediction that age effects are stable across closely related primate species. However, none of the associations reached the $p < 0.001$ significance level.

In contrast to the lack of sex differences in Extraversion among chimpanzees (King et al., 2008; Weiss and King, 2015), female bonobos were significantly more extraverted than males. However, this could be due to the fact that the traits associated with Extraversion differed between bonobos and chimpanzees. Specifically, unlike chimpanzees, for whom Extraversion is defined by traits related to gregariousness, activity and affect (King et al., 2008; Weiss and King, 2015; Weiss et al., 2007, 2009, 2015), the bonobo variant of Extraversion is defined by only a few traits related to social integration and affect (Weiss et al., 2015). Nonetheless, the sex differences in bonobos are consistent with the strength of social relationships in male and female bonobos, where male social bonds are relatively weak while relationships among females and between females and their adult sons are strong (Hohmann and Fruth, 2002; Kano, 1992; Parish, 1996; Stevens et al., 2015; Stumpf, 2007). One previous study reported that bonobo males behaved in a more extraverted manner than females (Schroepfer-Walker et al., 2015); however, the behavior recorded in this study was directed towards a human experimenter and not conspecifics. Furthermore, behavior towards caretakers or experimenters was not included in our study. As chimpanzees who are more extraverted attend more often to other chimpanzees than to humans (Pederson et al., 2005), and chimpanzees that had high levels of early human exposure scored lower on Extraversion (Freeman et al., 2016), these seemingly contradictory findings may, in fact, be consistent with the present findings. Further work that contrasts the association between Extraversion in bonobo- and human-directed interactions is needed to test this possibility.

For Conscientiousness, significant negative associations were found for frequencies of grooming given and received, for which the interpretation is more difficult. Individuals scoring high on Conscientiousness are more gentle and predictable and less manipulative, impulsive and bullying. As grooming exchanges serve an important function in bonobo society for the formation and maintenance of social bonds (Sakamaki, 2013; Surbeck and Hohmann, 2015; Stevens et al., 2015; Vervaecke et al., 2000b), Conscientiousness may be negatively related to being more strategic. In humans and chimpanzees, age-related increases in Conscientiousness are reported (King et al., 2008; McCrae et al., 2000). In bonobos, age differences were seemingly absent. However, this may reflect the fact that Conscientiousness in bonobos comprises only one aspect of that dimension in chimpanzees, with the other aspect being relegated to the bonobo Attentiveness dimension (Weiss et al., 2015). As such, these findings do not necessarily imply low levels of comparative validity.

For Agreeableness and Attentiveness there was little support for validity. Both factors showed few associations with behavioral variables, age or sex. In humans and chimpanzees age-related increases have been documented for Agreeableness (King et al., 2008; McCrae et al., 2000), which were not present in bonobos. Also, after Bonferroni correction, Agreeableness was not significantly correlated with coded behaviors. Because an Attentiveness factor was not identified in chimpanzees or humans, we were unable to make a priori hypotheses about age and sex effects for this dimension. Based on the item loadings, this dimension may reflect how vigilant individuals are to social and non-social cues in their environment. A similar dimension was related to focus during cognitive tasks in brown capuchin monkeys (Morton et al., 2013). Thus, the absence of significant correlations here and elsewhere may reflect shortcomings of the ethogram used in this study, and further development of the ethogram could reveal significant correlations between these dimensions and behaviors.

Developing valid measures of great ape personality is an important step in learning more about the evolutionary bases and functions of individual differences in personality. The present findings suggest that these dimensions derived from ratings are a promising beginning for studies focusing on how personality profiles influence fitness, or how personality variation is related to proximate mechanisms like candidate gene variation, not only in bonobos but also in their close relatives, the chimpanzees and humans (Staes et al. 2014; Staes et al., 2015, Staes et al., 2016).

Acknowledgements

We are grateful to the director and staff of the Royal Zoological Society of Antwerp (RZSA) for their support of this study and to the staff of the Centre for Research and Conservation (CRC) for their interesting suggestions and discussions. The CRC receives structural support from the Flemish Government. We thank all students involved in data collection: Adriana Solis (University of Groningen), Annemieke Podt, Sanne Roelofs, Wiebe Rinsma, Linda Jaasma, Marloes Borger and Martina Wildenburg (University of Utrecht) for their help in data collection. Special thanks go to all the zoological institutes who hosted us during behavioral data collection: Planckendael (Mechelen, Belgium), Apenheul (Apeldoorn, the Netherlands), Twycross Zoo World Primate Center (Twycross, United Kingdom),

Wuppertal Zoo (Wuppertal; Germany), Frankfurt Zoo (Frankfurt, Germany) and Wilhelma zoological and botanical garden (Stuttgart, Germany). We also thank all the zoological and research institutes who provided us with questionnaire data. Our thanks also to Brian Hare and an anonymous reviewer for comments on earlier drafts of this chapter.

References

Adams, M.J., Majolo, B., Ostner, J., Schulke, O., De Marco, A., Thierry, B., Engelhardt, A., Widdig, A., Gerald, M.S., and Weiss, A. (2015). Personality structure and social style in macaques. *Journal of Personality and Social Psychology*, 109, 338–53.

Andersson, A., Laikre, L., and Bergvall, U.A. (2014). Two shades of boldness: novel object and anti-predator behavior reflect different personality dimensions in domestic rabbits. *Journal of Ethology*, 32, 123–36.

Baker, K.R., Lea, E.G., and Melfi, V.A. (2015). Comparative personality assessment of three captive primate species: *Macaca nigra, Macaca sylvanus*, and *Saimiri sciureus*. *International Journal of Primatology*, 36, 625–46.

Bates, D., Maechler, M., Bolker, B., and Walker, S. (2015). Fitting linear mixed-effects models using lme4. *Journal of Statistical Software*, 67, 1–48.

Boesch, C. and Boesch-Acherman, H. (2000). *The Chimpanzees of the Taï Forest: Behavioural Ecology and Evolution*. Cambridge: Cambridge University Press.

Boesch, C., Hohmann, G., and Marchant, L.F. (2002). *Behavioural Diversity in Chimpanzees and Bonobos*. Cambridge, Cambridge University Press.

Buirski, P., Plutchik, R., and Kellerman, H. (1978). Sex differences, dominance, and personality in the chimpanzee. *Animal Behaviour*, 26, 123–9.

Capitanio, J.P. (1999). Personality dimensions in adult male rhesus macaques: prediction of behaviors across time and situation. *American Journal of Primatology*, 47, 299–320.

Carere, C. and Eens, M. (2005). Unravelling animal personalities: how and why individuals consistently differ. *Behaviour*, 142, 1149–57.

Carter, A.J., Marshall, H.H., Heinsohn, R., and Cowlishaw, G. (2012). Evaluating animal personalities: do observer assessments and experimental tests measure the same thing? *Behavioral Ecology and Sociobiology*, 66, 153–60.

Costa, P.T., Terracciano, A., and McCrae, R.R. (2001). Gender differences in personality traits across cultures: robust and surprising findings. *Journal of Personality and Social Psychology*, 81, 322–31.

D'amour, D.E., Hohmann, G., and Fruth, B. (2006). Evidence of leopard predation on bonobos (*Pan paniscus*). *Folia Primatologica*, 77, 212–17.

De Vries, H., Stevens, J.M.G., and Vervaecke, H. (2006). Measuring and testing the steepness of dominance hierarchies. *Animal Behaviour*, 71, 585–92.

Di Bitetti, M.S. (2000). The distribution of grooming among female primates: testing hypotheses with the Shannon–Wiener index. *Behaviour*, 137, 1517–40.

Dutton, D.M. (2008). Subjective assessment of chimpanzee (*Pan troglodytes*) personality: reliability and stability of trait ratings. *Primates*, 49, 253–9.

Franz, C. (1999). Allogrooming behavior and grooming site preferences in captive bonobos (*Pan paniscus*): association with female dominance. *International Journal of Primatology*, 20, 525–46.

Freeman, H.D. and Gosling, S.D. (2010). Personality in nonhuman primates: a review and evaluation of past research. *American Journal of Primatology*, 72, 653–71.

Freeman, H.D., Weiss, A., and Ross, S.R. (2016). Atypical early histories predict lower extraversion in captive chimpanzees. *Developmental Psychobiology*, 58, 519–27.

Furuichi, T. (2011). Female contributions to the peaceful nature of bonobo society. *Evolutionary Anthropology*, 20, 131–42.

Garai, C., Weiss, A., Arnaud, C., and Furuichi, T. (2016). Personality in wild bonobos (*Pan paniscus*). *American Journal of Primatology*, 78, 1178–89.

Godin, J.J. and Dugatkin, L.A. (1996). Female mating preference for bold males in the guppy, *Poecilia reticulata*. *Proceedings of the National Academy of Sciences*, 93, 10262–7.

Goldberg, L.R. (1990). An alternative 'description of personality': the big-five factor structure. *Journal of Personality and Social Psychology*, 59, 1216–29.

Goodall, J. (1986). *The Chimpanzees of Gombe: Patterns of Behavior*. Cambridge, MA: Harvard University Press.

Gosling, S.D. (1998). Personality dimensions in spotted hyenas (*Crocuta crocuta*). *Journal of Comparative Psychology*, 112, 107–18.

Gosling, S.D. (2001). From mice to men: what can we learn about personality from animal research? *Psychological Bulletin*, 127, 45–86.

Gosling, S.D. and John, O.P. (1999). Personality dimensions in nonhuman animals: a cross-species review. *Current Directions in Psychological Science*, 8, 69–75.

Gosling, S.D. and Vazire, S. (2002). Are we barking up the right tree? Evaluating a comparative approach to personality. *Journal of Research in Personality*, 36, 607–14.

Hare, B., Wobber, V., Wrangham, R. (2012). The self-domestication hypothesis: evolution of bonobo psychology is due to selection against aggression. *Animal Behaviour*, 83, 573–85.

Halekoh, U. and Hojsgaard, S. (2014). A Kenward–Roger approximation and parametric bootstrap methods for tests in linear mixed models—the R Package pbkrtest. *Journal of Statistical Software*, 59, 1–32.

Herrmann, E., Hare, B., Cissewski, J., and Tomasello, M. (2011). A comparison of temperament in nonhuman apes and human infants. *Developmental Science*, 14, 1393–405.

Hohmann, G. and Fruth, B. (2002). Dynamics in social organization of bonobos (*Pan paniscus*). In: Boesch, C., Hohmann, G., and Marchant, L.F. (eds). *Behavioral Diversity in Chimpanzees and Bonobos*. Cambridge: Cambridge University Press. 138–150.

Kano, T. (1992). *The Last Ape: Pygmy Chimpanzee Behavior and Ecology*. Stanford, CA: Stanford University Press.

King, A.J., Furtbauer, I., Mamuneas, D., James, C., and Manica, A. (2013). Sex differences and temporal consistency in stickleback fish boldness. *PLoS One*, 8, e81116.

King, J. E., Weiss, A., and Sisco, M.M. (2008). Aping humans: age and sex effects in chimpanzee (*Pan troglodytes*) and human (*Homo sapiens*) personality. *Journal of Comparative Psychology*, 122, 418–27.

Konečná, M., Lhota, S., Weiss, A., Urbanek, T., Adamova, T., and Pluhacek, J. (2008). Personality in free-ranging Hanuman langur (*Semnopithecus entellus*) males: subjective ratings and recorded behavior. *Journal of Comparative Psychology*, 122, 379–89.

Kuhar, C.W., Stoinski, T.S., Lukas, K.E., and Maple, T.L. (2006). Gorilla Behavior Index revisited: age, housing and behavior. *Applied Animal Behaviour Science*, 96, 315–26.

Martin, P.R. and Bateson, P.P.G. (1993). *Measuring Behaviour: An Introductory Guide*. Cambridge: Cambridge University Press.

Massen, J., Antonides, A., Arnold, A.-M., Bionda, T., and Koski, S.E. (2013). A behavioral view on chimpanzee personality exploration tendency, persistence, boldness and tool-orientation measured with group experiments. *American Journal of Primatology*, 75, 947–58.

McCrae, R.R., Costa, P.T., Hrebickova, M., Ostendorf, F., Angleitner, A., Avia, M.D., Sanz, J., and Sanchez-Bernardos, M.L. (2000). Nature over nurture: temperament, personality, and life span development. *Journal of Personality and Social Psychology*, 78, 173–86.

McCrae, R.R. and Costa, P.T. (1999). A five-factor theory of personality. In:Pervin, L. and John, O.P. (eds). *Handbook of Personality*. New York, NY: Guilford Press. 159–181.

McCrae, R.R., Costa, P.T., Pedroso De Lima, M., Simoes, A., Ostendorf, F., Angleitner, A., Caprara, G.V., Barbaranelli, C., Marusic, I., Bratko, D., and Chae, J. (1999). Age differences in personality across the adult life span: parallels in five cultures. *Developmental Psychology*, 35, 466–77.

Mehta, P.H. and Gosling, S.D. (2008). Bridging human and animal research: a comparative approach to studies of personality and health. *Brain, Behavior, and Immunity*, 22, 651–61.

Morton, B.F., Lee, P.C., Buchanan-Smith, H.M., Brosnan, S.F., Thierry, B., Paukner, A., De Waal, F.B.M., Widness, J., Essler, J.L., and Weiss, A. (2013). Personality structure in brown capuchin monkeys (*Sapajus apella*): Comparisons with chimpanzees (*Pan troglodytes*), orangutans (*Pongo* spp.), and rhesus macaques (*Macaca mulatta*). *Journal of Comparative Psychology*, 127, 282–98.

Murray, L. (2011). Predicting primate behavior from personality ratings. In: Weiss, A., King, J.E., and Murray, L. (eds). *Personality and Temperament in Nonhuman Primates*. New York, NY: Springer. 129–167.

Nakagawa, S. (2004). A farewell to Bonferroni: the problems of low statistical power and publication bias. *Behavioral Ecology*, 15, 1044–5.

Paoli, T. and Palagi, E. (2009). What does agonistic dominance imply in bonobos? In: Furuichi, T. and Thompson, J. (eds). *The Bonobos: Behavior, Ecology and Conservation*. Springer. 39–54.

Parish, A.R. (1996). Female relationships in bonobos (*Pan paniscus*)—Evidence for bonding, cooperation, and female dominance in a male-philopatric species. *Human Nature*, 7, 61–9.

Paunonen, S.V. (2003). Big five factors of personality and replicated predictions of behavior. *Journal of Personality and Social Psychology*, 84, 411–24.

Pederson, A.K., King, J.E., and Landau, V.I. (2005). Chimpanzee (*Pan troglodytes*) personality predicts behavior. *Journal of Research in Personality*, 39, 534–49.

Prado-Martinez, J., Sudmant, P.H., Kidd, J.M., Li, H., Kelley, J.L., Lorente-Galdos, B., Veeramah, K.R., Woerner, A.E., O'connor, T.D., Santpere, G., Cagan, A., Theunert, C., Casals, F., Laayouni, H., Munch, K., Hobolth, A., Halager, A.E., Malig, M., Hernandez-Rodriguez, J., Hernando-Herraez, I., Prufer, K., Pybus, M., Johnstone, L., Lachmann, M., Alkan, C., Twigg, D., Petit, N., Baker, C., Hormozdiari, F., Fernandez-Callejo, M., Dabad, M., Wilson, M.L., Stevison, L., Camprubi, C., Carvalho, T., Ruiz-Herrera, A., Vives, L., Mele, M., Abello, T., Kondova, I., Bontrop, R.E., Pusey, A., Lankester, F., Kiyang, J.A., Bergl, R.A., Lonsdorf, E., Myers, S., Ventura, M., Gagneux, P., Comas, D., Siegismund, H., Blanc, J., Agueda-Calpena, L., Gut, M., Fulton, L., Tishkoff, S.A., Mullikin, J.C., Wilson, R.K., Gut, I.G., Gonder, M.K., Ryder, O.A., Hahn, B.H., Navarro, A., Akey, J.M., Bertranpetit, J., Reich, D., Mailund, T., Schierup, M.H., Hvilsom, C., Andres, A.M., Wall, J.D., Bustamante, C.D., Hammer, M.F., Eichler, E.E., and Marques-Bonet, T. (2013). Great ape genetic diversity and population history. *Nature*, 499, 471–5.

Réale, D., Reader, S.M., Sol, D., Mcdougall, P.T., and Dingemanse, N.J. (2007). Integrating animal temperament within ecology and evolution. *Biological Reviews of the Cambridge Philosophical Society*, 82, 291–318.

Sakamaki, T. (2013). Social grooming among wild bonobos (*Pan paniscus*) at Wamba in the Luo Scientific Reserve, DR Congo, with special reference to the formation of grooming gatherings. *Primates*, 54, 349–59.

Schroepfer-Walker, K., Hare, B., and Wobber, V. (2015). Experimental evidence that grooming and play are social currency in bonobos and chimpanzees. *Behaviour*, 152, 545–62.

Staes, N., Koski, S.E., Helsen, P., Fransen, E., Eens, M., and Stevens, J.M.G. (2015). Chimpanzee sociability is associated with vasopressin (Avpr1a) but not oxytocin receptor gene (OXTR) variation. *Hormones and Behavior*, 75, 84–90.

Staes, N., Stevens, J.M.G., Helsen, P., Hillyer, M., Korody, M., and Eens, M. (2014). Oxytocin and vasopressin receptor gene variation as a proximate base for inter- and intraspecific behavioral differences in bonobos and chimpanzees. *PLoS One*, 9, e113364.

Staes, N. The role of vasopressin (*Avpr1a*) and oxytocin (*OXTR*) receptor gene variation as a proximate base for inter- and intraspecific differences in personality in bonobos (*Pan paniscus*) and chimpanzees (*Pan troglodytes*). PhD thesis, Universiteit Antwerpen (2016).

Staes, N., Weiss, A., Helsen, P., Korody, M., Eens, M., and Stevens, J.M.G. (2016). Bonobo personality traits are heritable and associated with vasopressin receptor gene 1a variation. *Scientific Reports*, 6, 38193.

Stevens, J.M.G., De Groot, E., and Staes, N. (2015). Relationship quality in captive bonobo groups. *Behaviour*, 152, 259–83.

Stevens, J.M.G., Vervaecke, H., Vries, H., and Elsacker, L. (2007). Sex differences in the steepness of dominance hierarchies in captive bonobo groups. *International Journal of Primatology*, 28, 1417–30.

Stumpf, R.M. (2007). Chimpanzees and bonobos: diversity within and between species. In: Campbell, C.J., Fuentes, A., Mackinnon, K.C., Panger, M., and Bearder, S.K. (eds). *Primates in Perspective*. Oxford: Oxford University Press. 321–344.

Surbeck, M. and Hohmann, G. (2013). Intersexual dominance relationships and the influence of leverage on the outcome of conflicts in wild bonobos (*Pan paniscus*). *Behavioral Ecology and Sociobiology*, 67, 1767–80.

Surbeck, M. and Hohmann, G. (2015). Social preferences influence the short-term exchange of social grooming among male bonobos. *Animal Cognition*, 18, 573–9.

Uher, J. and Asendorpf, J.B. (2008). Personality assessment in the Great Apes: Comparing ecologically valid behavior measures, behavior ratings, and adjective ratings. *Journal of Research in Personality*, 42, 821–38.

Van Oers, K. (2005). Context dependence of personalities: risk-taking behavior in a social and a nonsocial situation. *Behavioral Ecology*, 16, 716–23.

Vazire, S., Gosling, S.D., Dickey, A.S., and Shapiro, S.J. (2007). Measuring personality in nonhuman animals. In: Robins, R.W., Fraley, R.C., and Krueger, R. (eds). *Handbook of Research Methods in Personality Psychology*. New York, NY: Guilford Press. 190–206.

Vervaecke, H., De Vries, G.J., and Van Elsacker, L. (2000a). Dominance and its behavioral measures in a captive group of bonobos (*Pan paniscus*). *International Journal of Primatology*, 21, 47–68.

Vervaecke, H., De Vries, G.J., and Van Elsacker, L. (2000b). The pivotal role of rank in grooming and support behavior in a captive group of bonobos (*Pan paniscus*). *Behaviour*, 137, 1463–85.

Weiss, A., Adams, M.J., Widdig, A., and Gerald, M.S. (2011). Rhesus macaques (*Macaca mulatta*) as living fossils of hominoid personality and subjective well-being. *Journal of Comparative Psychology*, 125, 72–83.

Weiss, A., Inoue-Murayama, M., Hong, K–W., Inoue, E., Udono, T., Ochiai, T., Matsuzawa, T., Hirata, S., and King, J.E. (2009). Assessing chimpanzee personality and subjective well-being in Japan. *American Journal of Primatology*, 71, 283–92.

Weiss, A., Inoue-Murayama, M., King, J.E., Adams, M.J., and Matsuzawa, T. (2012). All too human? Chimpanzee and orang-utan personalities are not anthropomorphic projections. *Animal Behaviour*, 83, 1355–65.

Weiss, A. and King, J.E. (2015). Great ape origins of personality maturation and sex differences: a study of orangutans and chimpanzees. *Journal of Personality and Social Psychology*, 108, 648–64.

Weiss, A., King, J.E., and Hopkins, W.D. (2007). A cross-setting study of chimpanzee (*Pan troglodytes*) personality structure and development: zoological parks and Yerkes National Primate Research Center. *American Journal of Primatology*, 69, 1264–77.

Weiss, A., Staes, N., Pereboom, J.J.M., Inoue-Murayama, M., Stevens, J.M.G., and Eens, M. (2015). Personality in bonobos. *Psychological Science*, 26, 1439–30.

Wobber, V., Wrangham, R., Hare, B. (2010). Bonobos exhibit delayed development of social behavior and cognition relative to chimpanzees. *Current Biology*, 20, 226–30.

CHAPTER 14

Social cognition and brain organization in chimpanzees (*Pan troglodytes*) and bonobos (*Pan paniscus*)

William D. Hopkins, Cheryl D. Stimpson and Chet C. Sherwood

Abstract Bonobos and chimpanzees are two closely relates species of the genus *Pan*, yet they exhibit marked differences in anatomy, behavior and cognition. For this reason, comparative studies on social behavior, cognition and brain organization between these two species provide important insights into evolutionary models of human origins. This chapter summarizes studies on socio-communicative competencies and social cognition in chimpanzees and bonobos from the authors' laboratory in comparison to previous reports. Additionally, recent data on species differences and similarities in brain organization in grey matter volume and distribution is presented. Some preliminary findings on microstructural brain organization such as neuropil space and cellular distribution in key neurotransmitters and neuropeptides involved in social behavior and cognition is presented. Though these studies are in their infancy, the findings point to potentially important differences in brain organization that may underlie bonobo and chimpanzees' differences in social behavior, communication and cognition.

Les bonobos et les chimpanzés sont deux espèces du genus Pan prochement liées, néanmoins ils montrent des différences anatomiques, comportementales et cognitives marquées. Pour cette raison, les études comparatives sur le comportement social, la cognition et l'organisation corticale entre ces deux espèces fournissent des idées sur les modèles évolutionnaires des origines humaines. Dans ce chapitre, nous résumons des études sur les compétences socio-communicatives et la cognition sociale chez les chimpanzés et les bonobos de notre laboratoire en comparaison avec des rapports précédents. En plus, nous présentons des données récentes sur les différences et similarités d'organisation corticale du volume et distribution de la matière grise entre espèces. Nous présentons plus de résultats préliminaires sur l'organisation corticale microstructurale comme l'espace neuropile et la division cellulaire dans des neurotransmetteurs clés et les neuropeptides impliqués dans le comportement social et la cognition. Bien que ces études sont dans leur enfance, les résultats montrent des différences d'organisation corticale importantes qui sont à la base des différences de comportement social, la communication et la cognition entre les bonobos et les chimpanzés.

The human brain is roughly three-times larger than that of our closest living relatives, the great apes, being considerably larger then would be expected for a primate of human body size (Sherwood et al., 2012). A fundamental interest in neuroscience, psychology and anthropology has been determining what factors selected for the increased brain size in modern humans. A number of different theoretical accounts have been proposed to explain the enlargement of the brain in human evolution which can be broadly divided into those that emphasize social and non-social mechanisms. For instance, some have postulated that non-social factors such as such as tool use and tool-making, throwing, problem-solving skills, foraging and home-range sizes, to name a few (Byrne, 1995; Deaner et al., 2007; Gibson and Ingold, 1993; Tallerman and Gibson, 2012). In contrast, others have suggested that variation in social and cultural factors were critical such as group size and complexity, social learning, language and communication and different aspects of social cognition (Byrne, 1995; Byrne and Corp, 2004; Dunbar, 1903; Reader and Laland, 2002). Of specific interest to this chapter is the comparison in cognitive performance and brain structure between the two species of *Pan*, chimpanzees (*Pan troglodytes*) and bonobos (*Pan paniscus*). Comparative studies in these two species are ideal because they are genetically quite similar and exhibit similar brain sizes (Hopkins et al., 2009; Rilling et al., 2012) but they differ in a number of behavioral traits that lend themselves to tests of evolutionary models of human socio-cognitive specializations and the brain.

Genetic data indicate that bonobos and chimpanzees are the two species that are the closest living primates to humans (Consortium, 2005; Prufer et al., 2012). Modern humans shared a last common ancestor with the *Pan* lineage approximately 6 to 8 million years ago. The *Pan* genus split into the two species approximately 2 million years ago that currently reside on the continent of Africa. Chimpanzees (*Pan troglodytes*) are found throughout most of eastern and western-central Africa and inhabit a variety of different habitats such as dense rainforest and open savannah. In contrast, bonobos (*Pan paniscus*) are found only in a small region within the Democratic Republic of Congo and reside primarily in rainforest. Despite the fact that chimpanzees and bonobos are closely related, evidence to-date suggest they exhibit marked differences in anatomy, behavior and cognition (Tuttle, 2014).

Of specific interest to this chapter are the claims of differences in socio-communicative and cognitive abilities that have been reported between bonobos and chimpanzees. In terms of communication, bonobos have been reported to have a more sophisticated and flexible gestural (Pollick and de Waal, 2007) and vocal communication system than chimpanzees (Bermejo and Omedes, 1999; de Waal, 1988). Acoustically, bonobo vocalizations are more high-pitched and tonal than chimpanzees and some have attributed this to factors related to feeding ecology (Hohman and Fruth, 1994; Malenky and Stiles, 1991). The differences in vocal and gestural communication have led some to suggest that this accounts for bonobos' greater ease in acquiring and using artificial and augmentative communication system for interspecies communication, such as the remarkable abilities reported for Kanzi and other bonobos (Brakke and Savage-Rumbaugh, 1995; Brakke and Savage-Rumbaugh, 1996; Savage-Rumbaugh, 1984; Savage-Rumbaugh and Lewin, 1994). Comparative psychology studies have reported that bonobos perform significantly better than chimpanzees on measures of social cognition, such as gaze following and comprehension of communicative signalling, whereas chimpanzees perform better on physical cognition tasks, like tool use (Herrmann et al., 2010a) and spatial memory (Rosati and Hare, 2012). It is also of note that, although both chimpanzees and bonobos are capable of using tools in captivity, evidence of the variety and diversity in forms of tool use that are widespread in wild chimpanzees are essentially absent in wild bonobos (Furuichi et al., 2014; Gruber et al., 2010).

Beyond communication and tool use, it has also been reported that bonobos are more prosocial and tolerant than chimpanzees as manifest by higher levels of food sharing and other measures of cooperative behavior such as token or tool exchange, though this remains controversial (de Waal, 1992; Hare et al., 2007; Heilbronner et al., 2008; Jaeggi et al., in press; 2010; Jaeggi and Van Schaik, 2011; Wobber et al., 2010). For example, Hare and Kwetuenda (2010) reported that bonobos tolerate sharing of food, even among unfamiliar individuals. This is an intriguing finding but the apes (at least those seen in the video) were quite young and it is unclear whether they opened a door with the motivation to play with the other individual or for the purposes of sharing the food. With respect to prosocial behaviors, it is worth noting that a recent study on polymorphic variation in the vasopressin receptor *AVPR1A* gene found that bonobos showed no indel deletions in the RS3 portion of this gene while approximately two-thirds of chimpanzees show this deletion (Donaldson et al., 2008; Staes et al., 2014). The neuropeptides vasopressin and oxytocin play a prominent role in a variety of social behaviors in mammals (Goodson and Bass, 2001) and there are some reports that polymorphisms in the *AVPR1A* gene explain some individual differences in social behavior and cognition among captive chimpanzees (Anestis et al., 2014; Hopkins et al., 2012; Hopkins et al., 2014a; Latzman et al., 2014; Staes et al., 2015).

As a means of providing new insights into differences between chimpanzees and bonobos in social cognition, one aim of this chapter is to present previously unpublished data. In particular, data are presented on chimpanzees and bonobos in socio-communicative abilities for several tasks assessing receptive joint attention as well as physical and social cognition. A second aim of this chapter is to present some preliminary data on macro- and microstructural brain organization in chimpanzees and bonobos. We present data on cortical organization data from a variety of sources and approaches including voxel-based morphometry (VBM), sulcal surface area and depth and at the microstructural level. In contrast to traditional region-of-interest approaches used to compare species (Hopkins et al., 2009), VBM is an analytic technique that allows for whole-brain comparisons in either grey or white matter without the need for defining specific brain regions using cortical landmarks that may or may not differ between species.

Cognitive studies

Gaze and point following

Unpublished behavioral data from our laboratory are fairly consistent with previous reports on species differences in social cognition and prosocial behavior. Specifically, we tested 22 adolescent or adult bonobos (10 males, 12 females) on two tests of receptive joint attention that assessed their responses to gaze and pointing cues presented by a human experimenter. One measure was based on a previous study by Dawson and colleagues (2004) in children and was designed to assess the number of socio-communicative cues needed to elicit an orientation response from an experimenter (see Hopkins et al., 2014a for description). Briefly, a human experimenter would engage in some husbandry behavior with the focal subject and at some point would break engagement with the ape and look over their shoulder as if something were behind them. If the ape failed to respond, they would repeat the engagement and subsequently look over their shoulder and also point to a hypothetical object behind them. If the subject still did not respond, they would repeat the husbandry activities and then again attempt to get the subject to orient behind them by looking over their shoulder, pointing and saying their name. Thus, depending on the subject's performance within a trial, the experimenter would progressively increase the number of cues needed to elicit an orienting response. On a given trial, scores of 1, 2, 3 and 4 were assigned to the ape depending on whether they responded to gaze alone (1), gaze plus pointing (2), gaze plus pointing plus vocalization, (3) or never responded (4). Thus higher scores reflected poorer performance. Each subject underwent four test trials and the mean performances for the bonobos and chimpanzees are shown in Figure 14.1a. Overall, the bonobos performed significantly better than the chimpanzees $t(21) = 4.969, p < 0.01$.

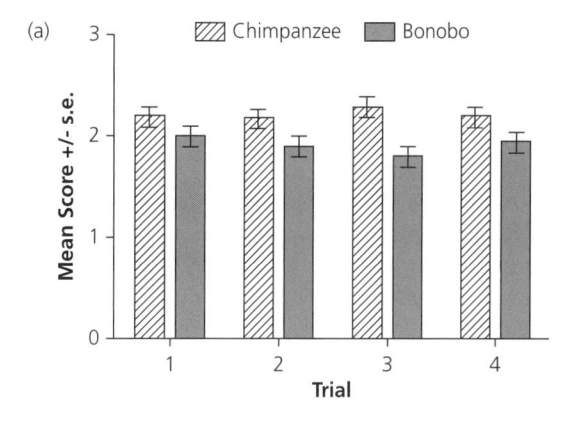

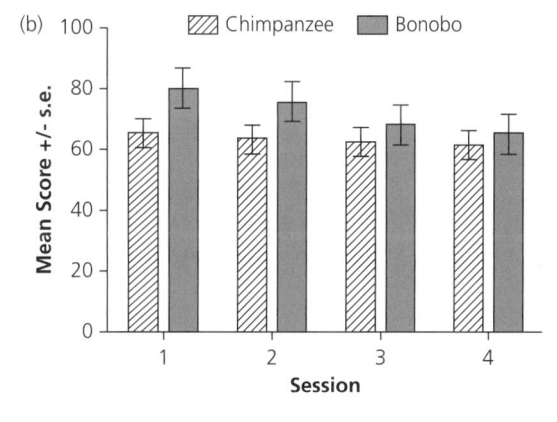

Figure 14.1 (a) Mean performance (+/− s.e.) on the DAWSON task for chimpanzees and bonobos for each of the four tests. Note, higher scores indicate worse performance on the task. (b). Mean performance (+/− s.e.) on the Mundy task for chimpanzees and bonobos for each test block of six trials. Note, higher values indicate better performance on the task. (a) Résultats moyens (+/− s.e.) de la tâche DAWSON chez les chimpanzés et les bonobos pour chacun des quatre examens. Noter qu'une marque haute indique un mauvais résultat dans la tâche. (b) Résultats moyens (+/− s.e.) de la tâche Mundy chez les chimpanzés et les bonobos pour chacun des six essais d'examens. Noter qu'une marque haute indique un bon résultat dans la tâche.)

The second task was developed by Mundy and colleagues to assess visual orienting responses to arbitrary declarative pointing cues by an experimenter in developing human children (Mundy et al., 2000). For this task, each of 21 chimpanzees and bonobos (10 male, 11 female) received 24 test trials, divided over 4 test sessions. Each session consisted of six test trials and only one session per day was performed. Subjects were tested alone. Prior to beginning the task, the experimenter placed two polyvinyl-choride (PVC) stations as high and far laterally apart on the cage mesh as possible. The experimenters positioned themselves equal distance between the two PVC stations and in front of the subject and subsequently engaged in some basic husbandry task. When the experimenter felt the subject was actively engaged, they stopped interacting with the subject and pointed (full arm extended) and looked towards one of the PVC stations (the cued PVC) and said the ape's name with increasing emphasis. If the subject looked, oriented towards, or touched the PVC station during this time, a correct response was recorded. If the subject did not look, orient or touch the cued PVC or looked, oriented towards or touched the non-cued PVP pipe, then an incorrect response was recorded. This process was repeated for all six trials, with each trial being separated by the experimenter re-engaging the subject with the basic husbandry task. The experimenter randomly alternated between which PVC station was the cued stimulus (left or right). The dependent measure was the proportion of correct responses divided by the incorrect responses across the 24 trials (see Figure 14.1b). We found that the bonobos performed significantly better than the chimpanzees, particularly in the first two blocks of 24 trial test sessions $t(20) = 2.46$, $p < 0.05$.

Social and physical cognition: the Primate Cognition Test Battery (PCTB)

In addition to the receptive joint attention tasks we have also collected data on a variety of social and physical cognition tasks in chimpanzees and bonobos. Herrmann and colleagues (2007) developed the Primate Cognitive Test Battery (PCTB) to measure cognitive abilities in a relatively large samples of chimpanzees, orangutans and two-and-a-half-year-old human children. The PCTB has subsequently been used to test physical and social cognition abilities in several primate species, including humans, chimpanzees, bonobos and macaques (Herrmann et al., 2010a,b; Schmitt et al., 2011). Here, we administered the PCTB to a sample of 26 bonobos residing at the Milwaukee Zoo, Jacksonville Zoo and Ape Cognition and Conservation Initiative. For comparison, we used the descriptive published data from 99 chimpanzees housed at the Yerkes National Primate Research Center that were also administered the PCTB (Hopkins et al., 2014c). The chimpanzee and bonobo subjects were tested on 12 of the original PCTB tasks using a somewhat modified version of the tasks originally described by Hermann and colleagues (2007) and presented in detail elsewhere (Russell et al., 2011a). Eight tasks were utilized in the Physical Cognition portion of our test including tasks exploring the apes' spatial memory and understanding of spatial relationships, ability to differentiate between quantities, understanding of causality in the visual and auditory domains and their understanding of tools. The four sets of tasks designated as Social Cognition in the PCTB that we utilized are as follows: the first two are designed to test the apes' ability to understand and to produce communicative signals. The third set assesses their sensitivity to the attentional state of an experimenter and their ability to use appropriate communicative modalities based on this information. The last social cognition task is designed to assess rudimentary aspects of Theory of Mind by testing their ability to follow gaze.

For the initial analysis, we compared the chimpanzee and bonobo PCTB performance for the average physical and social cognition tasks using a mixed model analysis of variance. Task (physical, social) was the repeated measure while species (chimpanzee, bonobo) and sex (male, female) were the between group factors. A significant interaction effect between task and species was found $F(1, 121) = 13.17$, $p < 0.01>$. The mean performance on each task and species are shown in Table 14.1. As can be seen, chimpanzees performed better on the physical compared to social cognition tasks where bonobos showed the opposite pattern. We next considered between-species performance for each task. For statistical comparisons, we performed one sample of t-tests with the mean value for the chimpanzees serving as the population estimate against which the performance of the

Table 14.1 Mean performance on the PCTB tasks by chimpanzees and bonobos.
(Performance moyenne sur les tâches PCTB par les chimpanzés et les bonobos.)

Task	Chimpanzee (n = 99)	Bonobo (n = 26)	T
Spatial Memory	60.9 (2.8)	57.9 (5.0)	−0.64
Object Permanence	63.0 (1.8)	59.2 (4.1)	−0.92
Rotation	45.5 (1.9)	45.7 (2.8)	+0.10
Transpose	61.2 (2.4)	51.1 (4.1)	−2.45 *
Quantity	67.5 (1.6)	72.3 (3.3)	+1.70
Noise—Causality	51.7 (1.5)	49.0 (3.6)	−0.73
Visual—Causality	66.9 (1.6)	64.3 (2.8)	−0.84
Tool Use	38.4 (4.9)	32.0 (9.4)	−0.67
Tool Properties	66.2 (2.0)	59.3 (3.9)	−1.73+
Comprehension	47.1 (3.1)	67.0 (6.7)	+2.94*
Production	39.9 (3.9)	28.1 (6.7)	−1.74+
Attention State	42.0 (2.7)	47.0 (6.1)	+0.80
Gaze	60.3 (3.7)	52.0 (7.0)	−1.16

Standard errors are shown in parentheses below each mean.
*$p < 0.05$,
+$p < 0.10$

bonobos was compared. Chimpanzees performed significantly better than bonobos on the transposition task while the bonobo performed significantly better on the comprehension of gaze and pointing cues. No other significant differences were found.

In summary, our findings suggest that bonobos perform significantly better than chimpanzees on some social cognition tasks whereas chimpanzees perform better on physical cognition tasks. These findings are broadly consistent with previous studies (MacLean, 2016).

Studies on brain organization

The behavioral and cognitive differences between chimpanzees and bonobos presumably are manifest in variation in brain organization and connectivity, an area of research that is beginning to receive some attention. For instance, Hopkins and colleagues (2014b) found that chimpanzees and bonobos differ in the relative size of the motor-hand area of the precentral gyrus, corresponding to the primary motor cortex, and these findings might explain the species differences in handedness and tool use (Hopkins et al., 2014b). Additionally, Hopkins and team (2009) reported that bonobos and chimpanzees differ in the volume and lateralization of the basal ganglia, cerebellum and hippocampus, which they hypothesized may underlie differences in spatial cognition and motor-learning capacities. Using a neuroimaging method that allows for measurement of cortical connectivity, Rilling and colleagues (2012) reported that bonobos have a larger amygdala volume and increased connectivity with the anterior cingulate cortex compared to chimpanzees. Rilling and coworkers (2011) suggest that these differences may explain species differences in social cognition.

Here, we present two sets of data on chimpanzees and bonobo brain organization. First, we present data on grey matter volume and organization using a method called voxel-based morphometry. Second, we present some previous and preliminary finding on microstructural differences (and similarities) between chimpanzees and bonobos.

Voxel-Based Morphometry (VBM)

For the VBM analysis, several pre-processing steps were taken to place the brains of all apes in the same stereotaxic space and to allow for direct comparison in cerebral cortical grey matter. Initially, each individual T1 and T2 scan was segmented into grey matter, white matter and CSF using the FAST function within FSL software (Smith et al., 2004). Partial volumes reflecting grey matter were used in all subsequent analyses. During step 1, each grey matter volume was registered and saved to an average chimpanzee template brain using a 12-degree of freedom linear transformation. The linear registered scans were then merged into a single 4D volume, averaged, flipped on the left–right axis and saved as a new 4D volume. The normal and flipped 4D volumes were then averaged together to create a single study-specific grey matter template volume. During step 2, each linear transformed volume was non-linearly registered to the study-specific grey matter template. The individual non-linear registered brains were then merged, as a single 4D volume, averaged, then the 4D volume was flipped on the left–right axis and saved as a separate volume. The normal and flipped 4D volumes were then averaged to create the final grey matter template brain. Finally, each individual non-linear registered brain was re-registered using a non-linear registration (FNIRT) and the volume and associated Jacobin warping field were saved as output. The individual non-linear registered volume was multiplied by the Jacobian warping field and saved as a volume. The final 12 bonobo and 12 chimpanzee grey matter volumes were appended into a single 4D volume and smoothed with a 2 mm sigma Gaussian kernel. Voxel-wide independent samples t-tests (bonobo > chimpanzee; chimpanzee > bonobo) were performed using the GLM function in FSL. Voxel clusters larger than 75 mm^3 with a $p < 0.05$ (uncorrected for multiple comparisons) were considered significantly different between the species.

For the analysis of bonobos compared to chimpanzees, 16 regions differed significantly at $p < 0.05$ (see Figure 14.2a). Grey matter density that was greater in bonobos than chimpanzees was found bilaterally in the postcentral gyrus, lateral precentral gyrus, orbital prefrontal cortex and the regions inferior to the putamen, which generally corresponds to the nucleus accumbens and basal forebrain region (reward regions of the brain). Leftward asymmetries were found in the dorsal prefrontal cortex and putamen while rightward biases in the lateral and medial grey matter around the intraparietal sulcus, medial precentral gyrus, medial postcentral gyrus and superior parietal sulcus. In contrast, chimpanzees showed higher grey matter density in 12 regions (see Figure 14.2b). Bilateral differences were found in the hippocampus and preoccipital sulcus. Left hemisphere differences were found for the intraparietal sulcus, temporal pole and ventrolateral precentral gyrus. Right hemisphere differences were found in the dorsolateral prefrontal cortex, fronto-orbital gyrus, middle inferior temporal gyrus and inferior frontal gyrus (Table 14.2).

Microstructural studies

In this section, we review data on the microstructural organization of brain regions that might be involved in social cognition differences between bonobos and chimpanzees. These include the insular cortex, hypothalamus and amygdala.

In humans, it has been shown that the insular cortex is important for self-recognition, awareness of emotions, time perception, empathy and decision-making in the face of uncertainty (Craig, 2009). The insula is a highly interconnected hub in cortical networks, sharing projections with anterior cingulate cortex, rostral and dorsolateral prefrontal cortex, regions of the parietal and temporal lobes, as well as entorhinal cortex, amygdala, hypothalamus and dorsal thalamus (Critchley, 2005; Price, 1999). The cellular composition of the insula in primates displays a gradient, transitioning from granular neocortex in the posterior and dorsal insula to agranular neocortex in the anterior and ventral insula, with an extensive intermediate zone of dysgranularity (Bauernfeind et al., 2013). In addition to the three sectors of insula that can be recognized in all primates, great apes and humans also exhibit a distinctive subdivision of agranular insula cortex, known as frontoinsular cortex (FI), located adjacent to the orbital cortex. The FI region is defined by the presence of clusters of von Economo neurons in layer Vb (Allman et al., 2010; von Economo and Koskinas, 1925). The von Economo neurons, which can also be found in the anterior cingulate and dorsolateral prefrontal cortex of humans and great apes, are large spindle-shaped projection neurons with a single basal dendrite (Fajardo et al., 2008; Nimchinsky et al., 1999). The anterior agranular insular cortex, which includes area FI, is particularly involved in social cognitive functions such as intersubjective perspective-taking, empathy and cooperation.

We previously analysed the volumes of insular cortex regions in a sample 30 different primate species based on cytoarchitecture (Bauernfeind et al., 2013). This sample included bonobos ($n = 3$, 1 male, 2 females) and chimpanzees ($n = 3$, 2 males, 2 female). We found that the anterior insular cortex (i.e. agranular insular cortex plus FI) is significantly larger in bonobos (mean = 0.500 cm^3) than chimpanzees (mean = 0.230 cm^3) (Mann–Whitney U test: Z = 2.80, $p < 0.05$) despite similar brain volumes in the sample (bonobos = 339 cm^3; chimpanzees = 330 cm^3). We also calculated the percentage by which the observed anterior insular cortex volumes in bonobos and chimpanzees differ from allometric scaling expectations based on brain volume (minus anterior insula) using the equation calculated from the total

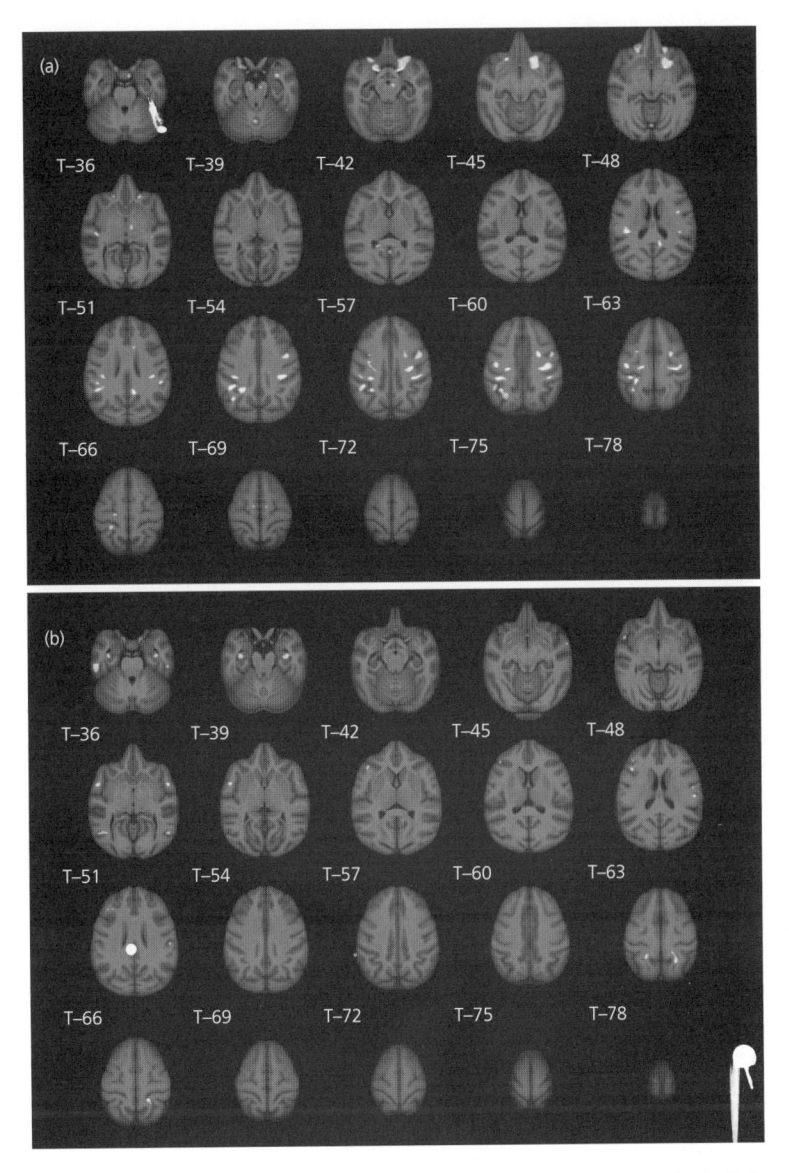

Figure 14.2 (a) Clusters with higher grey matter values in bonobos compared to chimpanzees. Clusters are thresholded at p < 0.05, uncorrected for multiple comparisons (Plate 14a). (b) Clusters with higher grey matter values in chimpanzees compared to bonobos. Clusters are thresholded at p < 0.05, uncorrected for multiple comparisons (Plate 14b).
(a) Grappes de plus hautes valeurs de matière grise chez les bonobos en comparaison avec les chimpanzés. Les grappes sont limitées à p < 0.05, non corrigé pour plusieures comparaisons (Plate 14a). (b) Grappes de avec plus hautes valeurs de matière grise chez les chimpanzés en comparaison avec les bonobos. Les grappes sont limitées à p < 0.05, non corrigé pour plusieures comparaisons (Plate 14b).

sample of 30 primate species ($y = 1.18x - 3.40$; $R^2 = 0.92$, $p < 0.05$). Based on this analysis, the anterior insular cortex is 22 per cent larger in bonobos and 90 per cent smaller in chimpanzees than what would be expected by allometric scaling. Thus, within the genus *Pan* the anterior insula is larger than would be expected for a species on their brain size while this area is smaller in chimpanzees. The expanded anterior insula region in bonobos may explain, in part, some of the differences in socio-emotional and socio-communicative behavior between bonobos and chimpanzees.

Table 14.2 Significant grey matter clusters differentiating bonobos and chimpanzees.
(Importants amas de matière grise différenciant les bonobos et les chimpanzés.)

	Cluster Size .mm³	T
Bonobo > Chimpanzee		
Left Post Central Gyrus (Dorsal)	247	2.84
Left Precentral Gyrus (Lateral)	509	2.86
Left Dorsal Prefrontal Cortex	375	2.99
Left Orbital Frontal Gyrus	144	2.83
Left Putamen	731	2.63
Left Dorsal Medial Prefrontal Cortex	258	2.62
Left Nucleus Accumbens	358	3.14
Right Precentral Gyrus (Medial)	320	2.62
Right Precentral Gyrus (Lateral)	180	2.61
Right Post Central Gyrus (Dorsal)	318	3.86
Right Post Central Gyrus (Ventral)	131	2.58
Right Intraparietal Sulcus (Lateral)	308	2.95
Right Intraparietal Sulcus (Medial)	162	2.79
Right Superior Parietal Sulcus	212	2.81
Right Orbital Prefrontal Cortex	213	2.80
Right Nucleus Accumbens	389	2.79
Chimpanzee > Bonobo		
Left Intraparietal Sulcus	190	2.38
Left Hippocampus	195	2.52
Left Temporal Pole	621	2.53
Left Preoccipital Sulcus	62	2.49
Left Ventrolateral Precentral Gyrus	79	2.38
Right Hippocampus	187	2.37
Right Ventral Precentral Gyrus	162	2.36
Right Preoccipital Sulcus	110	2.50
Right Inferior Frontal Gyrus	211	2.17
Right Dorsal Lateral Prefrontal Cortex	87	2.16
Right Fronto-orbital Sulcus/Gyrus	76	2.15
Right Middle Inferior Temporal Gyrus	273	2.43

Oxytocin (OXT) and arginine vasopressin (AVP) are peptides that have key roles in the regulation of social cognition and behaviors, such as attachment, social exploration, recognition and aggression, as well as anxiety and fear (reviewed in Meyer-Lindenberg et al., 2011). Several studies have shown that the function of these neuropeptides is impaired in psychological disorders associated with social deficits in human patients. OXT and AVP are synthesized in magnocellular and parvocellular neurons in the paraventricular nucleus (PVN) and supraoptic nucleus (SON) of the hypothalamus. From the hypothalamus, axon projections of the magnocellular neurons lead to the posterior lobe of the pituitary where these neuropeptides are released into the peripheral circulation. The smaller parvocellular neurons in the PVN and SON that produce OXT and AVP project directly to other regions in the brain, including the amygdala, hippocampus, striatum, suprachiasmatic nucleus, bed nucleus of stria terminalis and brainstem. In these brain structures, OXT and AVP act as neuromodulators or neurotransmitters. For example, OXT and AVP have been shown to modulate the activity of neural populations in the central amygdala (Huber et al., 2005). A number of studies report that intranasal delivery of OXT and AVP in humans increases prosocial behaviors that can provide buffering of positive social interactions on stress responsiveness, as well as enhance the decoding of social signals such as facial expression, increase memory for social information and promote interpersonal trust (Baumgartner et al., 2008; Kosfeld et al., 2005). This suggests that central brain levels of these neuropeptides might serve to modulate functions associated with social cognition.

We used immunohistochemistry to label AVP- and OXT-containing neurons in the PVN and SON of the hypothalamus in bonobos ($n = 5$, 3 males, 2 females) and chimpanzees ($n = 7$, 4 males, 3 females) (Figure 14.3). The counts of the total number of neurons in each nucleus that expressed AVP and OXT did not reveal statistically significant species differences (Figure 14.4). When the sample was divided by sex, there were also no significant differences between bonobos and chimpanzee. It is important to note that this analysis did not evaluate possible species-specific differences in the distribution of receptors to AVP and OXT. Eutherian mammals have one receptor for OXT (*OXTR*) and three for AVP (*AVPR1A, AVPR1B* and *AVPR2*). Notably, studies on variation in the genes encoding OXT and AVP in relation to social behavior in humans have not demonstrated strong effects, whereas a substantial body of evidence now implicates the genes that encode their receptors as playing a major role in regulating social behavior (reviewed in Meyer-Lindenberg et al., 2011). The polymorphisms in *OXTR* and *AVPR1A* that are most strongly implicated in social behavior do not change the amino acid sequence of the encoded protein but rather are located in upstream microsatellite regions that exert their effects on the distribution of receptors in the brain. Future studies of bonobos and chimpanzees that investigate the distribution of *OXTR* and *AVPR1A* in the brain might yield additional insight into the neural mechanisms of social behavior.

The amygdala is another area of the brain that is important in regulating social behavior through processing of aversive emotional stimuli that elicit fear and anxiety (LeDoux, 1996). Amygdala responsiveness is further regulated by the neurotransmitter serotonin, as it contains a high density of serotonin (5-HT) receptors (Albert et al., 2014). Data from a variety of mammalian species demonstrate that low levels of 5-HT in the cerebral spinal fluid and various brain regions are associated with increased aggression, anxiety, fearfulness and depression (Lesch et al., 1996; Wang et al., 2013). Given the importance of both the serotonergic system and amygdala in social behavior, we tested whether the amygdala of bonobos ($n = 7$) and chimpanzees ($n = 7$) differs in its neuroanatomical organization and serotonergic innervation (Stimpson et al., 2016). Quade's rank analysis controlling for brain mass as a covariate revealed a trend for whole amygdala volume to be relatively larger in chimpanzee than bonobos ($F_{1,11} = 3.95$, $p = 0.072$). Using immunohistochemistry to label serotonergic axons, we found that the amygdala of bonobos contained a higher density of serotonergic fibres (0.00521 ± 0.0019 μm/μm³) than

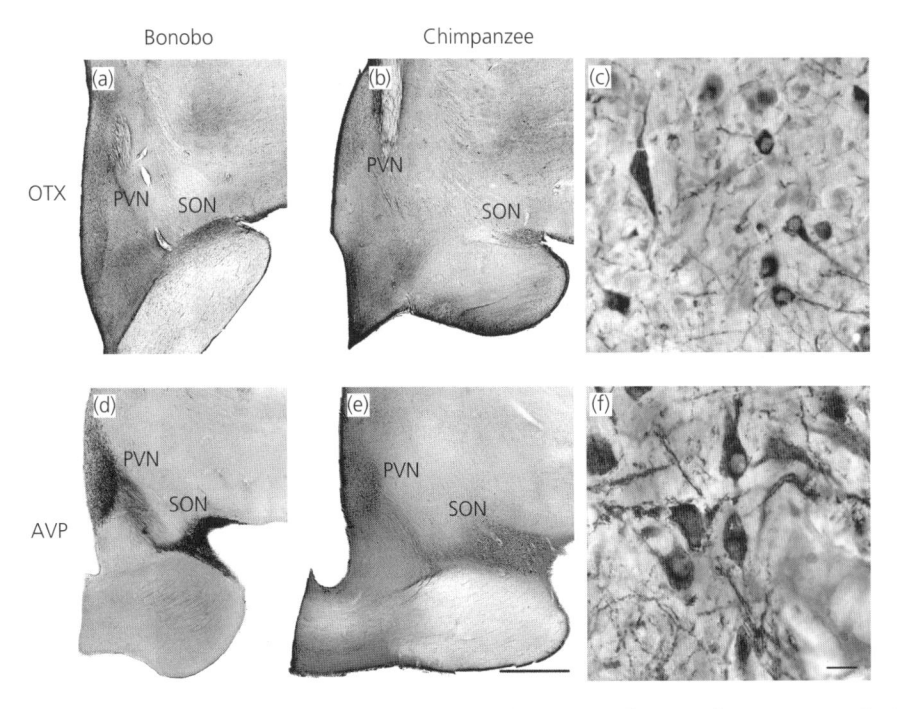

Figure 14.3 OTX and AVP immunostaining in the hypothalamus of bonobos and chimpanzees from coronal sections. OTX staining is shown at low magnification in bonobo (a) and chimpanzee (b), and higher magnification from a chimpanzee (c) in the PVN. AVP staining is shown at low magnification in bonobo (d) and chimpanzee (e), and higher magnification from a chimpanzee (f) in the PVN. Scale bar in E = 250 μm and scale bar in F = 25 μm.
(Immunocoloration OTX et AVP dans l'hypothalamus des bonobos et des chimpanzés dans des sections coronaires. La coloration OTX est montrée dans un faible grossissement chez le bonobo (a) et le chimpanzé (b), et plus fort grossissement d'un chimpanzé (c) dans le PVN. La coloration AVP est montrée dans un faible grossissement chez le bonobo (d) et le chimpanzé (e), et plus fort grossissement d'un chimpanzé (f) dans le PVN. La barre échelle de E = 250 μm et la barre échelle en F = 25 μm.)

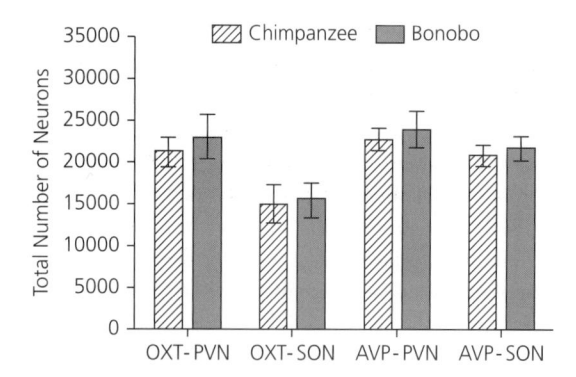

Figure 14.4 Total number of OTX- and AVP-immunostained neurons in the PVN and SON of bonobos and chimpanzees. Bars represent mean and error bars show standard error.
(Nombre total de neurones immunocolorés par OTX et AVP dans le PVN et SON des bonobos et des chimpanzés. Les bandes représentent la moyenne et les bandes d'erreur montrent l'erreur standard.)

in chimpanzees (0.00255 ±0.0008 μm/μm^3) (Mann–Whitney U test: $Z = -2.86$, $P = 0.004$). In particular, the density of serotonergic axons in the basal nucleus of the amygdala in bonobos (0.0543 ±0.0013 μm/μm^3) was nearly twofold greater than in chimpanzees (0.00279 ±0.0003 μm/μm^3; $Z = -3.00$, $P = 0.003$). The density of serotonergic innervation of the central nucleus was also significantly greater in bonobos (0.00481 ±0.0010 μm/μm^3) than in chimpanzees (0.00288 ±0.0005 μm/μm^3; $Z = -3.00$, $P = 0.003$).

Broca's area

Broca's area is located in the inferior frontal gyrus and numerous clinical and experimental studies have implicated this region as critical for language and speech functions in humans (among many other functions). Comparative studies in chimpanzees and bonobos on the volume of the Broca's area homolog have failed to find significant differences (Hopkins et al., 2009). Recently, we have begun to explore whether potential species differences may be evident in microstructure. Brodmann's areas 44 and 45 (BA44, BA45) comprise the constituent cytoarchitectural regions that define Broca's area in the human and chimpanzee brain (Schenker et al., 2010). In our preliminary studies, we have quantified the neuropil fraction (NF) within BA44 and BA45 in a sample of chimpanzee and bonobo post-mortem brains. The neuropil is space surrounding cell bodies of neurons and glia occupied by dendrites, synapses, axons and blood vessels surrounding as observed in a Nissl stain. The NF measures the percentage of unstained neuropil space between neurons and other cells in the grey matter. Accordingly, the NF provides an estimate of the local interconnectivity within a region of the cerebral cortex. We have examined NF in BA44 and BA45 from chimpanzee ($N = 6$) and bonobos ($N = 5$) brain specimens. The preliminary data indicate that overall NF is greater in bonobos as compared to chimpanzees in both hemispheres of BA44 and BA45, and that the difference between species is especially pronounced in left hemisphere BA45 (Figure 14.5). Thus, based on these preliminary data, bonobos may have increased localized interconnectivity in BA44 and BA45 compared to chimpanzees, which might be associated with species differences in communication systems.

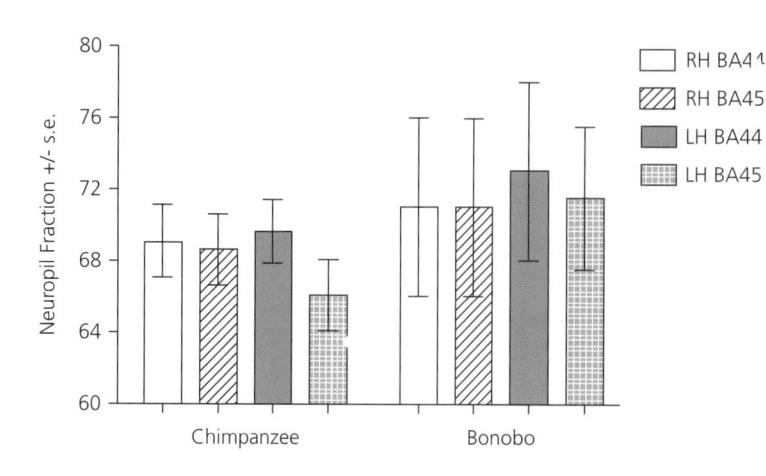

Figure 14.5 Mean neuropil fraction (+/− s.e.) for BA44 and B45 in the left and right hemispheres of chimpanzee and bonobo post-mortem brains.
(Moyenne de fraction neuropile (+/− s.e.) pour BA44 et B45 dans les hémisphères gauche et droite dans les cerveaux des chimpanzés et des bonobos après le décès.)

Discussion

In many ways, comparative studies between chimpanzees and bonobos on social behavior, cognition and the brain are in their infancy. With respect to social behavior and cognition, the data from our laboratory are somewhat, though not entirely, consistent with the current narrative regarding species differences between bonobos and chimpanzees (MacLean, 2016). Specifically, with respect to social cognition, our findings are consistent with the notion that bonobos are more sensitive to social cues than chimpanzees, such as following gaze and using social cues in discrimination tasks. However, we would suggest that some caution be used to interpret these results because the differences are relatively small (though consistent) and the issue of divergent rearing experiences of the subjects when comparing chimpanzees and bonobos is difficult to control. We emphasize the rearing of the subjects as a concern because we have previously found that chimpanzees with different early social rearing experiences differ significantly on social cognition tasks, and the effects in these studies rival or exceed the between species variation reported here (Lyn et al., 2010; Russell et al., 2011b). Though we used a match design in comparing the chimpanzees and bonobo in our analyses based on whether the apes were conspecific- or human-reared, it can certainly be argued that the complexity of the apes' social experience can vary considerably beyond the mother–infant dyad. Likewise, the physical environments of the apes can also vary considerably and this may have some influence on both cognitive performance and cortical organization. For instance, do the physical environments of a zoo enclosure, a research laboratory and a sanctuary all offer the same level of stimulation to the apes? Even within these different environs, there can be considerable variation in the space and level of complexity of the environment and this is difficult to control for when comparing species.

Caution aside, the fact that bonobos and chimpanzees differ with respect to genetic polymorphisms in genes coding for receptor distribution in neuropeptides strongly implicated in social behavior (i.e. *AVPR1A*) provides a possible genetic explanation for species differences. The challenge with this interpretation, at this point in time, is that we lack a full understanding of the function of these genes in individual variation in behavior and where these genes are expressed in brains of chimpanzees and bonobos. In chimpanzees, it is increasingly clear that the indel deletion of the RS3 portion of the *AVPR1A* is linked to individual differences in personality, anxiety-related behaviors and social cognition (Anestis et al., 2014; Hopkins et al., 2012; Hopkins et al., 2014a, Latzman et al., in press; 2014; Staes et al., 2015). Bonobos do not have the deletion in the RS3 region, thus, from a genetic polymorphic standpoint, bonobos are similar to chimpanzees without the RS3 deletion. Thus, it might be hypothesized that, all things being equal, RS3$^+$ chimpanzees and bonobos should not differ behaviorally or cognitively compared to the RS3$^-$ chimpanzees. It should also be noted that, within chimpanzees, allele frequencies of the RS3 deletion vary between subspecies, potentially providing a possible explanation for variation in social behavior between communities (Anestis et al., 2014). Finally, as already noted, an important line of inquiry for future studies would be assessing gene expression in chimpanzee and bonobos brains. It is one thing to differ in genetic polymorphisms, but a key mediating factor is also where these genes are expressed in terms of anatomy and how this mediates function.

The findings showing increased serotonergic innervation of the amygdala in bonobos relative to chimpanzees may play a role in reducing their emotional reactivity to potential threats in the social environment through a relatively lower neuronal excitability in major output nuclei of the amygdala (i.e. basal and central nuclei) that send projections to cortical and autonomic centres. Consequently, greater serotonergic innervation of these amygdala regions in bonobos compared to chimpanzees may reduce activation of targets that control the sympathetic nervous system in response to stimuli that potentially arouse fear or aggression (Asan et al., 2013). Prospectively, if this explanation is correct, it would also suggest that bonobos would be less reactive or aroused in response to certain classes of social stimuli such as unfamiliar conspecific faces or vocalizations, stimuli that evoke significant responses in chimpanzees (Hopkins et al., 2006). This explanation is also consistent with the notion that bonobos may be more inclined or tolerant of unfamiliar individuals under certain circumstances (Hare and Kwetuenda, 2010). Finally, the difference in amygdala serotonergic innervation between chimpanzees and bonobos leads to the question of whether differences in polymorphic variation in genes linked to serotonin may be implicated. For instance, the *5-HTTLPR* gene has been studied extensively in a number of primates (Dobson and Brent, 2013) and some have suggested that species differences in social behavior among species of *Macaca* may be linked to allele frequencies (Wendland et al., 2005). We know very little about genetic polymorphism associated with serotonin in chimpanzees and bonobos (Hong et al., 2011; Inoue-Murayama et al., 2000) and additional studies on this neurotransmitter system and others would be useful.

Neuroanatomically, chimpanzees and bonobos are quite similar in terms of total brain size, surface area and gyrification. The VBM analyses show that there is some species-specific variation in grey matter concentration and location in certain cortical regions between chimpanzees and bonobos. Though we identified a number of previously unreported differences between chimpanzees and bonobos in grey matter morphology, there was some overlap in regions with the previous report by Rilling and colleagues (2012) and Hopkins and his team (2009). For instance, the evidence that chimpanzees had higher grey matter clustering in the hippocampus and parahippocampus region is consistent with previous results from volumetric region-of-interest analyses. Given the role of the hippocampus and surrounding medial temporal lobe regions in spatial memory formation, this may explain why chimpanzees perform better than bonobos on spatial memory tasks.

Moreover, at least with respect to the insula and amygdala size, the findings described here are also consistent with both the grey and white matter VBM and tract-based connectivity data reported by Rilling and colleagues where it was reported that bonobos have increased white matter FA values and connectivity in the insula, ventral posterior frontal cortex and amygdala. As is the case with the behavioral data, we would echo a word of caution in interpreting the between-species differences in cortical organization and connectivity. Many of these studies are underpowered due to a limited amount of post-mortem tissue availability, particularly in bonobos, thereby limiting sample sizes. Further, species comparisons are done largely by convenience based on sample availability and not carefully based on subject characteristics. For instance, the DTI-based comparisons in the paper by Rilling and colleagues (2011) were based on a samples sizes of three subjects in each species and neither sex, age or rearing histories of the apes were controlled for in the analyses. While these findings are certainly valuable, nonetheless it is important to be cautious that inferences or claims of species differences are not confounded with other variables.

In sum, behavioral, neuroanatomical and genomic data point to important species differences between chimpanzees and bonobos. These two species are the closest living primate relatives to humans and continued studies are needed to identify the role of social and ecological factors on the evolution of cognition and the brain. These studies are important for understanding human behavior from a translational and evolutionary perspective. Further, given the threats to bonobos and chimpanzees in their native habitats that are influencing their sustainability for future generations, additional data from both captive and wild populations are critical for promoting scientific and conservation efforts of these amazing animals. Now, more than ever, it is vital that scientists working in field and captive setting combine their efforts and work collectively together to advance studies of these species.

Acknowledgement

This work was supported by National Science Foundation grant BCS-0824531 and National Institutes of Health grants NS-92988, HD-56232, NS-42867 and NS-73134. We would like to thank all the research assistants who assisted in behavioral data collection and the zoo staff and veterinarians who aided in the collection of post-mortem brains. We also appreciate the assistance of Brittany Moore in carrying out the neuropil studies in the chimpanzees and bonobos. Reprint requests can be sent to Dr William Hopkins, Neuroscience Institute, Georgia State University, PO Box 5030, Atlanta, GA 30302-5030 Email: whopkins4@gsu.edu or whopkin@emory.edu

References

Albert, P.R., Vahid-Ansari, F., and Luckhart, C. (2014). Serotonin-prefrontal cortical circuitry in anxiety and depression phenotypes: pivotal role of pre- and post-synaptic 5-HT1A receptor expression. *Frontiers in Behavioral Neuroscience*, 8, 199.

Allman, J.M., Tetreault, N.A., Hakeem, A.Y., Manaye, K.F., Semendeferi, K., Erwin, J.M., Goubert, V., and Hof, P.R. (2010). The von Economo neurons in the frontoinsular and anterior cingulate cortex in great apes and humans. *Brain Structure and Function*, 214, 495–517.

Anestis, S.F., Webster, T.H., Kamilar, J.M., Fontenot, M.B., Watts, D.P., and Bradley, B.J. (2014). AVPR1A variation in chimpanzees (*Pan troglodytes*): Population differences and association with behavioral style. *International Journal of Primatology*, 35, 305–24.

Asan, E., Steinke, M., and Lesch, P. (2013). Serotonergic innervation of the amygdala: targets, receptors, and implications for stress and anxiety. *Histochemistry and Cell Biology*, 139, 785–813.

Bauernfeind, A.A., Sousa, A.M.M., Avashti, T., Dobson, S.D., Raghanti, M.A., Lewandowski, A.H., Zilles, K., Semendeferi, K., Allman, J.M., Craig, A.D., Hof, P.R., and Sherwood, C.C. (2013). A volumetric comparison of the insular cortex and its subregions in primates. *Journal of Human Evolution*, 64, 263–79.

Baumgartner, T., Heinrichs, M., Vonlanthen, A., and Fischbacher, U. (2008). Oxytocin shapes the neural circuitry of trust and trust adaptation in humans. *Neuron*, 58, 639–50.

Bermejo, M. and Omedes, A. (1999). Preliminary vocal repertoire and vocal communication of wild bonobos (*Pan paniscus*) at Lilungu (Democratic Republic of Congo). *Folia Primatologica*, 70, 328–57.

Brakke, K.E. and Savage-Rumbaugh, E.S. (1995). The development of language skills in bonobo and chimpanzee—i. Comprehension. *Language and Communication*, 15, 121–48.

Brakke, K.E. and Savage-Rumbaugh, E.S. (1996). The development of language skills in *Pan*—ii. production. *Language and Communication*, 16, 361–80.

Byrne, R.W. (1995). *The Thinking Ape: Evolutionary Origins of Intelligence*. Oxford: Oxford University Press.

Byrne, R.W. and Corp, N. (2004). Neocortex size predicts deception rate in primates. *Proceedings of the Royal Society B*, 271, 1393–699.

Chang, L., Fang, Q., Zhang, S., Poo, M., and Gong, N. (2015). Mirror-induced self-directed behaviors in rhesus monkeys after visual-somatosensory training. *Current Biology*, 25, 212–17.

Consortium, T.C.S.A.A. (2005). Initial sequence of the chimpanzee genome and comparison with the human genome. *Nature*, 437, 69–87.

Craig, A.D. (2009). How do you feel-now? The anterior insula and human awareness. *Nature Neuroscience Reviews*, 10, 59–70.

Critchley, H.D. (2005). Neural mechanisms of autonomic, affective, and cognitive integration. *Journal of Comparative Neurology*, Vol. 393, 154–66.

Dawson, G., Toth, K., Abbott, R., Osterling, J., Munson, J., Estes, A., and Liaw, J. (2004). Early social attention impairments in autism: social orienting, joint attention and attention to distress. *Developmental Psychology*, 40, 271–83.

De Waal, F.B.M. (1988). The communicative repertoire of captive bonobos (*Pan paniscus*) compared to that of chimpanzees. *Behaviour*, 106, 183–251.

De Waal, F.B.M. (1992). Appeasement, celebration, and food sharing in the two *Pan* species. In: Nishida, T., Mcgrew, W. C., Marler, P., Pickford, M., and De Waal, F.B.M. (eds). *Topic in Primatology: Human Origins*, Vol. 1. Tokyo: University of Tokyo Press.

Deaner, R.O., Isler, K., Burkart, J., and Van Schaik, C.P. (2007). Overall brain size, and not encephalization quotient, best predicts cognitive ability across non-human primates. *Brain, Behavior and Evolution*, 70, 115–24.

Dobson, S.D. and Brent, L J.N. (2013). On the evolution of the serotonin transporter linked polymorphic region (*5-HTTLPR*) in primates. *Frontiers in Human Neuroscience*, 7, 588.

Donaldson, Z.R., Bai, Y., Kondrashov, F.A., Stoinski, T.L., Hammock, E.A.D., and Young, L.J. (2008). Evolution of a behavior-linked microsatellite-containing element of the 5′ flanking region of the primate avpr1a gene. *BMC Evolutionary Biology*, 8, 180–8.

Dunbar, R.I.M. (1998). The social brain hypothesis. *Evolutionary Anthropology*, 6, 178–90.

Fajardo, C., Escobar, M.I., Buritica, E., Arteaga, G., Umbarila, J., Casanova, M.F., and Pimienta, H. (2008). Von Economo neurons are present in the dorsolateral (dysgranular) prefrontal cortex of humans. *Neuroscience Letters*, 435, 215–18.

Furuichi, T., Sanz, C.M., Koops, K., Sakamaki, T., Ryu, H., Tokuyama, N., and Morgan, D. (2014). Why do wild bonobos not use tools like chimpanzees do? *Behaviour*, 152, 1–35.

Gibson, K.R. and Ingold, T. (1993). *Tools, Language and Cognition in Human Evolution*. Cambridge: Cambridge University Press.

Goodson, J.L. and Bass, A.H. (2001). Social behavior functions and related anatomical characteristics of vasotocin/vasopressin systems in vertebrates. *Brain Research Reviews*, 35, 246–65.

Gruber, T., Clay, Z., and Zuberbuhler, K. (2010). A comparison of bonobo and chimpanzee tool use: evidence for a female bias in the *Pan* lineage. *Animal Behaviour*, 80, 1023–33.

Hare, B. and Kwetuenda, S. (2010). Bonobos voluntarily share their own food with others. *Current Biology*, 20, R230–R231.

Hare, B., Melis, A.P., Woods, V., Hastings, S., and Wrangham, R. (2007). Tolerance allows bonobos to outperform chimpanzees on a cooperative task. *Current Biology*, 17, 619–23.

Heilbronner, S.R., Rosati, A.G., Stevens, J.R., Hare, B., and Hauser, M.D. (2008). A fruit in the hand or two in the bush? Divergent risk preferences in chimpanzees and bonobos. *Biology Letters*, 4, 246–9.

Herrmann, E., Call, J., Hernandez-Lloreda, M.V., Hare, B., and Tomasello, M. (2007). Humans have evolved specialized skills of social cognition: The cultural intelligence hypothesis. *Science*, 317, 1360–6.

Herrmann, E., Hare, B., Call, J., and Tomasello, M. (2010a). Differences in the cognitive skills of bonobos and chimpanzees. *PLoS One*, 5, e12438.

Herrmann, E., Hernandez-Lloreda, M.V., Call, J., Hare, B., and Tomasello, M. (2010b). The structure of individual differences in the cognitive abilities of children and chimpanzees. *Psychological Science*, 21, 102–10.

Hohman, G. and Fruth, B. (1994). Structure and use of distance calls in wild bonobos, (*Pan paniscus*). *International Journal of Primatology*, 15, 767–82.

Hong, K.W., Weiss, A., Morimura, N., Udono, T., Hayasaka, I., Humle, T., Murayama, Y., Ito, S., and Uinoue-Murayama, M. (2011). Polymorphism of the tryptophan hydroxylase 2 (*TPH2*) gene is associated with chimpanzee neuroticism. *PloS One*, 6, e22144.

Hopkins, W.D., Donaldson, Z.R., and Young, L.Y. (2012). A polymorphic indel containing the RS3 microsatellite in the 5′ flanking region of the vasopressin V1a receptor gene is associated with chimpanzee (*Pan troglodytes*) personality. *Genes, Brain and Behavior*, 11, 552–8.

Hopkins, W.D., Keebaugh, A.C., Reamer, L.A., Schaeffer, J., Schapiro, S.J., and Young, L.J. (2014a). Genetic influences on receptive joint attention in chimpanzees (*Pan troglodytes*). *Scientific Reports*, 4, 1–7.

Hopkins, W.D., Lyn, H., and Cantalupo, C. (2009). Volumetric and lateralized differences in selected brain regions of chimpanzees (*Pan troglodytes*) and bonobos (*Pan paniscus*). *American Journal of Primatology*, 71, 988–97.

Hopkins, W.D., Meguerditchian, A., Coulon, O., Bogart, S.L., Mangin, J.F., Sherwood, C.C., Grabowski, M.W., Bennett, A.J., Pierre, P. J., Fears, S.C., Woods, R.P., Hof, P.R., and Vauclair, J. (2014b). Evolution of the central sulcus morphology in primates. *Brain, Behavior and Evolution*, 84, 1930.

Hopkins, W.D., Russell, J.L., Freeman, H., Reynolds, E.A.M., Griffis, C., and Leavens, D.A. (2006). Lateralized scratching in chimpanzees (*Pan troglodytes*): Evidence of a functional asymmetry in arousal. *Emotion*, 6, 553–9.

Hopkins, W.D., Russell, J.L., and Schaeffer, J. (2014c). Chimpanzee intelligence is heritable. *Current Biology*, 24, 1–4.

Huber, D., Veinante, P., and Stoop, R. (2005). Vasopressin and oxytocin excite distinct neuronal populations in the central amygdala. *Science*, 308, 245–8.

Inoue-Murayama, M., Niimi, Y., Takenaka, O., Okada, K., Matsuzaki, I., Ito, S., and Murayama, Y. (2000). Allelic variation in serotonin transporter gene polymorphic region in apes. *Primates*, 41, 267–73.

Jaeggi, A.V., Stevens, J.M.G., and Van Schaik, C.P. (2010). Tolerant food sharing and reciprocity is precluded by despotism among bonobos but not chimpanzees. *American Journal of Physical Anthropology*, 143, 41–51.

Jaeggi, A.V. and Van Schaik, C.P. (2011). The evolution of food sharing in primates. *Behavioral Ecology and Sociobiology*, 65, 2125–40.

Kosfeld, M., Heinrichs, M., Zak, P.J., Fischbacher, U., and Fehr, E.M. (2005). Oxytocin increases trust in humans. *Nature*, 435, 673–6.

Latzman, R.D., Hopkins, W.D., Keebaugh, A.C., and Young, L.J. (2014). Personality in chimpanzees (*Pan troglodytes*): Exploring the hierarchical structure and associations with the asopressin *V1A* receptor gene. *PLoS One*, 9, e95741.

Ledoux, J. (ed.) (1996). *The Emotional Brain*. New York, NY: Simon & Schuster.

Lesch, K.P., Bengel, D., Heils, A., Sabol, S. Z., Greenberg, B.D., Petri, S., Benjamin, J., Muller, C.R., Hamer, D.H., and Murphy, D.L. (1996). Association of anxiety-related traits with a polymorphism in the serotonin transporter gene regulatory region. *Science*, 274, 1527–31.

Lyn, H., Russell, J.L., and Hopkins, W.D. (2010). The impact of environment on the comprehension of declarative communication in apes. *Psychological Science*, 21, 360–5.

Maclean, E.L. (2016). Unraveling the evolution of uniquely human cognition. *Proceedings of the National Academy of Sciences*, 113, 6348–54.

Malenky, R.K. and Stiles, E.W. (1991). Distribution of terrestrial herbaceous vegetation and its consumption by *Pan paniscus* in the Lomako Forest, Zaire. *American Journal of Primatology*, 23, 153–69.

Meyer-Lindenberg, A., Domes, G., Kirsch, P., and Heinrichs, M. (2011). Oxytocin and vasopressin in the human brain: social neuropeptides for translational medicine. *Nature Neuroscience Reviews*, 12, 524–38.

Mundy, P., Card, J., and Fox, N. (2000). EEG correlates of the development of infant joint attention skills. *Developmental Psychology*, 36, 325–38.

Nimchinsky, E.A., Gilissen, E., Allman, J.M., Perl, D.P., Erwin, J.M., and Hof, P.R. (1999). A neuronal morphology type unique to humans and great apes. *Proceedings of the National Academy of Sciences*, 96, 5268–73.

Pollick, A.S. and De Waal, F.M.B. (2007). Ape gestures and language evolution. *Proceedings of the National Academy of Sciences*, 104, 8184–9.

Price, J.L. (1999). Prefrontal cortical networks related to visceral function and mood. *Annals of the New York Academy of Sciences*, 877, 383–96.

Prufer, K., Munch, K., Hellmann, I., Akagi, K., Miller, J.R., Walenz, B., Koren, S., Sutton, G., Kodira, C., Winer, R., Knight, J.R., Mullikin, J.C., Meader, S.J., Ponting, C.P., Lunter, G., Higashino, S., Hobolth, A., Dutheil, J., Karakoc, E., Alkan, C., Sajjadian, S., Catacchio, C.R., Ventura, M., Marques-Bonet, T., Eichler, E.E., Andre, C., Atencia, R., Mugisha, L., Junhold, J., Patterson, N., Siebauer, M., Good, J.M., Fischer, A., Ptak, S.E., Lachmann, M., Symer, D.E., Mailund, T., Schierup, M.H., Andres, A.M., Kelso, J., and Paabo, S. (2012). The bonobo genome compared with the chimpanzee and human genomes. *Nature*, 486, 527–31.

Reader, S.M. and Laland, K.N. (2002). Social intelligence, innovation, and enhanced brain size in primates. *Proceedings of the National Academy of Sciences*, 99, 4436–41.

Rilling, J.K., Scholz, J., Preuss, T.M., Glasser, M.F., Errangi, B.V., and Behrens, T.E.J. (2012). Differences between chimpanzees and bonobos in neural systems supporting social cognition. *Social Cognitive and Affective Neuroscience*, 7, 369–79.

Rosati, A.G. and Hare, B. (2012). Chimpanzees and bonobos exhibit divergent spatial memory development. *Developmental Science*, 15, 840–53.

Russell, J.L., Lyn, H., Schaeffer, J.A., and Hopkins, W.D. (2011). The role of socio-communicative rearing environments in the development of social and physical cognition in apes. *Developmental Science*, 14, 1459–70.

Savage-Rumbaugh, E.S. (1984). *Pan paniscus* and *Pan troglodytes*: Contrast in preverbal communicative competence. In: Susman, R.L. (ed.). *The Pygmy Chimpanzee: Evolutionary Biology and Behavior*. New York, NY: Plenum Press.

Savage-Rumbaugh, E.S. and Lewin, R. (1994). *Kanzi: The Ape at the Brink of the Human Mind*. New York, NY: John Wiley.

Schenker, N.M., Hopkins, W.D., Spocter, M.A., Garrison, A.R., Stimpson, C.D., Erwin, J.M., Hof, P.R., and Sherwood, C.C. (2010). Broca's area homologue in chimpanzees (*Pan troglodytes*): probabilistic mapping, asymmetry, and comparison to humans. *Cerebral Cortex*, 20, 730–42.

Schmitt, V., Pankau, B., and Fischer, J. (2011). Old World monkeys compare to apes in the primate cognition test battery. *PLoS One*, 7, e32024.

Sherwood, C.C., Baurernfeind, A.L., Bianchi, S., Raghanti, M.A., and Hof, P.R. (2012. Human brain evolution writ large and small. In: Hofman, M.A. and Falk, D. (eds). *Progress in Brain Research*. New York, NY: Elsevier.

Smith, S.M., Jenkinson, M., Woolrich, M.W., Beckmann, C.F., Behrens, T.E.J., Johansen-Berg, H., Bannister, P.R., De Luca, M., Drobniak, I., Flitney, D. E., Niazy, R., Saunders, J., Vickers, J., Zhang, Y., De Stafano, N., Brady, J.M., and Matthews, P.M. (2004). Advances in functional and structural MR image analysis and implementation of FSL. *NeuroImage*, 23 (S1), 208–19.

Staes, N., Kkoski, S.E., Helsen, P., Fransen, E., Eens, M., and Stevens, J.M.G. (2015). Chimpanzee sociability is associated with vasopressin (Avpr1a) but not oxytocin receptor gene (OXTR) variation. *Hormones and Behavior*, 75, 84–90.

Staes, N., Stevens, J.M.G., Helsen, P., Hillyer, M., Korody, M., and Eens, M. (2014). Oxytocin and vasopressin receptor gene variation as a proximate base for inter- and intraspecific behavioral differences in bonobos and chimpanzees. *PLoS One*, 9, e113364.

Stimpson, C.D., Hopkins, W.D., Taglialatela, J., Barger, N., Hof, P.R., and Sherwood, C.C. (2016). Differential serotonergic innervation of the amygdala in bonobos and chimpanzees. *Social, Cognitive and Affective Neuroscience*, 11, 413–22.

Tallerman, M. and Gibson, K. (eds) (2012). *The Oxford Handbook of Language Evolution*. Oxford: Oxford University Press.

Tuttle, R.H. (2014). *Apes and Human Evolution*. Cambridge, MA: Harvard University Press.

Von Economo, C. and Koskinas, G. (1925). *Die cytoarchitectonik der hirnrinde des erwachsenen menschen*. Berlin: Springer.

Wang, C.C., Lin, H.C., Chan, Y.H., Gean, P.W., Yang, Y.K., and Chen, P.S. (2013). 5-HT1A-receptor agonist modified amygdala activity and amygdala-associated social behavior in a valproate-induced rat autism model. *The International Journal of Neuropsychopharmacology*, 16, 2027–39.

Wendland, J.R., Lesch, K.-P., Newman, T.K., Timme, A., Gachot-Neveu, H., Thierry, B., and Suomi, S.J. (2005). Differential functional variability of serotonin transporter and monoamine oxidase A genes in macaque species displaying contrasting levels of aggression-related behavior. *Behavior Genetics*, 36, 163–72.

Wobber, V., Hare, B., Maboto, J., Lipson, S., Wrangham, R., and Ellison, P.T. (2010). Differential changes in steroid hormones before competition in bonobos and chimpanzees. *Proceedings of the National Academy of Sciences*, 107, 12457–62.

Jaeggi, A.V., De Groot, E., Stevens, J.M.G., and Van Schaik, C.P. (2013). Mechanisms of reciprocity in primates: testing for short-term contingency of grooming and food sharing in bonobos and chimpanzees. *Evolution and Human Behavior*. Vol. 34, 69–77.

Latzman, R.D., Young, L.J., and Hopkins, W.D. (2016). Displacement behaviors in chimpanzees (*Pan troglodytes*): A neurogenomics investigation of the RDoC Negative Valence Systems domain. *Psychophysiology*. Vol. 53, 355–363.

CHAPTER 15

Cognitive comparisons of genus *Pan* support bonobo self-domestication

Brian Hare and Vanessa Woods

Abstract The self-domestication hypothesis (SDH) suggests bonobo psychology evolved due to selection against aggression and in favour of prosociality. This hypothesis was formulated based on similarities between bonobos and domesticated animals. This chapter reviews the first generation of quantitative research that supports the predictions of the SDH. Similar to domestic animals, bonobos are prosocial towards strangers, are more flexible with cooperative problems, are more responsive to social cues and show expanded windows of development compared to their closest relatives, chimpanzees. A preliminary comparison of bonobo and chimpanzee infants suggests that when hearing a stranger, bonobos have a xenophilic response while chimpanzees have a xenophobic response. The chapter explores why the research with bonobos has implications for theories of both human and animal cognitive evolution, and why bonobos will be central in studying evolutionary processes that lead to cognitive change.

L'hypothèse d'auto-domestication (SDH) suggère que la psychologie bonobo a évolué grâce à la sélection contre l'agression et en faveur de la prosocialité. Cette hypothèse fut formulée à partir de similarités entres les bonobos et les animaux domestiqués. Nous révisons la première génération de recherche quantitative qui soutient les prédictions du SDH. Comme les animaux domestiques, les bonobos sont prosociaux envers les étrangers, plus flexibles avec les problèmes de coopération, plus sensibles aux signaux sociaux, et montrent des fenêtres étendues de développement relativement à leur plus proche parent, le chimpanzé. Nous présentons une comparaison préliminaire des bébés bonobos et chimpanzés. Quand ils entendent un étranger, les bonobos ont une réaction xénophilique alors que les chimpanzés ont une réaction xénophobique. Nous expliquons pourquoi le travail des bonobos est impliqué dans les théories d'évolution cognitive humaine et animale, et pourquoi les bonobos seront au centre des études évolutionnaires des procès menants aux changements cognitives.

If the deduction that the bonobo is highly derived with respect to the other chimpanzees is correct, then the similarities of bonobos to humans become even more significant: the original human environment has long been exceeded, but we may suppose that bonobos still occupy the niche to which they are best adapted. By examining this niche we may learn something about the early human niche, or at least aspects of it. (Groves, 1981).

How does cognition evolve? Why are some species more flexible problem-solvers than others? What are the processes that lead to shifts in cognitive abilities? An ecological approach to cognitive evolution suggests that different domains of intelligence have evolved in animals and that these domains can vary independently within and between species. Different cognitive profiles are hypothesized to cause individual and species variation in problem-solving. Ecological differences are thought to have generated the specific selective pressures favouring differences in problem-solving. For example, primates rely on visual acuity while cetaceans use echolocation to locate food. Neither taxa is 'smarter'. Instead, each has evolved the cognitive tools necessary to solve the problems relevant to their survival. This echoes the logic used to understand the evolution of almost any trait.

A powerful way to generate and initially test hypotheses for cognitive evolution is through comparisons of closely related species. Ultimately,

Hare, B. and Woods, V., *Cognitive comparisons of genus Pan support bonobo self-domestication*. In: *Bonobos: Unique in Mind, Brain, and Behavior*. Edited by Brian Hare and Shinya Yamamoto: Oxford University Press (2017). © Oxford University Press.
DOI: 10.1093/oso/9780198728511.003.0015

large-scale phylogenetic studies are needed to test evolutionary hypotheses rigorously, but in cognitive research, large-scale comparisons must be built on the previous success of intensive, small-scale comparisons (i.e. valid and capable of being replicated; MacLean et al., 2011). Moreover, when or where an animal eats can be observed by watching its behavior, but uncovering the internal mechanisms of their perception and memory during these feeding events requires careful experimentation (Tomasello and Call, 2008). Natural observations will always inform the most valid experiments but the most powerful tests of cognitive hypotheses will always involve experimentation (Hare, 2001).

In this chapter, we review research over the past decade designed to compare the cognition of bonobos and chimpanzees as a test of the self-domestication hypothesis. The self-domestication hypothesis suggests that selection for prosociality and against aggression leads to changes in cognition as a result of shifts in development. Natural selection acting on temperament leads to more tolerance by acting on physiological mechanisms responsible for emotional reactivity. Shifts in development alter the physiology mediating social behavior and then allows for tolerance. These same shifts also cause a cascade of incidental changes across the phenotype known as the 'domestication syndrome'. It is suspected that changes in early neurohormonal development also alter the development of these other traits as a by-product of selection for prosociality (Trut, 1999; 2009). The first large-scale experimental comparisons of bonobos and chimpanzees were designed to test the predictions of the self-domestication hypothesis based on previous work with experimentally domesticated animals. The bonobo has served as the first candidate species for self-domestication through natural selection. The potential implications of these studies apply not only to ideas about human cognitive evolution but evolution in animal cognition more generally (Hare, 2017). Tolerant primates, island populations and urban populations are all candidates for self-domestication since each may experience selection against aggression or for increased prosociality. As we continue to test the self-domestication hypothesis with a wider range of species, it is possible that selection on social emotions is an important evolutionary force that provides selection the opportunity to act on 'revealed variance' in cognitive abilities (West-Eberhard, 2003), the idea being that variance in cognitive abilities are masked by constraints on emotional reactivity. Selection that favours more prosocial interactions creates new forms of social interactions where selection might then act on any unmasked cognitive variance expressed in these novel social contexts (Hare and Tomasello, 2005). For example, once a species can share food, variance in the ability to synchronize actions during cooperative foraging—that previously had no fitness consequence—now can pay reproductive dividends. We start by explaining the origin of the self-domestication hypothesis based on the experimental domestication of foxes. We then lay out the predictions of the hypothesis as it applies to bonobos as well as evidence suggesting that the predictions are largely supported.

The experimental discovery of self-domestication

Intensive, paired comparisons between domesticated animals and their wild progenitors have led to the hypothesis that selection for friendliness leads to changes in social cognition. Central to this hypothesis is the work of Dmitry Belyaev and colleagues who experimentally selected a population of foxes to be interested and prosocial towards humans as opposed to aggressive and fearful towards them (Trut, 1999; 2009). The experimental line of foxes was compared to a line of control foxes who were randomly selected for their behavior towards humans. The two fox populations were otherwise maintained identically. Compared to the control line of foxes, the experimental foxes showed the majority of physiological, morphological and behavioral elements observed in the domestication syndrome seen in most domesticated animals. This included a higher frequency of feminized crania (i.e. shortened muzzles), smaller teeth, floppy ears, shortened tails, more adult play, more sexual behavior and, not surprisingly, more prosocial behavior towards humans than the control line. Prosociality in the case of the foxes, and throughout this review, is defined as positive and affiliative behavior and does not imply whether these behaviors are selfishly or unselfishly motivated (i.e. Eisenberg et al., 1983). This is opposed to increased gregariousness that can lead to increases in both prosocial and antisocial behavior.

Developmental comparisons of the two populations suggest that many of the differences are a result of heterochrony (also see Gariepy et al., 2001). For example, the experimental foxes have both an earlier and extended socialization period (Trut, 1999). Selection against aggression and for prosociality seems to have acted on individual differences in developmental timing to produce a population of foxes that has not only become attracted to humans but also shows a suite of correlated by-products that were not directly under selection (i.e. morphology, social cognition, etc.; Trut, 2009).

The foxes have provided a powerful test for the link between selection for friendliness and cooperative communication. Domestic dogs show unusual skill at using human social gestures in relation to chimpanzees and wolves (Hare and Woods, 2013). They show basic skills for reading human gestures in the first months of life even with relatively little exposure to humans. This suggested that these skills evolved in dogs during domestication, but cognitive tests of the experimental foxes show these skills probably evolved as a *result* of domestication, or selection for friendliness. When tested on two spontaneous measures, experimental fox kits were also more skilled at using human gestures than control kits (Hare et al., 2005). Even though the experimental foxes were never selected based on their ability to communicate with humans, they showed the same level of skill at using human gestures as age-matched dog puppies. Hare and colleagues (2005) proposed that selection on the foxes' temperament unintentionally allowed them to apply their inherited skills at reading conspecific cues to humans once they began reacting positively to humans as though humans were conspecifics. The potential for further evolution is then greatly enhanced. Any revealed variance in these skills can then potentially come directly under selection and lead to dramatic cognitive evolution. Survival of the friendliest can lead to changes in social cognitive abilities.

Based on this experimental population, where highly controlled artificial selection on emotional reactivity had sweeping changes on everything from behavior, morphology, physiology and cognition, it has been suggested that a similar process occurred during dog domestication. In the case of dogs, evolution was the result of natural selection or 'self-domestication'. Wolves that were attracted to humans (i.e. were not fearful or aggressive) were able to take advantage of refuse in and around human habitations. Natural selection then favoured the 'friendliest' wolves—setting off a cascade of phenotypic changes in proto-dogs similar to those observed in the experimentally domesticated foxes (i.e. the domestication syndrome).

This raises the possibility that cognition can initially evolve without direct selection, as a correlated by-product of selection for other traits. Temperamental constraints (i.e. too much fear to have a social interaction where social cognition is needed) are removed to reveal variation in social abilities that selection can act on directly in a second wave of evolution (Wobber et al., 2009). Essentially, temperamental evolution can be the spark that leads to cognitive evolution.

The psychology of bonobo self-domestication

The next step of testing the self-domestication hypothesis requires comparisons with species that did not evolve as a result of interactions with humans. If natural selection can cause the domestication syndrome and correlated changes in cognition, it should be observable among wild species. Our closest relative, the bonobo, has recently been compared to chimpanzees as a candidate for self-domestication. While the two species are extremely similar, they are also more like humans than like each other genotypically and phenotypically across a number of evolutionarily significant traits (Hare and Yamamoto, 2015). This means both species are equally related to humans, to each other while also different from each other in substantial ways. This phylogenetic profile has made it difficult to infer whether our last common ape ancestor had a mosaic of traits or was more like a chimpanzee or bonobo (Hare and Wrangham, in press). Several hypotheses have been suggested regarding which ape our common ancestor more closely resembled. The *mosaic hypothesis* suggests that our common ancestor had a mix of traits and that no species is more representative than another. This hypothesis is essentially the null and is potentially supported by genomic comparisons that do not show that either bonobos or chimpanzees has experienced more positive selection or are genetically more human-like (Prufer et al., 2012). Surveys of phenotypic traits also suggest behavioral and cognitive traits for which each species of *Pan* is more similar to humans than to each other, with no clear pattern suggesting one species is more or less human-like (Hare and Yamamoto, 2015). The *bonobo-like hypothesis* suggests our common ancestor was overall more bonobo-like based on qualitative comparisons of behavior and morphology (de Waal and Lanting, 1997; Zilhmann, 1978). However, the bonobo-like hypothesis has not been formally proposed and no attempt has been made to account for patterns of trait change in bonobos, chimpanzees and humans. Finally, the *chimpanzee-like hypothesis* suggests that bonobos are more derived and potentially convergent with humans. The self-domestication hypothesis helps organize what seem to be unrelated derived traits in bonobos into a unified explanation. If bonobos are self-domesticated, chimpanzees are probably more representative of our last common ancestor. Recent arguments have been made that humans may also have experienced selection against aggression and this raises the possibility that bonobos and humans have converged in some ways both behaviorally and cognitively (Cieri et al., 2014; Hare, 2017).

Hare and colleagues (2012) formalized the *self-domestication hypothesis* (SDH) for bonobo evolution and reviewed the body of work testing its main predictions based on the initial proposal made by Wrangham and Pilbeam (2001). Bonobos have a host of traits that differ from chimpanzees and resemble those that differ between domesticated animals and their wild progenitors (see Figure 15.1). Because of the qualitative similarities between bonobos and domestic animals, quantitative tests were designed to examine the hypothesis that bonobos are self-domesticated. The SDH was informed by observational studies that led to predictions

Figure 15.1 Domesticated animals are thought to have expanded developmental windows with traits being expressed earlier and later in development relative to their wild ancestor. For example, dogs can be socialized with people earlier and for longer during development than wolves. Bonobos have been suggested to show earlier development of some traits (i.e. non-reproductive socio-sexual behavior) as well as a number of juvenile traits that continue to be expressed during adulthood (i.e. non-reproductive socio-sexual behavior, cranial morphology, play behavior, etc.). The self-domestication hypothesis explains why (a) juvenile chimpanzees and (b) adult bonobos are likely to share traits with each other that are not also shared with (c) adult chimpanzees. This may especially be the case in comparisons of males like the ones depicted in photos here. (Photo courtesy of Vanessa Woods).

(On pense que les fenêtres développementales des animaux domestiques se sont élargies pour les caractéristiques exprimées plus tôt et plus tard dans leur développement par rapport à leur ancêtre sauvage. Par exemple, on peut socialiser les chiens avec des personnes plus tôt et plus longtemps pendant leur développement que les loups. Apparemment, les bonobos présentent un développement précoce de certaines caractéristiques (à savoir le comportement socio-sexuel non reproductif) ainsi qu'un certain nombre de caractéristiques juvéniles qui continuent de s'exprimer à l'âge adulte (telles que le comportement socio-sexuel non reproductif, la morphologie crânienne, le comportement de jeu, etc.). L'hypothèse de l'autodomestication explique pourquoi (a) les jeunes chimpanzés et (b) les bonobos adultes sont susceptibles de partager des caractéristiques qui ne concernent pas (c) les chimpanzés adultes. Cela peut surtout être le cas dans les comparaisons de mâles tels que ceux présentés dans les photos ci-contre.)

about where cognitive differences should evolve in response to selection against aggression (Hare et al., 2012; Wrangham and Pilbeam, 2001). As a result, the majority of these a priori quantitative tests of the SDH involve comparing bonobo and chimpanzee cognition.

The main cognitive predictions are that bonobos should show greater prosociality than chimpanzees and more flexible social skills relating to cooperation and communication—similar to comparisons between domesticated animals and their wild types (e.g. Hare et al., 2002, 2010; Hernádi et al., 2012; Miklosi et al., 2003; Topal et al., 2009). In the following, work showing how bonobos react to strangers and work together to solve instrumental cooperative tasks is reviewed. Self-domestication also predicts a syndrome of changes resulting from increasing windows of development. Research testing whether bonobos show delays in their cognitive (and behavioral) development relative to chimpanzees is also reviewed. Since humans were not the driving factor in bonobo self-domestication, something ecological must have led to speciation. Comparisons between bonobo and chimpanzee foraging preferences are reviewed here that suggest that bonobo psychology is consistent with a more predictable feeding ecology that may have supported the predicted selection against aggression proposed in Hare et al. (2012).

Prosocial bonobos prefer strangers

The most obvious behavioral shift that is common to all domestic animals is an increase in prosociality. Domesticated animals are attracted to and interested in interacting with humans while their wild-type ancestors are fearful or even aggressive towards people (Hare et al., 2012). Following this pattern, a series of experimental studies suggest that bonobos are more prosocial to conspecifics than chimpanzees.

Bonobos are able to co-feed in a greater range of contexts than chimpanzees. When two individuals were released into a room with food, only bonobos consistently co-fed when food was placed in a single location. In the same context, one chimpanzee almost always monopolized the food. Almost any pairing of bonobos successfully co-fed, while chimpanzees were unable to share highly monopolizable food even when highly tolerant individuals were paired together. This allowed bonobos to cooperate in an instrumental task more flexibly than chimpanzees who had much more experience in the task (Hare et al., 2007).

Mechanistically, it appears that these different responses are partly hormonally mediated. In anticipation of sharing food, when males of the two species were compared, bonobos showed an increase in cortisol but not testosterone while chimpanzees had the opposite response (females were not assessed). Bonobos become socially anxious while chimpanzees are primed for competition before the same social stressor (Wobber et al., 2010c). Bonobos also engage in play and socio-sexual behavior during co-feeding while chimpanzees tend to space themselves to avoid contact (Hare et al., 2007). When crowded in captivity, bonobos increase socio-sexual and grooming behavior while chimpanzees avoid aggression by decreasing social interactions (Aureli and de Waal, 1997; Paoli et al., 2007). Bonobos appear to use social interactions to diffuse tension while chimpanzees often inhibit social interactions to accomplish the same outcome.

Bonobos and chimpanzees have been observed proactively sharing food in the wild (Yamamoto, 2015). Experiments have been designed to help understand the motivation and cognition behind this type of sharing. The differences between the two species become even more apparent in experiments that require some form of proactive sharing. Chimpanzees do proactively share with group members but only if they cannot access the food themselves (Melis et al., 2010; Warneken et al., 2007). Bonobos share food even if it is in their possession. They do this by opening a one-way door that releases another bonobo into the room with them. They repeatedly share this way even when the other bonobo does not beg for food and has no opportunity to reciprocate within the experiment (Hare and Kwetuenda, 2010). When paired with a stranger, bonobos will also open the one-way door to share food. Crucial to the self-domestication hypothesis, when given the choice of proactively sharing with a stranger or group-mate, bonobos *prefer* to share with strangers (Tan and Hare, 2013; also see Bullinger et al., 2013 for absence of sharing with group-mates). In strong support of the self-domestication hypothesis, bonobos, like domestic animals, are attracted to the same social partners that chimpanzees are either fearful or aggressive towards (Wilson et al., 2014). However, it is important to note where bonobo sharing is constrained. Similar to chimpanzees, bonobos have not been observed to share monopolizable food that is in their possession readily unless they the act of sharing results in a physical interaction as a result (Byrnit, et al., 2015; Cronin et al., 2015; Jaeggi et al., 2010; Tan and Hare, 2013; Tan et al., 2015). In addition, the unusual level of prosociality observed in bonobos may be limited to interactions involving food. Both bonobos and chimpanzees show similar tendencies to engage in joint activities with humans and conspecifics that involve playing with objects (MacLean and Hare, 2013). Both bonobos and chimpanzees have a strong preference to engage with others over an object rather than playing with the same object alone. While chimpanzees readily share objects upon request, bonobos do not, and they prefer to share food over objects (Krupenye et al., 2015; Warneken et al., 2007; Yamamoto et al., 2012).

Prosocial temperament leads to social cognition

Flexible cooperative communication is the main cognitive feature that has been observed as a result of domestication across species (Hare, 2011; although also see Lewejohann et al., 2010). Dogs are more skilled than wolves at spontaneously understanding human gestures (Hare et al., 2002, 2010). Dogs show similar flexibility in reading human pointing gestures as young human infants (Hare and Tomasello, 2005). Dogs are more likely to make eye contact with a human than a wolf when faced with an unsolvable task while also making socially mediated errors only seen in human infants (Miklosi et al., 2003; Topal et al., 2009). Exogenous oxytocin may also enhance the ability of dogs to use human gestures (Oliva et al., 2015). This unusual skill at using human gestures has been attributed to selection against aggression and selection for prosocial behaviors towards humans based on the work with Balyaev's foxes (Hare et al., 2005). Subsequently, a similar relationship between temperament and social skills has been observed (or inferred) in a variety of domesticated and non-domesticated species (e.g. Bray et al., 2015; Brust and Guenther, 2015; Lewejohann et al., 2010; Michelleta and Waller, 2012; Petit et al., 1992).

On a battery of temperamental tests, bonobos were most similar to human children who were shy and observant compared to other apes (Herrmann et al., 2011). On a standardized battery of cognitive tasks, bonobos and chimpanzees were identical except in areas consistent with the self-domestication hypothesis (Herrmann et al., 2010). Most significantly, when bonobos and chimpanzees were quantitatively compared across a range of cognitive tasks, bonobos were more sensitive than chimpanzees to social cues like gaze direction (Herrmann et al., 2010). When bonobos and chimpanzees were compared using an eye-tracker, only bonobos spent significant time examining the eyes of social stimuli such as conspecific and human faces. Chimpanzees spent the majority of each trial examining the mouths of the same set of stimuli. The response of bonobos is similar to normally functioning children while the response of the chimpanzees resembles people on the autistic spectrum (Kano et al., 2015). In corroboration of this interpretation, the 2D:4D ratios of bonobos are similar to normally functioning humans while the ratios of chimpanzees suggest high neonatal androgen levels. High neonatal androgen levels and concomitant 2D:4D levels have been linked to autism spectrum disorder and aggressive tendencies in humans (Liu et al., 2012; McIntyre et al., 2009). While chimpanzees are socially sophisticated, these first direct comparisons suggest that bonobos are more like humans in terms of their sensitivity to subtle social cues.

The relationship between oxytocin and levels of mutual gaze in dogs suggest that the oxytocin system may also be implicated in differences observed in bonobos. Dogs who make more eye contact with their owners have the highest levels of oxytocin. Their owners also have the highest levels of oxytocin (Nagasawa et al., 2009; 2015; Romero et al., 2014). When dogs are given oxytocin exogenously they make more eye contact and their owner's oxytocin level increases (Nagasawa et al., 2015). It is possible that the increased prosociality in bonobos is mediated by a similar oxytocin feedback loop that is less sensitive in chimpanzees. This idea is supported by genetic differences in bonobos related to the expression of oxytocin and vasopressin (Hammock and Young, 2005; Staes et al., 2015; see also Garai et al., 2014).

However, this idea remains largely untested. For example, while variability in chimpanzee social behavior has been linked to urinary oxytocin levels (Wittig et al., 2014), a similar test has not yet been conducted with bonobos which would allow for a comparison. In addition, while bonobos are more sensitive to spontaneous social cues like gaze direction, they perform the same as chimpanzees with human cooperative communicative gestures (MacLean et al., 2015). This may relate to the target of selection being conspecific interactions as opposed to human interactions as in domesticates.

Juvenilization: a mechanism for change

Heterochronic change remains the leading explanation for the domestication syndrome (Wilkins et al., 2014). Domestic animals tend to have expanded windows of development. They show traits that both emerge early or represent juvenile traits maintained throughout life (Hare, 2017). If correct, the domestication syndrome should be accompanied by developmental shifts that lead to juvenile traits being preserved into adulthood (Wobber et al., 2010b). Species comparisons are essential to test for developmental shifts. Table 15.1 summarizes a number of recent quantitative comparisons that reveal where bonobo cognition is developmentally delayed relative to chimpanzees. Wobber and colleagues (2010a) predicted that bonobos would show delay in the development of social inhibition since bonobo infants (with mothers) receive little aggression in comparison to infant chimpanzees. While it is crucial for an infant chimpanzee to understand quickly who they can and cannot forage near to avoid aggression and injury, the high level of feeding tolerance in bonobos, particularly towards infants makes this type of inhibition less vital. In a series of cognitive tasks, bonobos showed delayed development of social inhibition in comparison to age-matched chimpanzees. Bonobos were older than chimpanzees before they were able to show similar levels

Table 15.1 Cognitive and behavioral traits for which bonobos have been found to have expanded window of development or show juvenilization relative to chimpanzees. References refer to relevant quantitative comparisons that demonstrate development differences between bonobos and chimpanzees.
(Caractéristiques cognitives et comportementales qui étaient maintenues juvéniles relativement aux chimpanzés et la citation pertinente pour les comparaisons quantitatives qui montrent un développement en retard chez les bonobos.)

Juvenilized Traits in Bonobos	Descriptions of Delayed Development in Comparison to Chimpanzees	References
Play behavior	Adult bonobos show juvenile levels of play in as adults	Hare et al., 2007; Palagi & Cordoni, 2012; Palagi & Paoli, 2007; Wobber et al., 2010a
Non-conceptive sexual behavior	Emerges within first year of life during infancy with adult bonobos showing higher levels of non-conceptive social sexual behavior than chimpanzees	Hare et al., 2007; Woods & Hare, 2011
Xenophilia and Prosociality	Adult bonobos show juvenile levels of attraction to strangers (i.e. similar to immigrant females). And are prosocial, not aggressive, towards strangers in dyadic interactions.	Hare & Kwetuenda, 2010; Tan & Hare, 2013
Maternal dependence	Bonobo infants remain dependent on their mothers for longer; adult males remain socially dependent their mothers throughout life.	de Lathouwers and Van Elsacker, 2006; Surbeck et al., 2011
Food sharing	Adult bonobos show levels of food sharing observed in juvenile chimpanzees.	Hare et al., 2007; Wobber et al., 2010a
Social inhibition	Bonobo infants and juveniles show delays in development of social inhibition.	Wobber et al., 2010a
Spatial memory	Bonobos show delayed development in their ability to remember locations of hidden food.	Rosati and Hare, 2012b
Physical cognition	Bonobos show delayed development in understanding of the physical properties of the world (i.e. connectivity, transposition, etc.).	Wobber et al., 2014

of competence in the same tasks. This delay is probably specific to social inhibition since in a non-social reaching detour task, bonobos show the same amounts of inhibition as other non-human apes (Vlammings et al., 2010). Bonobos also show evidence for developmental delay in measures of spatial memory. Individuals watched as 20 pieces of food were hidden throughout a large enclosure. Once released after a 20-minute delay each subject could search for the hidden food. Juvenile bonobos and chimpanzees performed similarly in finding the food, however, while chimpanzee performance improved dramatically with age this was not the case in bonobos. Adult bonobos performed at very similar levels to juvenile chimpanzees (Rosati and Hare, 2012b). This difference might be hormonally mediated since individual differences in testosterone among male chimpanzees is strongly associated with performance on this same spatial navigation task but is not in bonobos (Wobber and Herrmann, 2015). Juvenile-like memory may also be specific to the spatial domain since bonobos and chimpanzees perform similarly in tasks requiring them to remember previous social interactions by preferring experimenters who had played with or groomed them in the past (Schroepfer et al., 2015). Bonobos also show episodic-like memory similar to chimpanzees (Martin-Ordas et al., 2010; Mulcahy and Call, 2006). Across a battery of cognitive tasks in which infant bonobos and chimpanzees were directly compared, bonobos show developmental delay in the tasks designed to measure understanding of space, tools and memory (Wobber et al., 2014). Finally, bonobos also show delayed offset of the expression of thyroid hormone relative to chimpanzees, which might be linked to lower levels of aggression (Beringher et al., 2014). There are other suspected cases of developmental delay, that still require additional experimentation to confirm (i.e. visual perspective-taking: MacLean and Hare, 2012; reputation formation: Herrmann et al., 2013; also, for potential example in orangutans see Schneider et al., 2012). Bonobos also show early expression in other traits similar to domestic animals that show overall expanded developmental windows. As infants, bonobos use non-reproductive socio-sexual behavior to reduce tension during co-feeding that is not observed in age-matched chimpanzees (Woods and Hare, 2011). Bonobo females also appear to reach sexual maturity several years before female chimpanzees (Beringher et al., 2014; Ryu et al., 2015). These differences may be related to the development of testosterone and follow the pattern in domesticates of early reproduction (Wilkins et al., 2014).

Direct developmental comparisons reveal a pattern of juvenilization as well as early emergence in bonobos relative to chimpanzees. While bonobos show delay on a number of cognitive tasks relative to chimpanzees, there is no cognitive skill where chimpanzees show delay relative to bonobos. This overall pattern of delay is consistent with bonobo self-domestication and the interpretation that chimpanzees represent the more conserved developmental pattern both morphologically and cognitively. Paramorphism, or extended cognitive development, has been suggested as unlikely in chimpanzees based on converging evidence from the development of the bonobo crania (Hare et al., 2012; Wrangham and Pilbeam, 2001). Quantitative comparison of bonobo and chimpanzee cranial growth also shows a peadomorphic pattern of development in bonobos, suggesting bonobo brain (and cognitive) development follows a similar pattern (Lieberman et al., 2007; Zollikofer and Ponce de Leon, 2010).

The developmental predictions of the SDH hypothesis account for features of bonobos that are difficult to explain as independent ecological adaptations. Juvenilization observed in bonobo physiology, morphology, behavior and cognition now needs to be explained beyond the phenotypic level. Wilkins and colleagues (2014) recently proposed that the domestication syndrome is the result of a delay in the migration of melanocytes early in the development of the neural crest. These cells are known to be responsible for the formation of morphological and physiological features that are altered through domestication (i.e. teeth, cartilage, melanin, etc.) as well as influencing brain development. Hare and colleagues (2012) raise the possibility that a similar developmental mechanism might account for the convergence between bonobos and other domestic animals. The neural crest hypothesis points to future tests that could reveal the heritable mechanism that created the domestication syndrome across a wide variety of species—perhaps even the bonobo.

Cognitive support for reduced scramble competition

The main scenario that has been proposed to explain the divergence of bonobos and chimpanzees is an ecological one (Wrangham, 2014). Over a million years of isolation on either side of the Congo River, ecological differences shaped the two species into their modern forms. Using current ecological differences as a proxy for past differences, it has been suggested that bonobos evolved with reduced scramble competition compared to chimpanzees. This reduction may have been due to the lack of another large ape species, like gorillas, a greater reliance on terrestrial herbaceous vegetation (THV) as fallback food, or a more productive forest south of the Congo River (reviewed in Wrangham, 2014; see also Wrangham and Peterson, 1997). However, novel quantitative tests of these hypotheses remain difficult since there is no fossil record from the Congo Basin to confirm the presence or absence of gorillas over evolutionary time and the relevant ecological comparison between THV or fruit availability on the north and south sides of the Congo River have not been conducted. It is unclear that the ecological differences that drove the evolution of each species' unique traits exist today.

A cognitive approach can help in testing ecological predictions (Rosati et al., 2014). If bonobos and chimpanzees were shaped by ecological differences that mirror those that exist today, we should see species differences in foraging strategies that reflect those differences. Previous comparisons between closely related species have shown links between feeding strategies, decision-making and spatial memory (Santos and Rosati, 2015). For example, insectivorous tamarins prefer small immediate rewards while gummivorous marmosets prefer larger delayed rewards (Rosati et al., 2005) but then reverse their preferences if travel is required to obtain the same rewards (Stevens et al., 2005). Likewise, frugivorous lemurs are more successful than more folivorous lemur species in a set of spatial memory tasks (Rosati et al., 2014).

The link between foraging cognition and the past ecologies that shaped them sets up a powerful test regarding potential differences in bonobos and chimpanzees (Hare et al., 2012). If bonobos and chimpanzees were partly shaped by differences in predictability or amount of fruit and access to fallback foods, then these differences should appear in their cognitive preferences (i.e. even though their teeth morphology largely overlaps because the two species' diet is so similar; although see Wrangham, 2014).

To test this, bonobos and chimpanzees were quantitatively compared across a range of foraging tasks designed to measure how they value time, risk and how they remember things in space. Bonobos and chimpanzees differ on each of these cognitive measures. Two different captive populations of bonobos and chimpanzees were measured for how they value a small reward immediately versus a large reward later. Both populations showed the same pattern; in two different tasks, chimpanzees were willing to wait longer to receive larger rewards than bonobos (Rosati et al., 2007; Rosati and Hare, 2013). When the same two populations were compared on two tasks that measure their preference for a fixed or variable reward, the two species showed the opposite preference from one another. Bonobos preferred the small fixed reward while chimpanzees preferred the riskier, but potentially higher pay-off variable reward (Heilbronner et al., 2008; Rosati and Hare, 2013; see also Haun et al., 2011). However, in follow-up comparisons related to risk-taking, where it was expected that bonobos would avoid ambiguity and loss frames more than chimpanzees, the two species performed similarly ((Krupenye et al., 2015); Rosati and Hare, 2012a).

When the two species were compared on a series of memory tasks they showed differences in their ability to maintain in working memory where food was hidden. Chimpanzees consistently outperformed bonobos in finding more hidden food items across longer delays than bonobos (Rosati, 2015; Rosati and Hare, 2012b). The differences observed in bonobos and chimpanzees may also be mediated by androgens since spatial performance was linked to basal testosterone levels in chimpanzees but not in bonobos (Rosati, 2015; Wobber and Herrmann, 2015). These findings support the ecological model of reduced scramble competition. It will be interesting to see if this pattern extends to memory in other non-foraging contexts in future tests.

The future: quantitative comparisons of reactions to strangers in *Pan*

Qualitative comparison overwhelmingly suggests that bonobos are xenophilic relative to xenophobic chimpanzees. However, if possible, it would be important to develop a method allowing for a direct quantitative comparison of the two species for their reaction to strangers versus group-mates.

Direct comparisons have mainly been prevented by ethical concerns. The main paradigm that shows bonobos prefer strangers to group-mates involves subjects voluntarily coming into physical contact in a relatively small test room (Hare and Kwetuenda, 2010). In the case of chimpanzees this could lead to serious injury and should not be attempted (Tan and Hare, 2013). Strange chimpanzees are integrated into groups in captive settings, but only for management purposes. While these types of integrations are often completed successfully without significant injury, they can take months, are carried out stepwise, provide escape routes that are not available in a controlled experimental setting and can still be highly stressful even if no injuries occur (Seres et al., 2001).

This does not rule out the possibility that chimpanzees are attracted to strangers. For example, sanctuary chimpanzees are attracted to strange humans in a temperament test (Herrmann et al., 2011) while adolescent female chimpanzees who decide to immigrate prefer strangers even though they risk serious injury and infanticide (Pusey and Schroepfer-Walker, 2013). While chimpanzees are generally xenophobic, there is a level of plasticity that allows for xenophilia depending on life stage, sex and reproductive state. This plasticity is likely the behavioral variance that selection acted on during bonobo evolution.

There may also be important variance in the xenophilic attraction expressed by bonobos. Males who do not show strong affiliative interactions with other males within and between groups may show little attraction to other males—males may even be xenophobic towards other adult males. Uncovering any plasticity in the xenophilic and xenophobic preferences of chimpanzees and bonobos can help uncover mechanisms that influence the evolution of these social behaviors in primates and other animals.

A hypothesis is only as strong as the tests used to falsify it. To encourage the development of ethical but powerful quantitative comparisons of behavior towards strangers in chimpanzees and bonobos, we conducted a pilot experiment with both species using the same method. Infants of both species were exposed to playbacks of screams from their own species made by strangers and group-mates. Their behavioral and physiological responses were measured. We predicted that the two species would have opposite reactions to the call of a stranger (i.e. regardless of call type). We expected bonobos to approach in the direction of the playback while chimpanzees would have a physiological response consistent with stress. Parr and Hopkins (2000) demonstrated asymmetrical ear temperatures in chimpanzees in response to observing stressful images and videos (videos of conspecific fighting, pictures of syringes, etc.). They conclude that stressful events with negative valence produce a higher temperature in the right rather than left ear as a result of the tympanic membrane heating up slightly as more haemodynamic activity occurs in the right hemisphere while processing things with negative emotional valence. We expected that infant chimpanzees but not bonobos would show an asymmetric pattern of ear temperature with the right ear hotter than the left after hearing a stranger calling. This would suggest these calls have negative valence even in the youngest chimpanzee infants.

Our subjects were 12 infant bonobos (0–5 years) at Lola ya Bonobo sanctuary outside of Kinshasa, Democratic Republic of Congo and 11 infant chimpanzees (0–5 years) at the Tchimpounga chimpanzee sanctuary outside of Pointe Noire, Congo–Brazzaville. All of the subjects were orphans of the bushmeat trade and were mother reared for two to three years before being raised by surrogate human mothers in a peer group at each sanctuary. Quantitative comparisons of mother-reared and orphan bonobos and chimpanzees showed no differences in cognition, behavior or hormonal profiles (Wobber et al., 2011).

Recordings used for the playbacks were made using a Seinhauser microphone and Edirol model R-1 24bit digital MP3 recorder and were chosen to match each other in amplitude and length as closely as possible. Screams were chosen for the comparison since it is the vocalization that most likely differs least between the two species (de Waal, 1988; although see Mitani and Gros-Louis, 1995).

All of the screams were made when the callers were clearly stressed (i.e. being chased by dominants or removed from their surrogate mother). Group-mate screams were obtained at the sanctuaries from individuals within the subject's group while stranger screams were obtained at the Leipzig Zoo in Germany.

The substitute mother providing the orphans with daily care separated subjects from their peer group. They were taken to a quiet but familiar room in the dormitory where they slept at night. They were then placed on their human substitute mother's lap or allowed to play nearby. A speaker was placed two metres away inside a plastic dog kennel. Subjects may have associated the kennel with strangers since it was used to bring new orphans to the sanctuary. This meant the speaker was not visible to the subject but they could potentially approach and explore the kennel in which it was hidden. In the *group-mate session*, subjects experienced the test where a recording of a group-mate scream was played back (and the kennel containing the speaker was surreptitiously shaken slightly by an experimenter) and a control condition where no playback occurred. The *stranger session* was identical except subjects experienced a test recording of a stranger scream. Half of subjects received either the group-mate or stranger session first. Within each session the test and the control condition were counterbalanced across subjects with half of all subjects either receiving the test or control condition first. Following each playback or neutral control period the substitute mother then measured the left and right ear temperature of each subject twice (i.e. following Parr and Hopkins, 2000) using a human ear thermometer (Bosotherm medical 0124). The order that the temperatures were taken was counterbalanced within and between subjects and across conditions. Most infants had their ear temperatures read daily by these same human mothers for veterinarian purposes, so recordings could be made with relative ease. However, due to the preliminary nature of the study only a single exemplar was available for playback in each condition with each species (i.e. in each test condition subjects only heard the same scream from the same individual three times as opposed to multiple screams from multiple individuals). While the caller in each playback was always a juvenile, the use of a single exemplar in each condition means any finding is clouded by the likelihood that subjects' responses were specific to idiosyncrasies of the specific call used in each test as opposed to their reaction to the variable of interest (the identity of the caller). Even with its significant methodological limitation, this pilot is reported here to inspire future work in this direction.

All behavioral and temperature measurements were coded live. The analysis presented is based on those live measurements. All the experiments were videotaped in case further analysis is ever warranted. 2 × 2 repeated measures ANOVA (species: bonobo and chimpanzee; condition: before and after) were used to examine ear temperature in relation to species and condition when subjects heard either a group-mate or stranger scream. A difference score was calculated for each condition and species by subtracting the temperature of the right ear from that of the left after each playback. The change for each subject in each condition was established using the mean difference between the first measurement of each ear and that of the second after each playback. The idea being that the relative responses of the ear temperature to the different playbacks would give an estimate of whether subjects perceived the playback as negative by showing an increase in the temperature of the right ear in comparison to the left (Parr and Hopkins, 2000).

Our findings are consistent with the idea that bonobos are xenophilic and chimpanzees are xenophobic. Figure 15.1a shows there is no significant change in ear temperature in either species after hearing the scream of a familiar group-mate. Figure 15.1b, however, shows that when subjects heard the scream of a stranger there was a trend towards an effect of species [$F(1, 21) = 3.76$, $p < 0.066$] and a significant interaction between species and condition [$F(1, 21) = 4.8$, $p < 0.04$]. Chimpanzees had an increase in the temperature of their right ear compared to their left after hearing the call of a stranger ($t = 1.81$, $df = 10$, $p = 0.05$, one sample t-test, one tailed) while bonobos did not. The ear temperature response of the two species also only differed after

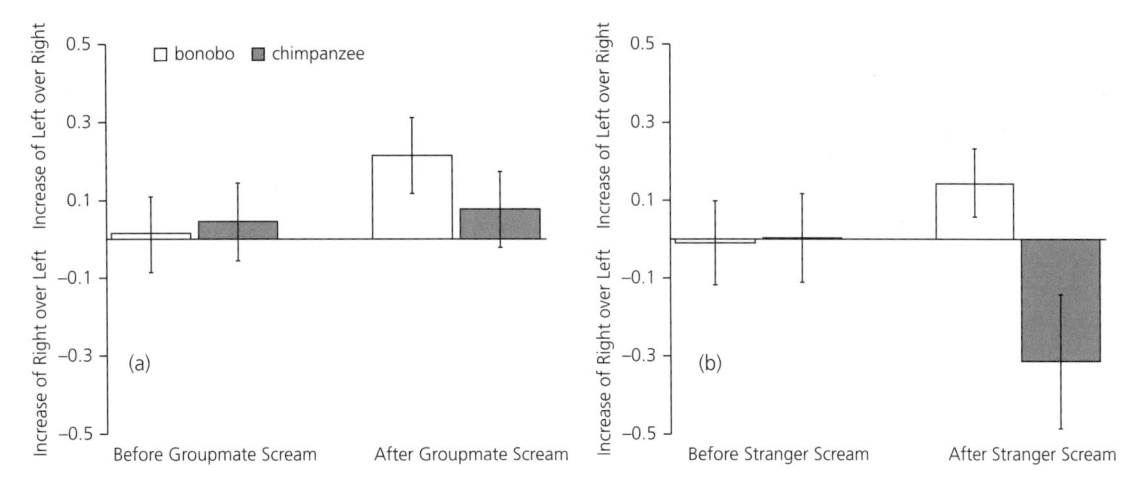

Figure 15.2 Comparison of the change in ear temperature of bonobos and chimpanzees before and after hearing the scream of A) a group-mate and B) a stranger. Only chimpanzees show evidence of negative valence with an increase in their right ear temperature after hearing a stranger scream. Most bonobo infants approached the playback speaker while most chimpanzee infants did not. Taken together, these pilot data support the idea that chimpanzees are fearful of out-group members while bonobos may even be attracted to strangers.
(Comparaison du changement de la température de l'oreille des bonobos et des chimpanzés après avoir entendu le cri de A) un partenaire et B) un étranger. Seuls les chimpanzés montrent des caractéristiques d'une valence négative avec une augmentation de la température de l'oreille droite après avoir entendu le cri d'un étranger. La plupart des enfants bonobos se sont rapprochés des hauts-parleurs alors que les enfants chimpanzés ne l'ont pas fait. Ces données soutiennent l'idée que les chimpanzés ont peur de membres hors-groupe alors que les bonobos peuvent même être attirés aux étrangers.)

they heard the stranger scream ($t = 2.34$, $df = 15.32$, $p = 0.033$, Welch independent t-test, two tailed). In fact, while not significant, the only change seen in bonobos was an increase in their left ear temperature after hearing each scream. Seven out of eight bonobo infants also approached the carrying crate containing the speaker during both playbacks while two chimpanzee infant approached the crate after the group-mate call and only one after the stranger scream. This suggests bonobos did not perceive either scream as threatening.

It is important to recognize that this is the first attempt to measure negative valence using ear temperatures in bonobos. It is possible that bonobos simply do not show the chimp-like right ear response in any situation or that this response has somehow become hemispherically reversed with higher left ear temperature indicating stress in bonobos. To rule this possibility out one would need to show an increase in right ear temperature in response to a non-social stressor in bonobos as seen in chimpanzees (Parr and Hopkins, 2000). However, given previous work with chimpanzees, one can feel more confident that the change observed here in the chimpanzees' right ear temperatures signals negative valence. The response of the infant chimpanzees is consistent with the fear that adult chimpanzees show towards strangers during playback experiments (Wilson et al., 2001) and their natural interactions (Wilson et al., 2014). The only question remains to which aspect of the playback the chimpanzees were actually responding, since only a single exemplar was used and it was always a juvenile scream. At the very least this pilot data should inspire future research aimed at exploring the possibility that bonobos and chimpanzees have opposite reactions to strangers and that this reaction is present even in the youngest infants who have had little opportunity to develop these preferences through direct interactions with adult out-group members. This pilot data is consistent with the characterization of chimpanzees as xenophobic and bonobos as xenophilic in response to strange conspecifics. More thorough investigations should now be possible using the latest methods to evaluate the emotional responses of non-human apes to others (i.e. Kano et al., in press).

Implications for cognitive evolution in humans and beyond

The first generation of developmental and cognitive comparisons between bonobos and chimpanzees support the SDH. Bonobos show a suite of traits that are seen in a range of species that have undergone selection for prosociality and against aggression. Cognitive comparisons support the idea that this process occurred due to natural selection caused by reduced scramble competition. Neither the mosaic nor the bonobo-like ancestor model predict or can explain the range of developmental differences reviewed here (Hare and Wrangham, submitted). Chimpanzees probably serve as the best (but still far from perfect) living representative of our species' last common ape ancestor. Meanwhile, the more derived bonobo gives us a powerful way to think about the evolution of the most derived ape of all, ourselves (Hare, 2017). The work with bonobos suggests that many species might have evolved due to natural selection targeting temperament. As a result, many wild species may express components of the domestication syndrome.

Research comparing humans with chimpanzees is often designed to demonstrate which human features are derived *hominin* features. The even bigger challenge is uncovering the evolutionary process that shaped features unique to human cognition (Hare, 2011). Bonobos will play a central role in testing ideas about how humans evolved. The flexible cooperation seen in bonobos relative to chimpanzees suggests tolerance is crucial to increases in cooperation between species (Hare et al., 2007). This observation and comparisons of domestic species suggests that the evolution of temperament is a powerful mechanism that can affect social problem solving abilities (Bray et al., 2015; Hare, 2011).

Children with shy temperament tend to develop sophisticated theory of mind skills earlier in development. This suggests that, even in humans, there may be an important link between temperament and social cognition that is crucial for unique forms of human cognition (Lane et al., 2013; Wellman et al., 2011; also see Rodrigues et al., 2009).

Direct comparisons of cognitive development across a range of cognitive tasks in bonobo, chimpanzee and human infants have also revealed that human infants have a set of early emerging social cognitive abilities. These same abilities are thought to be the cognitive foundation for participating in human culture (Herrmann et al., 2007). These findings support the idea that a shift in development led to the human condition (Wobber et al., 2014). This work inspired a comparison of fossil humans that suggests that late in our species' evolution we underwent selection against aggression that may have not only shaped our morphology but more importantly our whole way of life (Cieri et al., 2014; Nelson et al., 2011). Finally, an initial developmental comparison of gene expression in humans and chimpanzees suggests heterochronic shifts. Many genes that are switched off early in the development of chimpanzee prefrontal cortex are still switched on decades later in humans (Somel et al., 2009). This initial work points to the need for a new generation of tests to examine human self-domestication (Hare, 2017). Our more derived and potentially convergent relative the bonobo will be central to these tests.

Beyond humans, there are many species that can be considered candidates for selection against aggression and for prosociality. Island-living species and species invading urban environments typically live at high density and low levels of predation. Within the primate order, there are many species with more egalitarian social systems. Comparing these species to their mainland, rural, or more despotic relatives will help test the predictions of the self-domestication hypothesis (Hare, 2017; Hare et al., 2012). Do these species only show shifts in their social behavior or do they also present traits associated with the domestication syndrome in other animals? The ultimate goal of this work is large-scale phylogenetic comparisons that use the best phylogenetic techniques to examine how selection on temperament may have driven social cognitive evolution across different vertebrate orders (i.e. similar to MacLean et al., 2014). Only then can we know if the pattern seen in domesticated species and bonobos is an unusual evolutionary pattern or a larger story that might explain a significant portion of the cognitive variance observed in nature.

Summary

Bonobos have provided a powerful test of the self-domestication hypothesis. Comparisons between bonobo and chimpanzee cognition suggest that selection for prosociality has led to changes in the development and expression of cognition in bonobos since the two species diverged. These types of comparisons between species are the starting point for larger scale phylogenetic studies that can look for similar patterns across a larger range of species (i.e. MacLean et al., 2014). The comparisons between bonobos and chimpanzees suggest that natural selection can lead to the domestication syndrome, previously thought to be exclusively the result of artificial selection. The pattern observed in bonobos has implications for other animals including those living on islands, in urban habitats and may even include humans (Hare, 2017; Hare et al., 2012). The self-domestication hypothesis has already led to a number of tests regarding the evolution of human cooperation and communication as a result of selection against within group aggression. Both bonobos and chimpanzees are necessary to unlock the secrets of human evolution. Rather than standing in the shadow of their better-known cousins, bonobos will become the best friend of evolutionary anthropologists and psychologists alike.

References

Andre, C., Kamate, C., Mabonzo, P., Morel, D., and Hare, B. (2008). The conservation value of Lola ya Bonobo Sanctuary. In: Furuichi, T. and Thompson, J. (eds). *The Bonobos: Behavior, Ecology and Conservation*). New York, NY: Springer, pp. 303–22.

Aureli, F. and de Waal, F. (1997). Inhibition of social behavior in chimpanzees under high-density conditions. *American Journal of Primatology*, 41, 213–28.

Behringer, V., Deschner, T., Deimel, C., Stevens, J., and Hohmann, G. (2014). Age related changes in urinary testosterone levels suggest differences in puberty onset and divergent life history strategies in bonobos and chimpanzees. *Hormones and Behavior*, 66, 525–33.

Behringer, V., Deschner, T., Murtagh, R., Stevens, J., and Hohmann, G. (2014). Age-related changes in thyroid hormone levels of bonobos and chimpanzees indicate heterochrony in development. *Journal of Human Evolution*, 66, 83–8.

Bray, E., MacLean, E., and Hare, B. (2015). The effect of arousal on inhibitory control in dogs. *Animal Cognition*. Issue 6, pp 1317–1329.

Brust, V. and Guenther, A. (2015). Domestication effects on behavioral traits and learning performance: comparing wild cavies to guinea pigs. *Animal Cognition*, 18, 99–109.

Bullinger, A., Burkart, J., Melis, A., and Tomasello, M. (2013). Bonobos, chimpanzees, and marmosets prefer to feed alone. *Animal Behaviour*, 85, 51–60.

Byrnit, J., Hogh-Olesen, H., and Makransky, G. (2015). Share your sweets: Chimpanzee (Pan troglodytes) and bonobo (Pan paniscus) willingness to share highly attractive, monopolizable food sources. *Journal of Comparative Psychology*, 129(3), 218–228.

Cieri, R. L., Churchill, S. E., Franciscus, R. G., Tan, J., and Hare, B. (2014). Craniofacial feminization, social tolerance and the origins of behavioral modernity. *Current Anthropology*, 55(4), 419–43.

Clay, Z. and de Waal, F.B.M. (2015). Sex and strife: post conflict sexual contacts in bonobos. *Behaviour*. DOI:10.1163/1568539X-00003155.

Coolidge, H. (1933). *Pan paniscus*: pygmy chimpanzee from south of the Congo River. *American Journal of Physical Anthropology*, 18, 1–57.

Cronin, K., de Groot, E., and Stevens, J. (2015). Bonobos show limited social tolerance in a group setting: a comparison with chimpanzees and a test of the relational model. *Folia Primatologica*, 86, 164–77.

De Lathouwers, M. and Van Elsacker, L. (2006). Comparing infant and juvenile behavior in bonobos and chimpanzees: a preliminary study. *Primates*, 47, 287–93.

de Lathouwers, M. and Van Elsacker, L. (2006). Comparing infant and juvenile behavior in bonobos (*Pan paniscus*) and chimpanzees (*Pan troglodytes*): a preliminary study. *Primates*, 47, 287–93.

de Waal, F. (1988). The communicative repertoire of captive bonobo compared to that of chimpanzees. *Behaviour*, 106, 183–251.

de Waal, F.B.M. (1987). Tension regulation and non-reproductive functions of sex among captive bonobos (*Pan paniscus*). *National Geographic Research*, 3, 318–35.

de Waal, F. and Lanting, F. (1997). *Bonobo: The Forgotten Ape*. Berkeley and Los Angeles: California University Press.

Eisenberg N, Lennon R, and Roth K. (1983). Prosocial development: a longitudinal study. *Developmental Psychology*, 19: 846.

Feldblum, J.T., Wroblewski, E.E., Rudicell, R.S., Hahn, B.H., Paiva, T., Cetinkaya-Rundel, M., Pusey, A.E. and Gilby, I.C. Sexually coercive male chimpanzees sire more offspring. *Current Biology*, 24, 2855–60.

Furuichi, T. (2011). Female contributions to the peaceful nature of bonobo society. -*Evolutionary Anthropology*, 20, 131–142.

Furuichi, T., Sanz, C., Koops, K., Sakamaki, T., Ryu, H., Tokuyama, N., and Morgan, D. (2015). Why do wild bonobos not use tools like chimpanzees do? *Behaviour*. DOI:10.1163/1568539X-00003226

Garai, C., Furuichi, T., Kawamoto, Y., Ryu, H., and Inoure-Murayama, M. (2014). Androgen receptor and mono-

amine oxidase polymorphism in wild bonobos. *Meta Gene*, 2, 831–43.
Garièpy, J., Bauer, D. and Cairns, R. (2001). Selective breeding for differential aggression in mice provides evidence for heterochrony in social behaviours. *Animal Behaviour*, 61, 933–47.
Gould, S. (1977). *Ontogeny and Phylogeny*. Cambridge, MA: Harvard University Press.
Groves, C., (1981). (Comments) Bonobos: generalized hominid prototypes or specialized insular dwarfs? *Current Anthropology*, 22, 363–75.
Gruber, T., Clay, Z., and Zuberbühler, K. (2010). A comparison of bonobo and chimpanzee tool use: evidence for a female bias in the *Pan* lineage. *Animal Behaviour*, 80, 1023–33.
Hammock, E. and Young, L. (2005). Microsatellite instability generates diversity in brain and sociobehavioral traits. *Science*, 308. 1630–4.
Hare, B. (2001). Can competitive paradigms increase the validity of experiments on primate social cognition? *Animal Cognition*, 4, 269–80.
Hare, B. (2007). From nonhuman to human mind: what changed and why. *Current Directions in Psychological Science*, 16, 60–4.
Hare, B. (2009). What is the effect of affect on bonobo and chimpanzee problem solving? In: Berthoz, A. and Christen, Y. (eds) *The Neurobiology of the Umwelt: How Living Beings Perceive the World*. Springer Press, Berlin, pp. 89–102.
Hare, B. (2011). From hominoid to hominid mind: what changed and why? *Annual Review of Anthropology*, 40, 293–309.
Hare, B. (2017). Survival of the friendliest: *Homo sapiens* evolved due to selection for prosociality. *Annual Review of Psychology*, 68.
Hare, B., Brown, M., Williamson, C., and Tomasello, M. (2002). The domestication of social cognition in dogs. *Science*, 298, 1634–6.
Hare, B., Call, J., and Tomasello, M. (2001. Do chimpanzees know what conspecifics know? *Animal Behaviour*, 61, 139–51.
Hare, B. and Kwetuenda, S. (2010). Bonobos voluntarily share their own food with others. *Current Biology*, 20, 230–1.
Hare, B., Melis, A., Woods, V., Hastings, S., and Wrangham, R. (2007). Tolerance allows bonobos to outperform chimpanzees in a cooperative task. *Current Biology*, 17, 619–23.
Hare, B., Plyusnina, I., Ignacio, N., Schepina, O., Stepika, A., Wrangham, R., and Trut, L. (2005). Social cognitive evolution in captive foxes is a correlated by-product of experimental domestication. *Current Biology*, 15, 226–30.

Hare, B. Rosati, A. Breaur, J., Kaminski, J., Call, J., and Tomasello, M. (2010). Dogs are more skilled than wolves with human social cues: a response to Udell et al. (2008) and Wynne et al. (2008). *Animal Behaviour*, 79, e1–e6.
Hare, B. and Tomasello, M. (2005). Human-like social skills in dogs? *Trends in Cognitive Sciences*, 9, 439–44.
Hare, B., Wobber, T., and Wrangham, R. (2012). The self-domestication hypothesis: bonobo psychology evolved due to selection against male aggression. *Animal Behaviour*, 83, 573–85.
Hare, B. and Woods, V. (2013). *The Genius of Dogs*. Dutton, NY: Penguin.
Hare, B. and Wrangham, R. in press. Equal, similar but different: conserved chimpanzees and convergent bonobos. *Pancestor*. Muller, M. and Wrangham, R. (eds).
Hare, B. and Yamamoto, S. (2015). Moving bonobos off of the scientifically endangered list. *Behaviour*, 152, 545–62.
Hashimoto, C. and Furuich, T. (2006). Comparison of behavioral sequence of copulation between chimpanzees and bonobos. *Primates*, 47, 51–5.
Haun, D. (2011). Great apes' risk-taking strategies in a decision making task. *PLoS One*, 6, e28801–e28801.
Heilbronner, S., Rosati, A., Stevens, J., Hare, B., and Hauser, M. (2008). A fruit in the hand or two in the bush? Ecological pressures select for divergent risk preferences in chimpanzees and bonobos. *Proceedings of the Royal Society: Biology Letters*, 4, 246–9.
Hernádi, A., Kis, A., Turcsán, B. and Topál, J. (2012). Man's underground best friend: domestic ferrets, unlike the wild forms, show evidence of dog-like social-cognitive skills. *PLoS One*, 7, e43267.
Herrmann, E., Call, J., Hernández-Lloreda, M., Hare, B., and Tomasello, M. (2007). The cultural intelligence hypothesis: humans evolved specialized skills of social cognition. *Science*, 317, 1360–5.
Herrmann, E., Hare, B. Call, J., and Tomasello, M. (2010). Differences in the cognitive skills of bonobos and chimpanzees. *PLoS One*, 5, e12438.
Herrmann, E., Hare, B. Cisseski, J., and Tomasello, M. (2011). The origins of human temperament: children avoid novelty more than other apes. *Developmental Science*, 14, 1393–405.
Herrmann, E., Keupp, S., Hare, B., Vaish, A., and Tomasello, M. (2013). Direct and indirect reputation formation in great apes and human children. *Journal of Comparative Psychology*, 127, 63–75.
Hill, K. Boesch, C., Goodall, J., Pusey, A., Williams, J., and Wrangham, R. Mortality rates among wild chimpanzees. *Journal of Human Evolution*, 40, 437–50.
Hohmann, G. (2001). Association and social interactions between strangers and residents in bonobos (*Pan paniscus*). *Primates*, 42, 91–9.

Hohmann, G. and Fruth, B. (2000). Use and function of genital contacts among female bonobos. *Animal Behaviour*, 60, 107e120.

Hohmann, G. and Fruth, B. (2002). Dynamics in social organization of bonobos *(Pan paniscus)*. In: Boesch, C., Marchant, L., and Hohmann, G. (eds). *Behavioural Diversity in Chimpanzee*. Cambridge: Cambridge University Press, pp. 138–50.

Hohmann, G. and Fruth, B. (2003a). Culture in bonobos? Between-species and within- species variation in behavior. *Current Anthropology*, 44, 563–609.

Hohmann, G. and Fruth, B. (2003b). Intra- and inter-sexual aggression by bonobos in the context of mating. *Behaviour*, 140, 1389–413.

Hohmann, G. and Fruth, B. (2003c). Lui Kotal—a new site for field research on bonobos in Salonga National Park. *Pan African News*, 10(2). http://mahale.main.jp/PAN/10_2/10(2)_05.html.

Hopkins, W., Schaeffer, J., Russell, J., Bogart, S., Meguerditchian, A., and Coulon, O. (2015). A comparative assessment of handedness and its potential neuroanatomical correlates in chimpanzees and bonobos. *Behaviour*. DOI:10.1163/1568539X-00003204.

Idani, G. (1991). Social relationships between immigrant and resident bonobo *(Pan paniscus)* females at Wamba. *Folia Primatologica*, 57, 83–95.

Ihobe, H. (1992). Male-male relationships among wild bonobos at Wamba, Republic of Zaire. *Primates*, 33, 163–79.

Jaeggi, A., Stevens, J., and van Schaik, C. (2010). Tolerant food sharing and reciprocity is precluded by despotism among bonobos but not chimpanzees. *American Journal of Physical Anthropology*, 143, 41–51.

Kahlenberg, S., Emery-Thompson, M., Muller, M., and Wrangham, R. (2008). Immigration costs for female chimpanzees and male protection as an immigrant counterstrategy to intrasexual aggression. *Animal Behaviour*, 76, 1497–509.

Kano, T. (1992). *The Last Ape: Pygmy Chimpanzee Behavior and Ecology*. Stanford, CA: Stanford University Press.

Kano, F., Hirata, S., and Call, J. (2015). Social attention in the two species of Pan: bonobos make more eye contact than chimpanzees. *PLoS One*. DOI: 10.1371/journal.pone.0129684.

Kano, F., Hirata, S., Deschner, T., Behringer, V., and Call, J. (in press). Nasal temperature drop in response 2 to a playback of conspecific fights in chimpanzees: A thermo-imaging study. *Physiology & Behavior*.

Krupenye, C. and Hare, B. in preparation. Bonobos prefer hinderers over helpers.

Krupenye, C., Rosati, A. G. and Hare, B. (2015). Bonobos and chimpanzees exhibit human-like framing effects. *Biology letters*, 11, 20140527.

Kuroda, S. (1989). Developmental retardation and behavioral characteristics of pygmy chimpanzees. In: Heltne, P. and Marquardt, L. (eds). *Understanding Chimpanzees*. Cambridge, MA: Harvard University Press, pp. 184–93.

Lane, J., Wellman, H., Olson, S., Miller, A., Wang, L., and Tardif, T. (2013). Relations between temperament and theory of mind development in the United States and China: biological and behavioral correlates of preschoolers' false belief understanding. *Developmental Psychology*, 49, 825–36.

Leach, H., Groves, C., O'Connor, T., Pearson, O., and Zeder, M. (2003). Human domestication reconsidered. *Current Anthropology*, 44, 349–68.

Lewejohann, L., Pickel, T., Sachser, N., and Kaiser, S. (2010). Wild genius-domestic fool? Spatial learning abilities of wild and domestic guinea pigs. *Frontiers in Zoology*, 7, 1–8.

Lieberman, D., Carlo, J., Ponce de Leon, M., and Zollikofer, C. (2007). A geometric morphometric analysis of heterochrony in the cranium of chimpanzees and bonobos. *Journal of Human Evolution*, 52, 647–62.

Liu, J., Portnoy, J., and Raine, A. (2012). Association between a marker for prenatal testosterone exposure and externalizing behavior problems in children. *Development and Psychopathology*, 24(3), 771–82.

MacLean, E and Hare, B. (2012). Bonobos and chimpanzees infer the target of an actor's attention. *Animal Behaviour*, 83, 345–53.

MacLean, E. and Hare, B. (2013). Spontaneous triadic play in bonobos and chimpanzees. *Journal of Comparative Psychology*, 127, 245–55.

MacLean, E., and Hare, B. (2015). Bonobos and chimpanzees read helpful gestures better than prohibitive gestures. *Behaviour*. DOI:10.1163/1568539X-00003203.

MacLean, E. and Hare, B. (2015). Dogs hijack the human bonding pathway. *Science*, 348, 280–1.

MacLean, E., Hare, B., Nunn, C., Addessi, E., Amici, F., Anderson, R., Aureli, F., Baker, J., Bania, A., Barnard, A., Boogert, N., Brannon, E., Bray, E., Bray, J., Brent, L., Burkart, J., Call, J., Cantlon, J., Cheke, L., Clayton, N., Delgado, M., DiVincenti, L., Fujita, K., Hiramatsu, C., Jacobs, L., Jordan, K., Laude, J., Leimgruber, K., Messer, E., Moura, A., Ostojić, L., Picard, A., Platt, M., Plotnik, J., Range, F., Reader, S., Reddy, R., Sandel, A., Santos, L., Schumann, K., Seed, A., Sewall, K., Shaw, K., Slocombe, K., Su, Y., Takimoto, A., Tan, J., Tao, J., van Schaik, C., Virányi, Z., Visalberghi, E., Wade, J., Watanabe, A., Widness, J., Zentall, T., and Zhao, T. (2014). The evolution of self-control. *Proceedings of the National Academy of Sciences*. 111 (20), E2140-E2148.

MacLean, E., Matthews, L., Hare, B., Nunn, C., Anderson, R., Aureli, F., Brannon, E., Call, J., Drea, C., Emery, N., Haun, D., Herrmann, E., Jacobs, L., Platt, M., Rosati, A., Sandel, A., Schroepfer, K., Seed, A., Tan, J., van Schaik, C., and Wobber, V. (2011). How does cognition evolves?:

phylogenetic comparative psychology. *Animal Cognition*, 15, 223–38.

Martin-Ordas, G., Haun, D., Colmenares, F., and Call, J. (2010). Keeping track of time: evidence for episodic-like memory in great apes. *Animal Cognition*, 13, 331–40.

McIntyre, M., Herrmann, E., Wobber, V., Halbwax, M., Mohamba, C., De Sousa, N., Atencia, R., Cox, D., and Hare, B. (2009). Bonobos have a more human-like second-to-fourth finger length ratio (2D:4D) than chimpanzees: a hypothesized indication of lower prenatal androgens. *Journal of Human Evolution*, 56, 361–5.

Melis, A., Warneken, F., and Hare, B. (2010). Collaboration and helping in chimpanzees. In: Longsdorf, E., Ross, S., and Matsuzawa, T. (eds). *The Chimpanzee Mind*. Springer, Berlin, pp. 166–83.

Micheletta, J. and Waller, B. (2012). Friendship affects gaze following in a tolerant species of macaque. *Animal Behaviour*, 83, 459–67.

Miklosi, A., Kubinyi, E., Topal, J., Gacsi, M., Viranyi, Z., and Csanyi, V. (2003). A simple reason for a big difference: wolves do not look back at humans, but dogs do. *Current Biology*, 13, 763–66.

Mitani, J. and Gros-Louis, J. (1995). Species and sex differences in the screams of chimpanzees and bonobos. *International Journal of Primatology*, 16, 393–411.

Morimoto, N., Ponce de Leon, M., and Zolikofer, C. (2012). Phenotypic variation in infants, not adults, reflect genotypic variation among chimpanzees and bonobos. *PLoS One*. DOI: 10.1371/journal.pone.0102074.

Moscovice, L., Descher, T., and Hohmann, G. (2015). Welcome back: responses of female bonobos to fusions. *PLoS One*, 10, e0127305.

Mulcahy, N. and Call, J. (2006). Apes save tools for future use. *Science*, 312, 1038–40.

Muller, M., Emery-Thompson, M., Kahlenberg, S., and Wrangham, R. (2011). Sexual coercion by male chimpanzees shows that female choice can be more apparent than real. *Behavioral Ecology and Sociobiology*, 65, 921–33.

Muller, M., Kahlenberg, S., Emery-Thompson, M., and Wrangham, R. (2007). Male coercion and the costs of promiscuous mating for female chimpanzees. *Proceedings of the Royal Society B*, 274, 1009–14.

Muller, M. and Mitani, J., (2005). Conflict and cooperation in wild chimpanzees. *Advances in the Study of Behavior*, 35, 275–331.

Nagasawa, M., Mitsui, S., En, S., Ohtani, N., Ohta, M., Sakuma, Y., Onaka, T., Mogi, K., and Kikusui, T. (2015). Oxytocin-gaze positive loop and the co-evolution of human-dog bonds. *Science*, 348, 333–336.

Nagasawa, M., Kikusui, T., Onaka, T., and Ohta, M. (2009). Dog's gaze at its owner increases owner's urinary oxytocin during social interaction. *Hormones and Behavior*, 55, 434–41.

Nelson, E., Campbell, R., Cashmore, L., and Shultz, S. (2011). Digit ratios predict polygyny in early apes, Ardipithicus, Neanderthals and early modern humans but not in Australopithecus. *Proceedings of the Royal Society B*, 278, 1556–63.

Oliva, J., Rault, J., Appleton, B., and Lill, A. (2015). Oxytocin enhances the appropriate use of human social cues by the domestic dog (*Canis familiaris*) in an object choice task. *Animal Cognition*, 18, 767–75.

Palagi, E. and Cordoni, G. (2012). The right time to happen: play development divergence in the two *Pan* species. *PLoS One*. DOI: 10.1371/journal.pone.0052767.

Palagi, E. and Paoli, T. (2007). Play in adult bonobos (*Pan paniscus*): modality and potential meaning. *American Journal of Physical Anthropology*, 134, 219–25.

Paoli, T., Tacconi, G., Borgognini Tarli, S. and Palagi, E. (2007). Influence of feeding and short-term crowding on the sexual repertoire of captive bonobos (*Pan paniscus*). *Annals of Zoological Fennici*, 44, 81–8.

Parr, L., Hopkins, W. (2000). Brain asymmetries and emotional perception in chimpanzees. *Physiology and Behavior*, 71, 1–15.

Petit, O. Desportes, C., and Thierry, B. (1992). Differential probability of 'coproduction' in two species of macaque (Macaca tonkeana, M. mulatta). *Ethology*, 90, 107–20.

Prufer, K, Munch, K., Hellmann, I., Akagi, K. et al. (2012). The bonobo genome compared to the chimpanzee and human genome. *Nature*, 486, 527–31.

Pusey, A., Murray, C., Wallauer, W., Wilson, M., Wroblewski, E. and Goodall, J. (2008). Severe aggression among female *Pan troglodytes schweinfurthii* at Gombe National Park, Tanzania. *International Journal of Primatology*, 29, 949–73.

Pusey, A. and Schroepfer-Walker, K. (2013). Female competition in chimpanzees. *Philosophical Transactions of the Royal Society B*, 368. DOI:10.1098/rstb.2013.0077.

Rodrigues, S., Saslow, L., Garcia, N., Oliver, J., and Keltner, D. (2009). Oxytocin receptor genetic variation relates to empathy and stress reactivity in humans. *Proceedings of the National Academy of Sciences*, 106, 21437–41.

Romero, T., Nagasawa, M., Mogi, K., Hasegawa, T., and Kikusui, T. (2014). Oxytocin promotes social bonding in dogs. *Proceedings of the National Academy of Sciences*, 201322868.

Rosati, A. (2015). Context influences spatial frames of reference in bonobos (*Pan paniscus*). *Behaviour*. Volume 152, Issue 3–4, pp. 375–406.

Rosati, A. and Hare, B. (2012a). Decision-making across social contexts: competition increases risk-prone choices in chimpanzees and bonobos. *Animal Behaviour*, 84, 869–79.

Rosati, A. and Hare, B. (2012b). Chimpanzees and bonobos exhibit divergent spatial memory development. *Developmental Science*, 15, 840–53.

Rosati, A. and Hare, B. (2013). Chimpanzees and bonobos exhibit emotional responses to decision outcomes. *PLoS One*, 8 (5), e63058.

Rosati, A., Rodriguez, K., and Hare, B. (2014). The ecology of spatial memory in four lemur species. *Animal Cognition*. 4, 947–961.

Rosati, A., Stevens, J., Hare, B., and Hauser, M. (2007). The evolutionary origins of human patience: temporal preferences in chimpanzees, bonobos, and human adults. *Current Biology*, 17, 1663–8.

Rosati, A., Stevens, J., and Hauser, M. (2005). The effect of handling time on temporal discounting in two New World primates. *Animal Behavior*, 71, 1379–87.

Ryu, H., Hill, D., and Furuichi, T. (2015). Prolonged maximal sexual swelling in wild bonobos facilitates affiliative interactions between females. *Behaviour*. DOI:10.1163/1568539X-00003212.

Sakamaki, T., Behncke, I., Laporte, M., Mulavwa, M., Ryu, H., Takemoto, H., Tokuyama, N., Yamamoto, S., Yamagiwa, J., Aureli, F. and Furuichi, T. Intergroup transfer of females and social relationships between immigrants and residents in bonobo societies. In: Furuichi, T., et al. (eds) *Dispersing Primate Females*. DOI: 10.1007/978-4-431-55480-6_6.

Santos, L. and Rosati, A. (2015). The evolutionary roots of human decision making. *Annual Review of Psychology*, 66, 321–47.

Schneider, C., Call, J., and Liebal, K. (2012). Onset and early use of gestural communication in nonhuman great apes. *American Journal of Primatology*, 74, 102–113.

Schroepfer-Walker, K., Wobber, T., and Hare, B. (2015). Experimental evidence that grooming and play are social currency in bonobos and chimpanzees. *Behaviour*.

Schubert, G., Vigilant, L., Boesch, C., Klenke, R., Langergraber, K., Mundry, R., Surbeck, M., and Hohmann, G. (2013). Co-residence between males and their mothers and grandmothers is more frequent in bonobos than chimpanzees. *PLoS One*, DOI:10.1371/journal.pone.0083870.

Schultz, A. (1941). The relative size of the cranial capacity in primates. *American Journal of Physical Anthropology*, 28, 273–87.

Seres, M., Aureli, F., de Waal, F. (2001). Successful formation of a large chimpanzee group out of two preexisting subgroups. *Zoo Biology*, 20, 501–15.

Somel, M., Franz, H., Yan, Z., Lorenc, A., Guo, S., Giger, T., Kelso, J., Nickel, B., Dannemann, M., and Bahn, S. (2009). Transcriptional neoteny in the human brain. *Proceedings of the National Academy of Sciences*, 106, 5743–8.

Staes, N., Stevens, J., Helsen, P., Hillyer, M., Korody, M., and Eens, M. (2014). Oxytocin and vasopressin receptor gene variation as a proximate base for inter- and intraspecific behavioral differences in bonobos and chimpanzees. *PLoS One*. DOI: 10.1371/journal.pone.0113364.

Staes, N., Koski, S. E., Helsen, P., Fransen, E., Eens, M., and Stevens, J. M. (2015). Chimpanzee sociability is associated with vasopressin (Avpr1a) but not oxytocin receptor gene (OXTR) variation. *Hormones and behavior*, 75, 84–90.

Stevens, J., De Groot, E., and Staes, N. (2015). Relationship quality in captive bonobo groups. *Behaviour*.

Stevens, J., Rosati, A., Ross, K., and Hauser, M. (2005). Will travel for food: spatial discounting in two New World monkeys. *Current Biology*, 15, 1855–60.

Surbeck, M. and Hohmann, G. (2008). Primate hunting by bonobos at LuiKotale, Salonga National Park. *Current Biology*, 18, R906–R907.

Surbeck, M., Mundry, R., and Hohmann, G. (2011). Mothers matter! Maternal support, dominance status and mating success in male bonobos (*Pan paniscus*). *Proceedings of the Royal Society B*, 278, 590.

Tan, J., Ariely, S., and Hare, B. in preparation. Bonobos demonstrate evolution in weak ties.

Tan, J. and Hare, B. (2013). Bonobos share with strangers. *PLoS One*. 8(1): e51922.

Tan, J., Kwetuenda, S., and Hare, B. (2015). Preference or paradigm? Bonobos do not share in 'the' prosocial choice task. *Behaviour*. DOI:10.1163/1568539X-00003230.

Tomasello, M. and Call, J. (2008). Assessing the validity of ape-human comparisons: A reply to Boesch 2007. *Journal of Comparative Psychology*, 122, 449–52.

Tomasello, M., Hare, B., Call, J., and Leehman, H. (2007). Reliance on head versus eye gaze in great apes and human infants. *Journal of Human Evolution*, 52, 314–20.

Topal, J., Gergely, G., Erdohegyi, A., Csibra, G., Miklosi, A. (2009). Differential sensitivity to human communication in dogs, wolves, and human infants. *Science*, 325, 1269–72.

Trut, L. (1999). Early canid domestication: the farm-fox experiment. *American Science*, 87, 160–9.

Trut, L., Oskina, I., and Kharlamova, A. (2009). Animal evolution during domestication: the domesticated fox as a model. *Bioessays*, 31, 349.

Vlamings, P.H.J.M., Hare, B., and Call, J. (2010). Reaching around barriers: the performance of the great apes and 3–5-year-old children. *Animal Cognition*, 13, 273–85.

Warnaken, F., Hare, B., Melis, A., Hanus, D., and Tomasello, M. (2007). Spontaneous altruism by chimpanzees and children. *PLoS Biology*, **5**, 1–7.

Watts, D. and Mitani, J. (2002). Hunting behaviour of chimpanzees at Ngogo, Kibale National Park, Uganda. In: Boesch, C., Hohmann, G., and Marchant, L. (eds). *Behavioral Diversity in Chimpanzees and Bonobos.*. Cambridge: Cambridge University Press, pp. 244–57.

Wellman, H. M., Lane, J. D., LaBounty, J., and Olson, S. L. (2010). Observant, nonaggressive temperament predicts theory of mind development. *Developmental Science*, 14, 319–28.

West-Eberhard, M. (2003). *Developmental Plasticity and Evolution.* New York, NY: Oxford University Press.

Wilkins, A., Wrangham, R., and Fitch, T. (2014). The domestication syndrome in mammals: a unified explanation based on neural crest cell behavior and genetics. *Genetics,* 197, 795–808.

Wilson, M., Boesch, C., Fruth, B., Furuichi, T., Gilby, I. et al. (2014). Lethal aggression in *Pan* is better explained by adaptive strategies than human impacts. *Nature,* 513, 414–17.

Wilson, M., Hauser, M., and Wrangham, R. (2001). Does participation in intergroup conflict depend on numerical assessment, range location or rank for wild chimpanzees? *Animal Behaviour,* 61, 1203–16.

Wittig, R., Crockford, C., Deschner, T., Langergraber, K., Ziegler, T., and Zuberbuhler, K. (2014). Food sharing is linked to urinary oxytocin level and bonding in related and unrelated wild chimpanzees. *Proceedings of the Royal Society B,* 281, 20133096.

Wobber, T. and Hare, B. (2011). Psychological health of orphan bonobos and chimpanzees in African sanctuaries. *PLoS One,* 6, e17147.

Wobber, V., Hare, B., Koler-Matznick, J., Wrangham, R., and Tomasello, M. (2009). Breed differences in domestic dogs' (*Canis familiaris*) comprehension of human communicative signals. *Interaction Studies,* 10, 206–24.

Wobber, V., Hare, B., Maboto, J., Lipson, S. Wrangham, R., and Ellison, P. (2010c). Differential reactivity of steroid hormones in chimpanzees and bonobos when anticipating food competition. *Proceedings of the National Academy of Sciences,* 107, 12457–62.

Wobber, V. and Herrmann, E. (2015). The influence of testosterone on cognitive performance in bonobos and chimpanzees. *Behaviour,* 152, 407–23. DOI:10.1163/1568539X-00003202.

Wobber, T., Herrmann, E., Hare, B., Wrangham, R., and Tomasello, M. (2014). Differences in the early cognitive development of children and great apes. *Developmental Psychobiology,* 56, 547–73.

Wobber, V., Wrangham, R., and Hare, B. (2010a). Evidence for delayed development of social behaviour and cognition in bonobos relative to chimpanzees. *Current Biology,* 20, 226–30.

Wobber, V., Wrangham, R. and Hare, B. (2010b). Application of the heterochrony framework to the study of behaviour and cognition. *Communicative & Integrative Biology,* 3, 337.

Woods, V. and Hare, B. (2011). Bonobo but not chimpanzee infants use socio-sexual contact with peers. *Primates,* 52, 111–16.

Wrangham, R. (1993). The evolution of sexuality in chimpanzees and bonobos. *Human Nature,* 4, 47–79.

Wrangham, R. (1999). Evolution of coalitionary killing. *Yearbook of Physical Anthropology,* 42, 1–30.

Wrangham, R. (2002). The cost of sexual attraction: is there a trade-off in female *Pan* between sex appeal and received coercion? In: Boesch, C., Hohmann, G., and Marchant, L. (eds). *Behavioral Diversity in Chimpanzees and Bonobos.* Cambridge: Cambridge University Press, pp. 204–15.

Wrangham, R. (2014). Ecology and social relationships in two species of chimpanzees. In: Rubenstein, D. and Wrangham, R. (eds). *Ecological Aspects of Social Evolution: Birds and Mammals.* Princeton, NJ: Princeton University Press.

Wrangham, R. and Peterson, D. (1997). *Demonic Males: Apes and the Origins of Human Violence.* Boston, MA: Houghton Mifflin.

Wrangham, R. and Pilbeam, D. (2001). African apes as time machines. In:Galdikas, B., Briggs, N., Sheeran, L., Shapiro, G., and Goodall, J. (eds). *All Apes Great and Small.* New York, NY: Plenum Publishers, pp. 5–17.

Wrangham, R., Wilson, M., and Muller, M. (2006). Comparative rates of violence in chimpanzees and humans. *Primates,* 47, 14–26.

Yamamoto, S. (2015). Non-reciprocal but peaceful fruit sharing in the wild bonobos of Wamba. *Behaviour,* 152, 335–57.

Yamamoto, S., Humle, T., and Tanaka, M. (2012). Chimpanzees' flexible targeted helping based on an understanding of conspecific's goals. *Proceedings of the National Academy of Sciences,* 109, 3588–92.

Zihlman, A. and Cramer, D. (1978). Skeletal differences between pygmy (*Pan paniscus*) and common chimpanzees (*Pan troglodytes*). *Folia Primatologica,* 29, 86–94.

Zollikofer, C. and Ponce de Leon, M. (2010). The evolution of hominin ontogenies. *Seminars in Cell & Development Biology,* 21, 441–52.

PART VII

Evolution

CHAPTER 16

The formation of Congo River and the origin of bonobos: A new hypothesis

Hiroyuki Takemoto, Yoshi Kawamoto and Takeshi Furuichi

Abstract The Congo River functions as a strong geographical barrier for many terrestrial mammals in the Congo Basin, separating forest habitat into right and left banks of the river. However, there has been little discussion on the biogeography of the Congo Basin because the history of the river has been obscured. Based on the recent information of the sea-floor sediments near the mouth of the river and the geophysical survey on the continent, this chapter proposes a plausible hypothesis on the Congo River formation and presents a consequent hypothesis on the divergence of bonobos (*Pan paniscus*) from other *Pan* populations. The present hypothesis is also helpful for understanding the distribution of other primates and other mammals in the basin. Furthermore, this hypothesis suggests that all hominid clades, including human, chimpanzee and gorilla, except bonobo, evolved in the area north or east of the Congo River.

La rivière du Congo a la fonction d'une barrière géographique forte pour plusieurs mammifères dans le bassin du Congo, séparant l'habitat forêt dans les banques gauches et droites de la rivière. Cependant, il y a eu peu de discussions sur la biogégraphie du bassin du Congo, parce que l'histoire de la rivière a été voilée. Récemment, quelque données importantes qui peuvent avoir des liens avec la formation de la rivière du Congo ont été acquise, surtout par la recherche des sédiments du fond marin près de la bouche de la rivière et par l'enquête géographique du continent. À partir de cette nouvelle information, nous avons proposé une hypothèse plausible sur la formation de la rivière du Congo. Nous avons aussi présenté une hypothèse conséquente sur la divergence des bonobos (Pan paniscus) des autres populations Pan (voire Takemoto et al., 2015 pour la publication originale de cette étude). L'hypothèse présente nous aide aussi à comprendre la distribution des autre primates et des autres mammifères dans le bassin. De plus, cette hypothèse suggère que tous les hominidés clades, humains inclus, chimpanzés et gorilla à l'exception du bonobo, ont évolué dans la régions du nord ou de l'est de la rivière du Congo.

Previous hypotheses on the origin of bonobos

The ranges of both bonobos (*Pan paniscus*) and chimpanzees (*Pan troglodytes*) are clearly separated by the Congo River, a geographical barrier (Figure 16.1). Bonobos range in the forest area on the left bank of the Congo River (Coolidge, 1933) while two subspecies of chimpanzees (*P. t. schweinfurthii* and *P. t. troglodytes*) range on the right bank of the river.

We counted five relevant hypotheses concerning the origin of bonobos in previous studies (Table 16.1, Figure 16.2). Hypothesis C was drawn from hypothesis D as a derivative idea. Other hypotheses, except for hypothesis B, have been slightly modified by us because the original studies contained little or no discussion on the divergence of bonobos.

These hypotheses can be characterized by four main factors: (1) the proto-*Pan* (the *Pan* population prior to the division into two *Pan* clades) population ranged on both banks of the river (A–C) or on the right bank of the present river (D–F); (2) the separation of the two clades (bonobos and chimpanzees) is either tied to the formation of the Congo River

Takemoto, H., Kawamoto, Y., and Furuichi, T., *The formation of congo river and the origin of bonobos: A new hypothesis*. In: *Bonobos: Unique in Mind, Brain, and Behavior*. Edited by Brian Hare and Shinya Yamamoto: Oxford University Press (2017).
© Oxford University Press. DOI: 10.1093/oso/9780198728511.003.0016

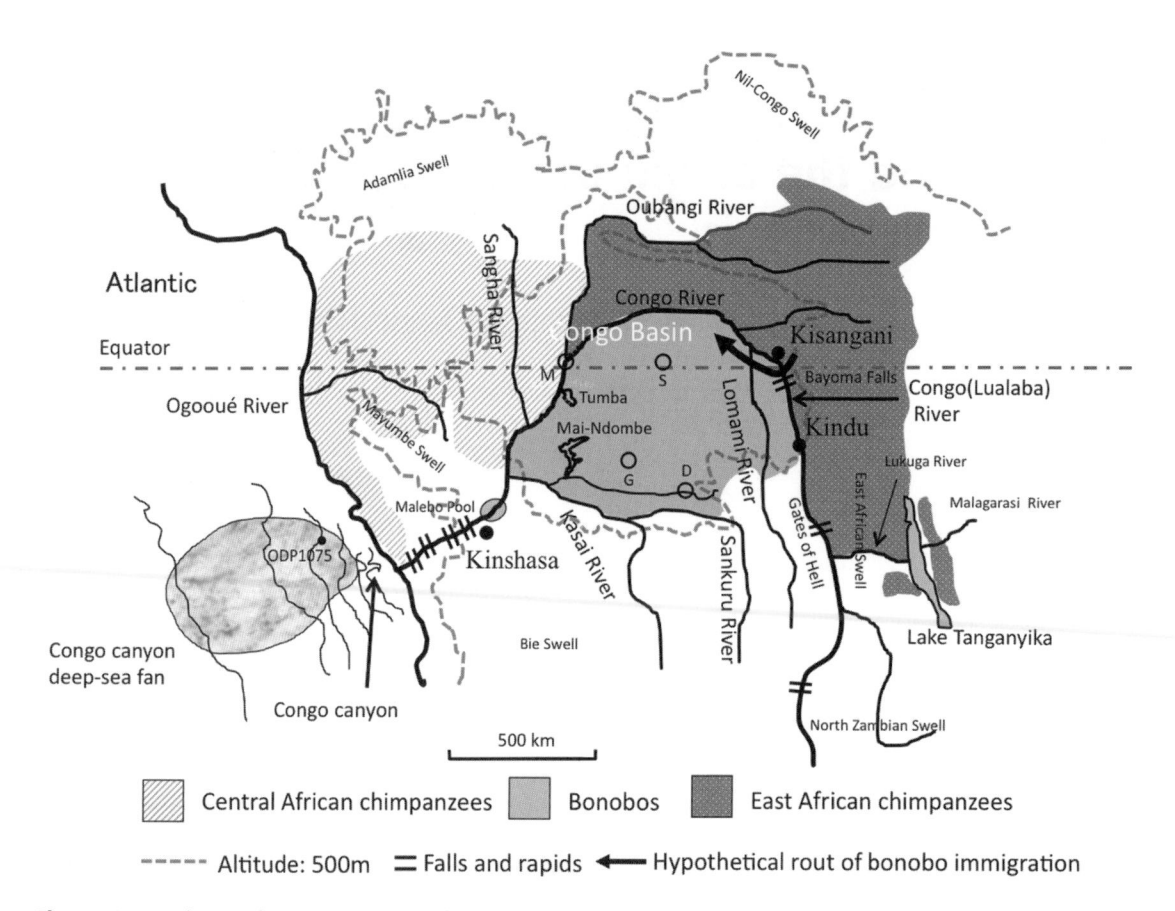

Figure 16.1 Distributions of *Pan* species in Central Africa and the geography around the Congo Basin. G, M, S and D indicate the locations of boreholes for geological survey. ODP 1075 is one of the sites where the sedimentary sequence of the Congo Fan was reported.
(Distributions des espèces Pan dans l'Afrique Centrale et la géographie autour du bassin du Congo. G, M, S et D indiquent l'emplacement des trous de forage pour l'enquête géologique. ODP 1075 est un des sites où les séquences sédimentaires du Congo Fan a été rapporté.)

(A, B) or is not (C–F); (3) separation by the Congo River as a geographical barrier was the main cause of the separation into two *Pan* clades (A–C, E, F) or was not (D); (4) proto-*Pan* was adapted to savannah vegetation and had a wider range than that of present day chimpanzees (C, D) or not (A, B, E, F).

These hypotheses are roughly classified into three groups. Hypotheses A (Caswell et al., 2008; Ferris et al., 1981; Prüfer et al., 2012; Stone et al., 2002; Yu et al., 2003), and B (Thompson, 2003) regard the formation of the Congo River as a key to their separation. Hypotheses C and D (Becquet and Przeworski, 2007) are based on an adaptation to savannah areas and habitats further beyond the forest areas inhabited by proto-*Pan*. Hypotheses E (Chaline et al., 2000) and F (Won and Hey, 2005) regard the moving of an ancestral population of bonobos to the left bank of the river as having occurred in a similar geographical setting for both the river course and the habitat occupied by *Pan* species today.

Among these, hypothesis A and B are the two major hypotheses because previous genetic studies discussing mechanism of the separation assume a simple model of allopatric speciation. Additionally, some previous literatures in paleoecology and limnology suggested that the current state of the Congo River appeared near to the Plio-Pleistocene boundary (1.8 Ma before 2009 and 2.6 Ma after 2009; see the caption of Figure 16.3) (Beadle, 1981; Moreau, 1963). It is quite natural, therefore, that some genetic studies, which estimated that the divergence time of two *Pan* clades was relatively close to the Plio-Pleistocene boundary, regarded the appearance of the Congo River as a cause of the separation of two *Pan* clades.

Table 16.1 Possible hypotheses on the origin of bonobos.
(Hypothèses sur l'origine des bonobos).

Hypothesis	The range of proto-*Pan* before the separation	How did bonobos come to range in the left bank of the river?	Is it linked to the river formation?	Congo River as a factor of the speciation	Savannah adaptation for proto-*Pan*	Main weakness	Source
A Divided by the Congo River formation	Both banks of the present Congo River in the basin.	Congo River formation divided the proto-*Pan* population into the right and left bank populations	Yes	Yes	No	The age of Congo River formation has been obscure	1
B Forest refugia	Both banks near the headwater of the assumed former river/ Forest refugium in the west to Lake Tanganyika, involving Marungu Mts	Bonobo ancestors descended to the southern basin without crossing the main stream of the Congo River and the appearance of the current river systems divided the populations	Yes	Yes	Partially Yes	Historical changes in the river course has also been obscure.	2
C Wider habitat	Both bank / Wider than present *Pan* populations exceeding the head water of the Congo River	The reduction of the habitat to the forest area in the basin spontaneously prevented the gene flow across the river	No	Yes	Yes	The wider habitat and/or the habitat reduction for proto-*Pan* are imaginary	-
D Speciation in a drier habitat	Right Bank to far from the river / Wider than present chimpanzees including semi-dry habitat	Speciation occurred outside of their current habitats such as semi-dry environment and moved to the forest area in the left bank	No	No	Yes	The mechanism of the speciation and re-adaptation to the forest habitat are unclear	3
E Detouring the Lualaba River	Right bank of the present Congo River in the basin.	Some populations detoured the headwater of the river to enter the left bank	No	Yes	No	It is unknown what condition allowed proto-*Pan* to detour the headwater	4
F Corridor on some locations	Right bank of the present Congo River in the basin	Bonobo ancestor crossed the river on the dispersal corridor in some locations along the river	No	Yes	No	It is unknown what corridor and where the corridor existed	5

1. Caswell et al., 2008; Ferris et al., 1981; Prüfer et al., 2012; Stone, et al., 2002; Yu et al., 2003.
2. Myers Thompson, 2003.
3. Bequet and Przeworski, 2007.
4. Chaline et al., 2000.
5. Won and Hey, 2005; this study.

Genetic information concerning the origin of bonobos

Studies of the divergence of bonobos from chimpanzees began with analyses of mitochondrial DNA (mtDNA). Most mtDNA studies dated the divergence of the two clades around 2.5 Ma to around 1.3 Ma (Adachi and Hasegawa, 1995; Bjork et al., 2011; Ferris et al., 1981; Gagneux et al., 1999; Horai et al., 1992; Pesole et al., 1992; Stone et al., 2010). The subsequent studies on nuclear DNA proposed dates ranging from 2.6 to 0.85 Ma (Bailey et al., 1992; Caswell et al., 2008; Gonder et al., 2011; Hey, 2010; Kaessmann et al., 1999; Langergraber et al., 2012; Prüfer et al., 2012; Stone et al., 2002; Wegmann and Excoffier, 2010; Yu et al., 2003). Most of these results, including Bayesian calculation methods, produced time periods concentrating around 1.8 Ma or 1 Ma. Recalculation using a new estimation of the generation interval for

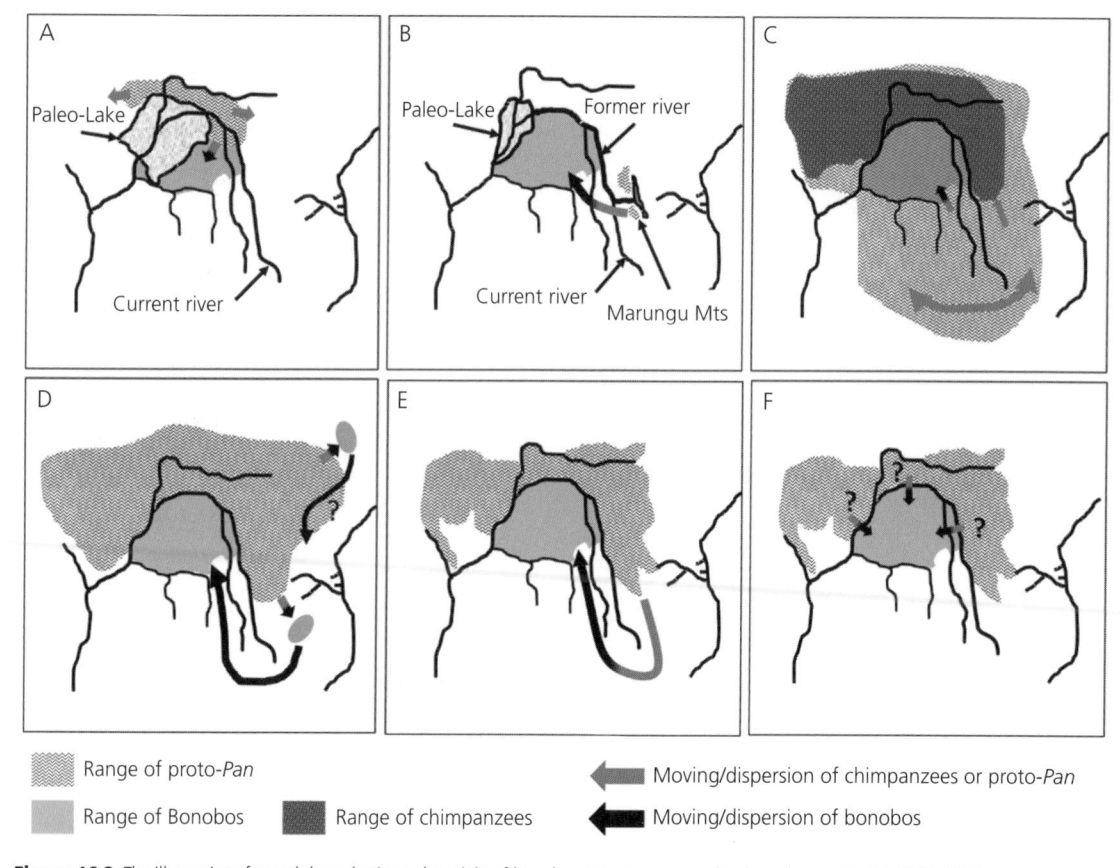

Figure 16.2 The illustrations for each hypothesis on the origin of bonobos. A to F correspond to hypotheses A to F in Table 16.1. (Les illustrations de chaque hypothèse sur l'origine des bonobos. A jusqu'à F correspondent aux hypothèses A jusqu'à F dans la Table 1.)

apes may give a slightly earlier origin of bonobos (Langergraber et al., 2012). In recent genetic studies, based on a divergence of human from *Pan* at 7 Ma to 6 Ma (Bjork et al., 2011; Caswell et al., 2008; Hey, 2010; Prüfer et al., 2012; Stone et al., 2002, 2010; Wegmann and Excoffier, 2010; Yu et al., 2003), the time of divergence between the two clades of *Pan* falls on approximately 2.1 Ma to 0.8 Ma, allowing for a margin of error. Based on this, bonobos would have branched off from other *Pan* clades sometime in the middle Pleistocene.

Genetic analyses indicate a monophyletic origin of bonobos (Fischer et al., 2011; Prüfer et al., 2012). Although the recent genome analysis indicated the possibility that some gene flow occurred from bonobos into chimpanzees during the late Pleistocene, the divergence of bonobos from chimpanzees was thought to be between 1.7 Ma and 2.1 Ma (de Manuel et al., 2016). The ancestral population of bonobos after divergence from chimpanzees was relatively small (Hey, 2010; Wegmann and Excoffier, 2010), though the population size might have remained constant at around 20,000 over the last 0.3 Ma (Hvilsom et al., 2014). These facts suggest that the major divergence event of ancestral bonobo populations from other *Pan* populations occurred only once and that bonobos might experience a 'bottleneck' sometime in the past.

Geological and geographical information concerning the origin of bonobos

The course of the Congo River follows a unique pattern in the centre of the basin (Runge, 2007). The upper stream, called Lualaba, flows to the north. The middle stream from Kisangani changes its course

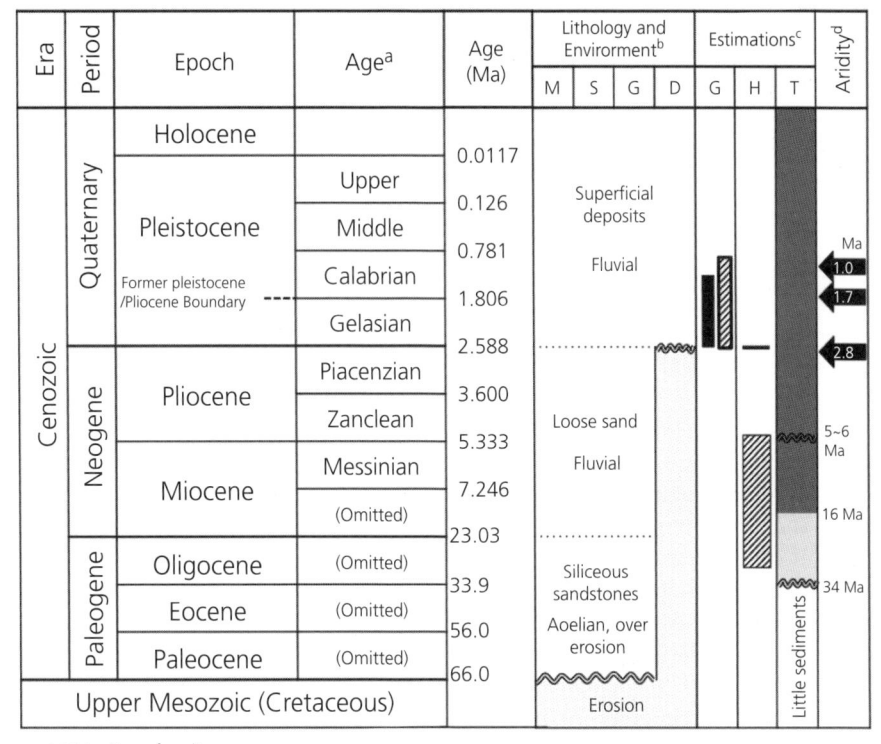

Era	Period	Epoch	Age[a]	Age (Ma)	Lithology and Environment[b] M S G D	Estimations[c] G H T	Aridity[d]
Cenozoic	Quaternary	Holocene		0.0117	Superficial deposits		
		Pleistocene	Upper	0.126			Ma
			Middle	0.781	Fluvial		1.0
		Former pleistocene /Pliocene Boundary	Calabrian	1.806			1.7
			Gelasian	2.588			2.8
	Neogene	Pliocene	Piacenzian	3.600	Loose sand		
			Zanclean	5.333			5~6 Ma
		Miocene	Messinian	7.246	Fluvial		
			(Omitted)	23.03			16 Ma
	Paleogene	Oligocene	(Omitted)	33.9	Siliceous sandstones		34 Ma
		Eocene	(Omitted)	56.0	Aoelian, over erosion		
		Paleocene	(Omitted)	66.0			
Upper Mesozoic (Cretaceous)					Erosion		

∼∼∼: Uncomformity

Figure 16.3 Geological time scale in the Cenozoic (IUGS, 2012) and geological events in the Congo Basin. a. Gelasian, which had been placed within in the Neogene since 1983, was moved into the Pleistocene in 2009 by IUGS. The term 'Tertiary,' which consisted of the Paleogene and Neogene, also disappeared (or became an 'unofficial' word). b. Lithostratigraphy and the supposed sedimentary environment in the central Congo Basin (Kadima et al., 2011; Giress, 2005). G, M, S and D are the boreholes for geological survey. c. Various estimates concerning bonobo divergence. G. The separation time between bonobos and chimpanzees estimated by genetic studies (2.6~0.85 Ma). The black rectangle is the estimate by mtDNA analysis and the striped one indicates the estimate by nuclear DNA analysis. H. The black bar indicates the old estimates of the time when the paleo-Congo Lake drained to the Atlantic Ocean (2.6 Ma) and the striped rectangle indicates the time of the Congo River formation based on geological and hydrological studies (30~5.3 Ma). T. Chronology of the depositing of Congo Fan sediments in the sea. Most of the studies showed terrigenous sediments in the Cenozoic beginning at 35 Ma to 30 Ma or 16 Ma. An unconformity was found at the Pliocene/Miocene boundary (6 Ma to around 5 Ma), but after that, deposits seemed to have accumulated continuously. d. Strong aridity in the Pleistocene indicated by the dust flux (deManocal, 2004).

(L'échelle de temps géologique dans la Cenozoic (IUGS, 2012) et les événements géologiques dans le bassin du Congo. a) Gelasian, qui a été placé dans le Néogène depuis 1983, a été mis dans le Pleistocene dans 2009 par IUGS. Le terme 'Tertiair', qui consistait du Paléogène et le Néogène, aussi a disparu (ou est devenu un mot 'non officiel'). b) Lithostratigraphie et l'environnement sédimentair supposé dans le bassin du Congo central (Kadime et al., 2011; Giress, 2005). G, M, S et D sont les trous de forage pour l'enquête géologique. c) Estimations variées concernant la divergence des bonobos. G) La séparation de temps entre les bonobos et les chimpanzés estimée par des études génétiques (2.6~0.85 Ma). Le rectangle noir est l'estimation par analyse mtDNA et les hachures indiquent l'estimation par l'analyse nucléaire de DNA. H) La bande noire indique les anciennes estimations du temps quand le lac paléo-Congo a drainé à l'Océan Atlantique (2.6 Ma) et le rectangle hachuré indique le temps de la formation de la rivière du Congo basé sur les études géologiques et hydrologiques (30~5.3 Ma). T) Chronologie du dépôt des sédiments Congo Fan dans la mer. La plupart des études ont montré des sédiments terriens au début Cenozoic en 35–30 Ma ou 16 Ma. Une non-conformité a été trouvé la frontière Pliocène/Miocène (6~5 Ma), mais après cela, les dépôts se sont accumulés continuellement. d) Acidité forte dans le Pleistocene indiqué par le flux de poussière (deManocal, 2004).)

to the west and then around to the south-west at the end, without interruption by falls or rapids. The lower stream from Malebo pool to the mouth of the river, containing rapids and the deepest part in the river, flows to the south-west. From the mouth of the river, the submarine canyon continues into the deep-sea floor, carrying with it enormous amounts of Congo Fan sediments.

Surmising the geoscience literatures concerning the deep-sea Congo Fan sediments and the topographical changes on the continent, we concluded that the unique form of the Congo River in Cenozoic appeared much earlier than previously considered: it seems to have been formed by 34 Ma, due to the sinking of the basin and the uplifts in the East African Rift (EAR) systems (see Takemoto et al., 2015). Although the formation of the Congo Basin is still debated, the uplift of EAR was confirmed to have occurred since 50 Ma by the estimation of the river profiles (Rudge et al., 2015) and it is considered that the thick sediments up to 9 km in the centre of the basin were a result of the subsidence of the basin since Neoproterozoic (800 Ma) (Crosby et al., 2010; Kadima et al., 2011; Moucha and Forte, 2011). The sinking of the centre of the basin seems to be a strong factor for the stability of the course of the main stream. The fact that the fluvial sediments are accumulated along the current main stream of the river (Kadima et al., 2011) suggests that the main stream of the Congo River flowed in the centre of the basin without changing the course dramatically since its formation.

This unique form and the old origin of the river must have affected the biogeographical pattern of animal species in the central African region. The forest area largely overlaps with the ranges of bonobos and chimpanzees (Figure 16.1). The majority of upper parts of the river from the headwaters to the Kindu area and downstream from the junction of Kasai River to the estuary flow through marshy or savannah vegetation. Only the middle part of the river flows through forest vegetation. The forest living mammals cannot easily reach the opposite bank of the main stream of this river nor have they been able to do so since 34 Ma. These patterns are very peculiar, being different from those of another big tropical river, the Amazon, which flows only through the tropical wetland and has many signs of old currents of the main stream. Contrasted with the Congo River, the instability of the course of the Amazon River may have affected the distribution of the forest animal species (Haffer, 1997).

Re-evaluation of the hypotheses on the origin of bonobos

The geological evidence in the Congo Basin and genetic information of bonobos provide us with quite an important suggestion for the divergence of two *Pan* clades. First, the time of divergence of two *Pan* clads (2.1–0.8 Ma) is at quite a temporal distance from the Congo River formation (34 Ma). Therefore, the hypotheses that link origin of bonobos to the formation of the Congo River (hypotheses A, B), a once dominant assumption for the divergence of these two *Pan* clades, are no longer plausible. Second, the assumptions that the ancestral population of bonobos was small and that the divergence between two *Pan* clades occurred only once historically (monophyletic origin of bonobos) suggest that the idea of a wider range for proto-*Pan* (hypothesis C) is not plausible.

Hypothesis D suggests that the original population of bonobos branched from chimpanzees while adapting to life in an isolated drier habitat and bonobos may have performed another adaptation to the present forested habitat from the previous drier habitat. However, the mechanism of speciation and re-adaptation to the forest habitat was not explained by this hypothesis. Although a fossil chimpanzee of 0.5 Ma in Kenya was found far from their current habitat (McBrearty and Jablonski, 2005), the range of proto-*Pan*, living between 7 Ma and 6 Ma and 2.1 Ma and 0.8 Ma, is totally unknown. Thus our current knowledge of ecology, genetics and fossil evidences of *Pan* species could not support hypothesis D.

Remaining possibilities

In conclusion, hypotheses E (Chaline et al., 2000) and F (Won and Hey, 2005) seem to be two remaining possibilities by which proto-*Pan*s could have reached the southern Congo Basin, under conditions similar to the present state of the river and the range of modern chimpanzees. These two hypotheses appear to be more consistent with the monophyletic origin (Fischer et al., 2011; Prüfer et al., 2012) and a small ancestral population (Hey, 2010; Wegmann and Excoffier, 2010) of bonobos than the other hypotheses.

Currently there is no corridor between the ranges of chimpanzees and bonobos either across the river or detouring near the headwaters. Therefore, two hypotheses require an exceptional reduction in the Congo River discharge. This might have enabled exposure of some rocky or sandy corridors somewhere in the river or some shortening of the headwaters. The 'normal' reduction in the river discharge occurred periodically during the Pleistocene, corresponding to the glacial–interglacial cycle. Among these reduction phases, an exceptional reduction in river discharge should have occurred only rarely during the Pleistocene, sometime near to the estimated *Pan* clade divergence time calculated by genetic studies, because the monophyletic origin of bonobos suggests that such movement to the left bank of the river occurred only once.

Based on these assumptions, we examined the plausibility of these two hypotheses as follows. One scenario involves detouring around the shortened headwaters during a very dry phase, such as a glacial period (the conditional detouring hypothesis E), and the other is crossing the Congo River at shallow points when the water level was at its lowest (the conditional corridor hypothesis F). Besides, the corridor hypothesis, without detouring around the headwaters in a savannah area and long-distance travel, is just simpler than the detouring hypothesis. Of course, if there are no shallow points in the main stream of the river, we may have to accept the detouring hypothesis.

A reduction of the Congo River discharge

Among Pleistocene dry periods, extensive aridity in the eastern to northern parts of Africa occurred around 2.8 Ma, 1.7 Ma and 1.0 Ma (Figure 16.2), as shown by analyses of terrigenous dust flux from marine sediments (deMenocal, 2004). Because pollen analyses in the Niger Delta showed decreases in the forest area in West Africa at nearly the same times (Bonnefille, 2010), these dry phases were probably on a pan-African scale. The Congo River discharge may have fluctuated in response to these pan-African arid phases, and the ancestral bonobo population(s) would have been able to move to the left bank of the Congo River during either of these periods.

Palynological records since 1.35 Ma from the ODP 1075 site (Figure 16.1) in the Congo Fan sediments have shown that the most prominent change in pollen assemblages occurred at 1.05 Ma (Dupont et al., 2001). An increase in *Podocarpus* pollen, coupled with decreases in tropical, woodland and mangrove tree pollen, suggests a shift to a drier phase in the Congo Basin. At the same time, dinoflagellate (phytoplankton) cyst associations changed, showing a reduction in river discharge (Dupont et al., 2001). The fresh water derived from the Congo River is relatively depleted of nutrients (Marret et al., 2008). However, the annual average water discharge of the Congo River is the second largest in the world, $1,250 \times 10^9$ m^3 (Wohl, 2007), which induces a nutrient-rich upwelling of coastal seawater. Therefore, a decrease in river discharge (i.e. a reduction in deep seawater upwelling) would result in an increase in salinity and a decrease in nutrients around the sea surface that lead to changes in the dinoflagellate flora. The sharp peaks in several dinoflagellate species during 1.1 Ma to around 0.9 Ma (Dupont et al., 2001) indicate great amplitude fluctuations in river discharge over a short time period. The peaks of *Ataxiodinium choane*, a dinoflagellate species from the fully marine environment with higher salinity (Marret and Zonneveld, 2003) during 1.05 Ma to around 0.95 Ma seem to suggest a dramatic reduction in Congo River discharge at this time.

At around 1.0 Ma, the Congo River may have nearly dried up. The headwater could have shortened and the water level of the river could have been very low. Regrettably, there is no evidence for fluctuation in the Congo River discharge before 1.35 Ma. As the estimate of 1.0 Ma for the reduction in Congo River discharge corresponds to a period of pan-African aridity, another period of pan-African aridity between 2.8 Ma and 2.5 Ma or 1.8 Ma to around 1.7 Ma might also have caused a reduction in the Congo River discharge. Two time periods (1.8 Ma to around 1.7 Ma, 1.0 Ma) are within the range of the molecular dating for two *Pan* clades (2.1 Ma to around 0.8 Ma). A reduction in the river discharge did occur at least at 1.0 Ma, as revealed directly by Congo River sediments. It is difficult to know to what extent the water level declined, but there is a possibility that proto-*Pan* could have crossed the Congo River to the opposite bank during these time periods.

Possible points to cross the river

Some points where the river might possibly be crossed exist east of Kisangani. The lower Lualaba

River has some rocky shallow points. For example, around the gently sloped waterfalls stretching for 100 km (Bayoma Falls, Figure 16.1), the water level is very low in the dry season (Kennis et al., 2011). Villagers can walk in the river to repair their fishing gear following stepping stones. The seasonal change in the discharge of the Congo River at Kinshasa is relatively small. The discharge of the Lualaba River, however, shows strong seasonal fluctuation, dropping down to one-quarter at Kisangani and one-tenth of the wet months discharge at Kindu in dry months (Runge, 2007). Thus, it seems plausible that the ancestors of bonobos could have crossed the river in the driest months of the driest phase without getting their bodies wet at some time in the Pleistocene, such as at 1.0 Ma. In the southern Congo Basin, forest refugia could be expected to exist along the river or in the centre of the basin during a dry phase, such as at the Last Glacial Maximum (Anhof et al., 2006; Maley, 2001). The ancestors of bonobos could have survived on the left bank of the river in such forest refugia.

Although the detouring hypothesis (E) cannot be rejected, we are inclined to support the conditional corridor hypothesis (F) since it is likely that shallow river crossing points existed within the forested area. The travel distance across the savannah area by the detouring hypothesis would still be quite long because forest area would also decrease in such a period (Trauth et al., 2007). Although other shallow points might have appeared along the Lualaba River at the same time, such as near Kindu area or the Gates of Hell area, only the area of Bayoma falls probably lay within the forest area (Figure 16.1).

Biogeographic implications of the history of the Congo River

The most important contribution of this study to biogeography would be that the older origin of the Congo River was delineated. Second, this study suggested that the corridor or detouring hypotheses focused on the Pleistocene is realistic, and that the exchange of populations between the opposite banks of the river for most forest mammals probably occurred in the driest phases, that is, 2.8 Ma to around 2.5 Ma, 1.8 Ma to around 1.7 Ma or 1.0 Ma, in the eastern area of the basin.

The number of primate species in each region of central Africa was counted using a recent report on the range of each species (Mittermeier et al., 2013) (Figure 16.4). The range of each region and its name code used here follow Grubb's criteria (Grubb, 2001). The southern Congo Basin (SC: South-Central) has the least number of primate species among three tropical forest regions. Four common species were found throughout three regions; Demidoff's dwarf galago (*Galagoides demidoff*), Thomas's dwarf galago (*G. thomasi*), Allen's swamp monkey (*Allenopithecus nigroviridis*) and De brazza's monkey (*Cercopithecus neglectus*). The two galago species range across a broad area including semi-dry vegetation. Allen's swamp monkey and De brazza's monkey are excellent swimmers (Butynski et al., 2013; Kingdon, 1989), therefore it is possible for them to have moved between both banks of the river.

When these four species were excluded, the findings regarding species common to the two regions were as follows: no common species between WC (West-Central) and SC (South-Central), four between WC and EC (East-Central) and four between SC and EC. Malbrouck monkey (*Chlorocebus cynosures*), Red-tailed monkey (*Cercopithecus ascanius*), Blue monkey (*Cercopithecus mitis*) and Angolian colobus (*Colobus angolensis*) are the common species occurring in SC and EC. The Malbrouck monkey mainly inhabits savannah areas in the northern Zambia highlands, but for the latter three species, the area around headwaters of the river might be a strong geographical barrier and shallow points on the Lualaba River in the forest area might be plausible corridors. Thus, the SC does not seem to have been a dispersion route between EC and WC for most of primate species (Grubb, 1982; Kingdon, 1989) and the primate species composition of SC seems to be closely connected with EC than WC (Colyn et al., 1991; Grubb, 1982).

Genetic analysis of primate populations between the left and the right banks of the Congo River are rare, except for *Pan* species. The divergence time of Lesula (*Cercopithecus lomamiensis*), ranging in the western area along the Lomami River, and the Owl-faced monkey (*C. hamlyni*), living in the right bank of the Congo River in Kisangani–Kindu region, was estimated at 1.7 Ma (3.2 Ma to 0.5 Ma) by the

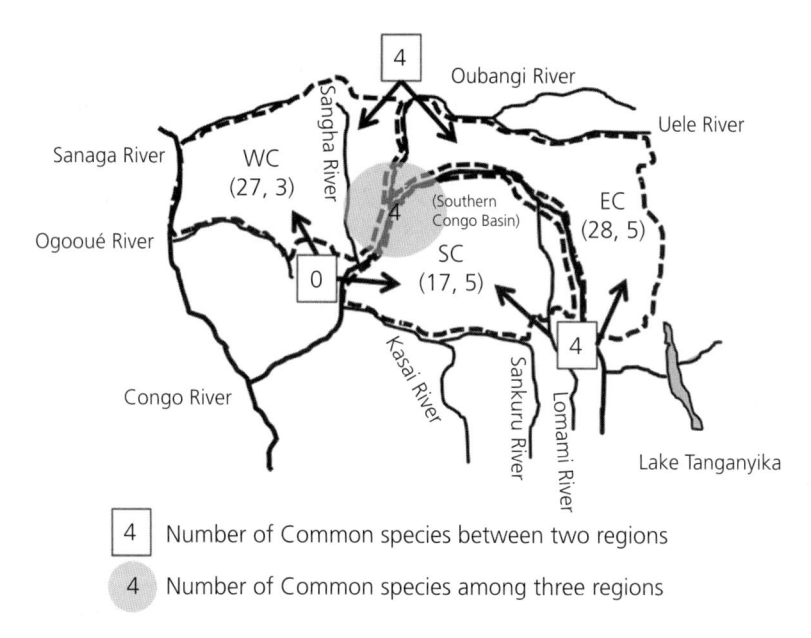

Figure 16.4 The species diversity and common species of primates in central African forest region. WC: West-Central region, SC: South-Central region, EC: East-Central region. Numbers in the parenthesis beneath the region name show numbers of all primate species (left) and endemic primate species (right) in each region. The number in the grey circle: the common species through the three regions. The numbers in a square: the common species between two regions except the four species commonly found in the three regions.
(La diversité des espèces et les espèces de primates communes dans la région de forêt Africaine centrale. WC: région Ouest-Centrale, SC: région Sud-Centrale, EC: région Est-Centrale. Les nombres entre parenthèses sous le nom de la région montrent les nombres de tous les espèces primates (gauche) et les espèces primates endémiques (droite) dans chaque région. Le nombre dans le cercle gris: les espèces communes à travers les trois régions. Les nombres dans le carré: les espèces communes entre les deux régions à l'exception de quatre espèces couramment trouvées dans les trois régions.)

Y chromosome and 2.8 Ma (4.3 Ma to 0.6 Ma) by the X chromosome (Hart et al., 2012). Two species of red colobus living in the southern Congo Basin (*Piliocolobus paramentieri* and *P. tholloni*) diverged from other clades at around 2.3 ±0.4 Ma or 0.8 ±0.2 Ma (Ting, 2008). These estimations are very close to the estimates for drought phases of the Congo River in the Pleistocene: 2.8~2.5 Ma, 1.8~1.7 Ma, 1.0 Ma.

Lower species diversity in the left bank of the river, the more number of common species between SC and EC and the estimated divergence times of primate species support our idea. Some literature supposed the existence of the paleo-Congo lake in the Pliocene (Beadle, 1981; Moreau, 1952), however, geological evidence could not support the paleo-Congo lake hypothesis (see Takemoto et al., 2015) and the lower species diversity in the southern half of the basin cannot be explained by the existence of the paleo-Congo lake, which might occupy over the both banks of the current Congo River (see hypotheses A and B in Figure 16.2).

Phylogeographical studies of other animals in the basin and the Congo River history

Recently, phylogeographical studies have provided new biogeographical perspectives on the southern Congo Basin. The forest elephants (*Loxodonta africana*) on the left bank of the river, having been thought common species along both banks of the Congo River, were regarded as being from the same lineage as the savannah elephants in Zambezi/Botswana populations (Ishida et al., 2013). This means that elephant populations could not have crossed the Congo River. The rocky corridor might not be suitable for their heavy body weight to walk on.

The phylogeny of cichlid fish (Schwarzer et al., 2011) and spiny eels (Alter et al., 2015) in the Lower Congo River (west of the Malebo Pool) suggests that the Lower Congo River continued to exist and was connected with the middle part of Congo River since around 5 Ma and some colonization events to the

Lower Congo River occurred in the Pleistocene such as 2.7 Ma and 1.6 Ma. These findings suggest that the formation of the Lower Congo River cannot be as recent as at the Plio-/Pleistocene boundary (2.6 Ma) as previous reports suggested. The global aridity phase of Messinian salinity crisis (around 5.3 Ma, see the unconformity in the Congo Fan sediments at 6 Ma to around 5 Ma in Figure 16.3) and strong dry phases in the Pleistocene (2.8 Ma to around 2.5 Ma, 1.8 Ma to around 1.7 Ma, 1.0 Ma) seemed to have affected the divergence of these fish species.

On the other hand, a corridor in the Kisangani region has been envisioned for some species or genetic clades within a species of rodents (Kennis et al., 2011; Nicolas et al., 2008). In the same region, the phylogeographical analysis of forest understorey specialist bird species indicates that a fording point appeared near 1 Ma (Voelker et al., 2013). Further phylogeographical studies would be useful in establishing the Congo River's history because geological evidence and paleogeographical studies are still limited.

The origin of bonobos

We proposed the conditional corridor hypothesis (F) as the most plausible answer to the question of how bonobos came to range on the left bank of the Congo River branching from chimpanzees. Several sources of evidence from geology, biogeography, molecular phylogeny and paleoecology were used to examine all possibilities, but the conclusion obtained was simple. There is no need to consider a major change in the river form in the Pleistocene or a significantly different lifestyle of proto-*Pan* from those of present chimpanzees and bonobos. The ancestral bonobos could have moved to the left bank of the river under similar geographical settings as today, while remaining in a forest habitat. Our hypothesis requires only an exceptional reduction of the Congo River discharge, which most likely occurred as an effect of global climatic changes. The conditional corridor hypothesis seems to be more convincing than the other hypotheses, owing to its simplicity.

After branched from chimpanzees, six major mitochondrial haplogroups diverged in bonobos (Kawamoto et al., 2013). There are two assumptions to explain the genetic variation of bonobo populations. One insists that the riverine barrier was the important factor for generating the internal genetic clades (Eriksson et al., 2004). The other maintained that the forest refugia through the Pleistocene affected the genetic structures of bonobos rather than current riverine barriers (Kawamoto et al., 2013). One still needs to investigate where the most recent common ancestor of bonobos lived and when the six major haplogroups diverged, from geographical distributions and phylogenetic tree of mtDNA haplotypes. This analysis can clarify whether our hypothesis on the origin of bonobos is truthful. It may be also useful for judging which effects, current riverine barriers or historical changes of the forest area, affected bonobos' dispersion over the southern Congo Basin.

Evolution of other hominid clades

Given that the Congo River was formed by 34 Ma, *gorilla* clade, which diverged from *Pan-Homo* clade at around 8.5 Ma to 12 Ma (Scally et al., 2012), seems to have evolved on the right bank of the Congo River. If so, why did gorillas not wade across the Congo River, as did the ancestral populations of bonobos? Genetic analyses of the Western and Eastern gorilla populations suggest that the two populations diverged at 1.75 Ma or earlier (Langergraber et al., 2012; Scally et al., 2012). They might have survived in eastern and western forest refugia in their current habitats, not in lowland forest refugia along the Congo River, through arid phases of the Pleistocene (Anthony et al., 2007). Therefore, gorillas might not have been ranging in an area along the Congo River when exceptional aridity made it possible to wade across the Congo River.

The lineage of apes diverged from the other catarrhines in the Oligocene in Afro-Arabia (30~25 Ma) (Stevens et al., 2013) and Proconsul lived across a wider range of habitat involving dense humid forests (Michel et al., 2014). Since then, most African ape fossils have been found in east Africa and Arabian Peninsula. Remarkable exceptions are Otavipithecus (13 Ma) from Namibia (Conroy et al., 1992) and the unnamed Ryskop hominoid (ca. 17 Ma) from South Africa (Senut et al., 1997). Although the phylogenetic relations of these fossils from south Africa with hominids are unknown, the fact that the extant African great apes, except bonobo, range in a wide area from west to east Africa on the right bank of the Congo River suggests that the common ancestor of extant African great apes too lived on the right bank of the Congo River after the formation of the river, until ancestor of bonobos moved to the left bank in the Pleistocene. This also suggests that the earliest human that diverged from *Pan* clade sometime from 6 Ma to 7 Ma (Prüfer et al., 2012; Stone et al., 2010) might also evolved in the area north or east to the Congo River.

Acknowledgements

This chapter includes a large part of the paper published in *Evolutionary Anthropology* 24:170–84 (2015). We thank the publisher Wiley for allowing us to republish the original material in this chapter. The idea for this article was born in the field at the many bonobo research sites we have visited. We thank the Centre de Recherche en Ecologie et Foresterie (CREF), the Ministère de la Recherche Scientifique (MIN), the African Wildlife Foundation (AWF), the World Wide Fund for Nature (WWF), the DRC staff of the Zoological Society of Milwaukee's Bonobo and Congo Biodiversity Initiative (ZSM), the Institut Congolais pour la Conservation de la Nature (ICCN), the ICCN Salonga National Park guards and the Tshuapa–Lomami–Lualaba (TL2) Project. We thank Dr Huffman and Dr Macintosh for comments on the English. This research was supported by the Environment Research and Technology Development Fund (D-1007) (to Furuichi), the JSPS Asia–Africa Science Platform Program (2009–11, 2012–14 to Furuichi); Japan Society for the Promotion of Science (JSPS) Grants-in-aid for Scientific Research (17255005 and 22255007, to Furuichi), the JSPS Grant-in-Aid for Challenging Exploratory Research (19657074 to Takemoto) and the JSPS International Training Program for Young Researchers: Primate Origins of Human Evolution (HOPE) (#8 to Primate Research Institute, Kyoto University).

References

Adachi, J. and Hasegawa, M. (1995). Improved dating of the human/chimpanzee separation in the mitochondrial DNA tree: heterogeneity among amino acid sites. *Journal of Molecular Evolution*, 40, 622–8.

Alter, S.E., Brown, B., and Stiassny, M.L. (2015). Molecular phylogenetics reveals convergent evolution in lower Congo River spiny eels. *BMC Evolutionary Biology*, 15(1), 224.

Anhuf, D., Ledru, M.P., Behling, H., Da Cruz, Jr, F.W., Cordeiro, R.C., Van der Hammen, T., Karmann, I., Marengo, J.A., De Oliveira, P.E., Pessenda, L., Siffedine, A., Albuquerque, A.L., and Da silva Dias, P.L. (2006). Paleo-environmental change in Amazonian and African rainforest during the LGM. *Palaeogeography, Palaeoclimatology, Palaeoecology*, 239(3), 510–27.

Anthony, N.M., Johnson-Bawe, M., Jeffery, K., Clifford, S.L., Abernethy, K.A., Tutin, C.E., Lahm, S.A., White, L.J., Utley, J.F., Wickings, E.J., and Bruford, M.W. (2007). The role of Pleistocene refugia and rivers in shaping gorilla genetic diversity in central Africa. *Proceedings of the National Academy of Science*, 104, 20432–6.

Bailey, W.J., Hayasaka, K., Skinner, C.G., Kehoe, S., Sieu, L.C., Slightom, J.L., and Goodman, M. (1992). Reexamination of the African hominoid trichotomy with additional sequences from the primate β-globin gene cluster. *Molecular Phylogenetics and Evolution*, 1, 97–135.

Beadle, L.C. (1981). *Inland Waters of Tropical Africa*. New York, NY: Longman.

Becquet, C. and Przeworski, M. (2007). A new approach to estimate parameters of speciation models with application to apes. *Genome Research*, 17, 1505—19.

Bjork, A., Liu, W., Wertheim, J.O., Hahn, B.H., and Worobey, M. (2011). Evolutionary history of chimpanzees inferred from complete mitochondrial genomes. *Molecular Biology and Evolution*, 28, 615–23.

Bonnefille, R. (2010). Cenozoic vegetation, climate changes and hominid evolution in tropical Africa. *Global Planet Change*, 72, 390–411.

Butynski, T.M., Kingdon, J., and Kalina, J. (eds). (2013). *Mammals of Africa*, Vol.II. *Primates*. London: Bloomsbury Publishing.

Caswell, J.L., Mallick, S., Richter, D.J., Neubauer, J., Schirmer, C., Gnerre, S., and Reich, D. (2008). Analysis of chimpanzee history based on genome sequence alignments. *PLoS Genetics*, 4, e1000057.

Chaline, J., Durand, A., Dambricourt Malassi, A., David, B., Magniez-Jannin, F., and Marchand, D. (2000). Were climatic changes a driving force in hominid evolution? In: Hart, M.B. (ed.). *Climates: Past and Present*. Special Publications, Vol. 181. London: Geological Society, pp. 185–98.

Conroy, G.C., Pickford, M., Senut, B., Van Couvering, J., and Mein, P. (1992). *Otavipithecus namibiensis*, first Miocene hominoid from southern Africa. *Nature*, 356, 144–7.

Coolidge, H.J. (1933). *Pan paniscus*, pygmy chimpanzee from south of the Congo River. *American Journal of Physical Anthropology*, 18, 1–59.

Colyn, M., Gautier-Hion, A., and Verheyen, W. (1991). A re-appraisal of palaeoenvironmental history in central Africa: evidence for a major fluvial refuge in the Zaïre Basin. *Journal of Biogeography*, 18, 403–7.

Crosby, A.G., Fishwick, S., and White, N. (2010). Structure and evolution of the intracratonic Congo Basin. *Geochemistry, Geophysics, Geosystems*, 11, Q06010. DOI:10.1029/2009GC003014.

de Manuel, M., Kuhlwilm, M., Frandsen, P., Sousa, V.C., Desai, T., Prado-Martinez, J., Hernandez-Rodriguez, J., Dupanloup, I., Lao, O., Hallast, P., Schmidt, J.M., Heredia-Genestar, J.M., Benazzo, A., Guido Barbujani, G., Peter, B.M., Kuderna, L.F.K., Casals, F., Angedakin, S., Arandjelovic, M., Boesch, C., Kuühl, H., Linda Vigilant, L., Langergraber, K., Novembre, J., Gut, M., Gut, I., Navarro, A., Carlsen, F., Andrés, A.M., Siegismund, H.R., Scally, A., Excoffier, L., Tyler-Smith, C., Castellano, S., Xue, Y., Hvilsom, C., and Marques-Bonet1, T. (2016). Chimpanzee genomic diversity reveals ancient admixture with bonobos. *Science*, 354(6311), 477–81.

deMenocal, P.B. (2004). African climate change and faunal evolution during the Pliocene-Pleistocene. *Earth and Planetary Science Letters*, 220, 3–24.

Dupont, L.M., Donner, B., Schneider, R., and Wefer, G. (2001). Mid-Pleistocene environmental change in tropical Africa began as early as 1.05 Ma. *Geology*, 29, 195–8.

Eriksson, J., Hohmann, G., Boesch, C., and Vigilant, L (2004) Rivers influence the population genetic structure of bonobos (*Pan paniscus*). *Molecular Ecology*, 13, 3425–35.

Ferris, S.D., Brown, W.M., Davidson, W.S., and Wilson, A.C. (1981). Extensive polymorphism in the mitochondrial DNA of apes. *Proceedings of the National Academy of Science*, 78, 6319–23.

Fischer, A., Prüfer, K., Good, J.M., Halbwax, M., Wiebe, V., Andre, C., Atencia, R., Mugisha, L., Ptak, S.E., and Pääbo, S. (2011). Bonobos fall within the genomic variation of chimpanzees. *PLoS One*, 6, e21605.

Gagneux, P., Wills, C., Gerloff, U., Tautz, D., Morin, P.A., Boesch, C., Fruth, B., Hohmann, G., Ryder, O.A., and Woodruff, D.S. (1999). Mitochondrial sequences show diverse evolutionary histories of African hominoids. *Proceedings of the National Academy of Science*, 96, 5077–82.

Giresse, P. (2005). Mesozoic-Cenozoic history of the Congo Basin. *Journal of African Earth Science*, 43, 301–15.

Gonder, M.K., Locatelli, S., Ghobrial, L., Mitchell, M.W., Kujawski, J.T., Lankester, F.J., Stewart, C.B., and Tishkoff, S.A. (2011). Evidence from Cameroon reveals differences in the genetic structure and histories of chimpanzee populations. *Proceedings of the National Academy of Science*, 108, 4766–4771.

Grubb, P. (1982). Refuges and dispersal in the speciation of African forest mammals. In: Prance, G.T. (ed.). *Biological Diversification in the Tropics*. New York, NY: Columbia University Press, pp. 537–53.

Grubb, P. (2001). Endemism in African rain forest mammals. In: Weber, W., White, L.J.T., Vedder, A. and Naughton-Treves, L. (eds). *African Rain Forest Ecology and Conservation*. New Haven, CT: Yale University Press, pp. 88–100.

Haffer, J. (1997). Alternative models of vertebrate speciation in Amazonia: an overview. *Biodiversity and Conservation*, 6, 451–76.

Hart, J.A., Detwiler, K.M., Gilbert, C.C., Burrell, A.S., Fuller, J.L., Emetshu, M., Hart, T.B., and Tosi, A.J. (2012). Lesula: a new species of Cercopithecus monkey endemic to the Democratic Republic of Congo and implications for conservation of Congo's Central Basin. *PloS One*, 7, e44271.

Hey, J. (2010). The divergence of chimpanzee species and subspecies as revealed in multipopulation isolation-with-migration analyses. *Molecular Biology and Evolution*, 27, 921–33.

Horai, S., Satta, Y., Hayasaka, K., Kondo, R., Inoue, T., Ishida, T., Hayashi, S., and Takahata, N. (1992). Man's place in Hominoidea revealed by mitochondrial DNA genealogy. *Journal of Molecular Evolution*, 35, 32–43.

Hvilsom, C., Carlsen, F., Heller, R., Jaffré, N., and Siegismund, H.R. (2014). Contrasting demographic histories of the neighboring bonobo and chimpanzee. *Primates*, 55, 101–102.

Ishida, Y., Georgiadis, N.J., Hondo, T., and Roca, A.L. (2013). Triangulating the provenance of African elephants using mitochondrial DNA. *Evolutionary Applications*, 6, 253–65. DOI:10.1111/j.1752–4571.2012.00286.x

IUGS (2012). http://www.stratigraphy.org/ICSchart/ChronostratChart2012.pdf

Kadima, E., Delvaux, D., Sebagenzi, S.N., Tack, L., and Kabeyaz, S.M. (2011). Structure and geological history of the Congo Basin: an integrated interpretation of gravity, magnetic and reflection seismic data. *Basin Research*, 23, 499–527.

Kaessmann, H., Wiebe, V., and Pääbo, S. (1999). Extensive nuclear DNA sequence diversity among chimpanzees. *Science*, 286, 1159–62.

Kawamoto, Y., Takemoto, H., Higuchi, S., Sakamaki, T., Hart, J.A., Hart, T.B., Tokuyama, N., Reinartz, G.E., Guislain, P., Dupain, J., Cobden, A.K., Mulavwa, M.N., Yangozene, K., Darroze, S., Devos, C., and Furuichi, T. (2013). Genetic structure of wild bonobo populations: diversity of mitochondrial DNA and geographical distribution. *PLoS One*, 8, e59660. DOI:10.1371/journal.pone.0059660.

Kennis, J., Nicolas, V., Hulselmans, J., Katuala, P.G., Wendelen, W., Verheyen, E., Dudu, A.M., and Leirs, H. (2011). The impact of the Congo River and its tributaries on the rodent genus Praomys: speciation origin or range expansion limit? *Zoological Journal of the Linnean Society—London*, 163, 983–1002.

Kingdon, J. (1989). Island Africa. *Journal of Geography*, 111(6), 798–815.

Langergraber, K.E., Prüfer, K., Rowney, C., Boesch, C., Crockford, C., Fawcett, K., Inoue, E., Inoue-Muruyama, M., Mitani, J.C., Muller, M.N., Robbins, M.M., Schubert, G., Stoinski, T.S., Viola, B., Watts, D., Wittig, R.M., Wrangham, R.W., Zuberbühler, K., Pääbo, S., and Vigilant, L. (2012). Generation times in wild chimpanzees and gorillas suggest earlier divergence times in great ape and human evolution. *Proceedings of the National Academy of Science*, 109, 15716–21.

Leakey, M., Ungars, P.S., and Walker, A. (1995). A new genus of large primate from the late Oligocene of Lothidok, Turkana District, Kenya. *Journal of Human Evolution*, 28, 519–31.

Maley, J. (2001). The impact of arid phases on the African rain forest through geological history. In: Weber, W., White, L.J.T., Vedder, A., and Naughton-Treves, L. (eds). *African Rain Forest Ecology and Conservation*. New Haven, CT: Yale University Press, pp. 68–87.

Marret, F. and Zonneveld, K.A.F. (2003). Figures of dinoflagellate cysts (relative abundance, temperature, salinity and nutrients). DOI:10.1594/PANGAEA.88307.

Marret, F., Scourse, J., Kennedy, H., Ufkes, E., and Jansen, J.H.F. (2008). Marine production in the Congo-influenced SE Atlantic over the past 30,000 years: a novel dinoflagellate-cyst based transfer function approach. *Marine Micro Paleontology*, 68, 198–222.

McBrearty, S. and Jablonski, N.G. (2005). First fossil chimpanzee. *Nature*, 437, 105–105.

Michel, L.A., Peppe, D.J., Lutz, J.A., Driese, S.G., Dunsworth, H.M., Harcourt-Smith, W.E., Hornor, W.H., Lehman, T., Nightingale, S., and McNulty, K.P. (2014). Remnants of an ancient forest provide ecological context for Early Miocene fossil apes. *Nature Communications*, 5, 3236 doi: 10.1038/ncomms4236.

Mittermeier, R.A., Rylands, A.B., and Wilson, D.E. (eds). (2013). *Handbook of the Mammals of the World*, Vol. 3. Barcelona: Lynx Editions.

Moreau, R.E. (1952). Africa since the Mesozoic: with particular reference to certain biological problems. *Proceedings of the Zoological Society of London*, 121, 869–913.

Moreau, R.E. (1963). Vicissitudes of the African biomes in the late Pleistocene. *Proceedings of the Zoological Society of London*, 141, 395–421.

Moucha, R. and Forte, A.M. (2011). Changes in African topography driven by mantle convection. *Nature Geoscience*, 4, 707–12.

Nicolas, V., Mboumba, J.F., Verheyen, E., Denys, C., Lecompte, E., Olayemi, A., Missoup, A.D., Katuala, P. and Colyn, M. (2008). Phylogeographic structure and regional history of Lemniscomys striatus (Rodentia: Muridae) in tropical Africa. *Journal of Biogeography*, 35, 2074–89.

Olson, D.M., Dinerstein, E., Wikramanayake, E.D., Burgess, N.D., Powell, G.V., Underwood, E.C., D'amico, J.A., Itoua, I., Strand, H.E., Morrison, J.C., Loucks, C.J., Allnutt, T.F., Ricketts, T.H., Kura, Y., Lamoreux, J.F., Wettengel, W.W., Hedao, P., and Kassem, K.R. (2001). Terrestrial Ecoregions of the World: A New Map of Life on Earth: A new global map of terrestrial ecoregions provides an innovative tool for conserving biodiversity. *BioScience*, 51, 933–8.

Pesole, G., Sbisa, E., Preparata, G., and Saccone, C. (1992). The evolution of the mitochondrial D-loop region and the origin of modern man. *Molecular Biology and Evolution*, 9, 587–98.

Prüfer, K., Munch, K., Hellmann, I., Akagi, K., Miller, J.R., Walenz, B., Koren, S., Sutton, G., Kodira, C., Winer, R., Knight, J.R., Mullikin, J.C., Meader, S.J., Ponting, C.P., Lunter, G., Higashino, S., Hobolth, A., Dutheil, J., Karakoc, E., Can Alkan, C., Sajjadian, S., Catacchio, C.R., Ventura, M., Marques-Bonet, T., Eichler, E.E., André, C., Atencia, R., Mugisha, L., Junhold, J., Patterson, N., Siebauer, M., Good, J.M., Fischer, A., Ptak, S.E., Lachmann, M., Symer, D.E., Mailund, T., Schierup, M.H., André A.M., Janet Kelso, J., and Pääbo, S. (2012).The bonobo genome compared with the chimpanzee and human genomes. *Nature*, 486, 527–31.

Rudge, J.F., Roberts, G.G., White, N., and Richardson, C.N. (2015). Uplift histories of Africa and Australia from linear inverse modeling of drainage inventories. *Journal of Geophysical Research: Earth Surface*, 120, 894–914. DOI: 10.1002/2014JF003297.

Runge, J. (2007). The Congo River, Central Africa. In: Gupta, A. (ed.). *Large Rivers*. Chichester: John Wiley & Sons Ltd., pp. 293–309.

Scally, A., Dutheil, J.Y., Hillier, L.W., Jordan, G.E., Goodhead, I., Herrero, J., Hobolth, A., Lappalainen, T., Mailund, T., Marques-Bonet, T., McCarthy, S., Montgomery, S.H., Schwalie, P.C., Tang, Y.A., Ward, M.C., Xue1, Y., Yngvadottir, B., Alkan, C., Andersen, L.N., Ayub, Q., Ball, E.V., Beal, K., Bradley, B.J., Chen, Y., Clee, C.M., Fitzgerald, S., Graves, T.A., Gu, Y., Heath, P., Heger, A., Karakoc, E., Kolb-Kokocinski, A., Laird, G.K., Lunter, G., Meader, S., Mort, M., Mullikin, J.C., Munch, K., O'Connor, T.D., Phillips, A.D., Prado-Martinez, J., Rogers, A.S., Sajjadian, S., Schmidt, D., Shaw, K., Simpson,

J.T., Stenson, P.D., Turner, D.J., Vigilant, L., Vilella, A.J., Whitener, W., Zhu, B., Cooper, D.N., de Jong, P., Dermitzakis, E.T., Eichler, E.E., Flicek, P., Goldman, N., Mundy, N.I., Ning, Z., Odom, D.T., Ponting, C.P., Michael A. Quail, M.A., Ryder, O.A., Searle, S.M., Warren, W.C., Wilson, R.K., Schierup, M.H., Rogers, J., Tyler-Smith, C., and Durbin, R. (2012). Insights into hominid evolution from the gorilla genome sequence. *Nature*, 483(7388), 169–75.

Schwarzer, J., Misof, B., Ifuta, S.N., and Schliewen, U.K. (2011). Time and origin of cichlid colonization of the lower Congo rapids. *PloS One*, 6, e22380.

Senut, B., Pickford, M., and Wessels, D. (1997). Panafrican distribution of Lower Miocene Hominoidea. *Comptes rendus de l'Academie des sciences, Paris, Series IIa*, 325, 741–6.

Stevens, N.J., Seiffert, E.R., O'Connor, P.M., Roberts, E.M., Schmitz, M.D., Krause, C., Goracak, E., Ngasala, S., Hieronymus, T.L., and Temu, J. (2013). Palaeontological evidence for an Oligocene divergence between Old World monkeys and apes. *Nature*, 497(7451), 611–14.

Stone, A.C., Griffiths, R.C., Zegura, S.L., and Hammer, M.F. (2002). High levels of Y chromosome nucleotide diversity in the genus Pan. *Proceedings of the National Academy of Science*, 99, 43–8.

Stone, A.C., Battistuzzi, F.U., Laura, S., Kubatko, L.S., Perry, G.H., Trudeau, E., Lin, H., and Kumar, S. (2010). More reliable estimates of divergence times in Pan using complete mtDNA sequences and accounting for population structure. *Philosophical Transactions of the Royal Society of London B*, 365, 3277–88.

Takemoto, H., Kawamoto, Y., and Furuichi, T. (2015). How did bonobos come to range south of the Congo river? Reconsideration of the divergence of *Pan paniscus* from other *Pan* populations. *Evolutionary Anthropology*, 24, 170–84. DOI:10.1002/evan.21456.

Thompson, J.A.M. (2003). A model of the biogeographical journey from *Proto-pan* to *Pan paniscus*. *Primates*, 44, 191–7.

Ting N. 2008. Mitochondrial relationships and divergence dates of the African colobines: evidence of Miocene origins for the living colobus monkeys. *Journal of Human Evolution*, 55, 312–325.

Trauth, M.H., Maslin, M.A., Deino, A., Bergner, A.G.N., Dühnforth, M., and Strecker, M.R. (2007). High- and low-latitude forcing of Plio-Pleistocene East African climate and human evolution. *Journal of Human Evolution*, 53, 475–86.

Yu, N., Jensen-Seaman, M.I., Chemnick, L., Kidd, J.R., Deinard, A.S., Ryder, O., Kidd, K.K., and Li W.H. (2003). Low nucleotide diversity in chimpanzees and bonobos. *Genetics*, 164, 1511–18.

Voelker, G., Marks, B.D., Kahindo, C., A'genonga, U., Bapeamoni, F., Duffie, L.E., Huntley, J.W., Mulotwa, E., Rosenbaum, S.A., and Light, J.E. (2013). River barriers and cryptic biodiversity in an evolutionary museum. *Ecology and Evolution*, 3, 536–45.

Wegmann, D. and Excoffier, L. (2010). Bayesian inference of the demographic history of chimpanzees. *Molecular Biology and Evolution*, 27, 1425–35.

Wohl, E.E. (2007). Hydrology and discharge. In: Gupta, A. (ed.). *Large Rivers*. Chichester: John Wiley & Sons Ltd., pp. 29–43.

Won, Y.J. and Hey J. (2005). Divergence population genetics of chimpanzees. *Molecular Biology and Evolution*, 22, 297–307.

PART VIII

Conservation and Captive Care

CHAPTER 17

Geospatial information informs bonobo conservation efforts

Janet Nackoney, Jena Hickey, David Williams, Charly Facheux, Takeshi Furuichi and Jef Dupain

Abstract The endangered bonobo (*Pan paniscus*), endemic to the Democratic Republic of Congo (DRC), is threatened by hunting and habitat loss. Two recent wars and ongoing conflicts in the DRC greatly challenge conservation efforts. This chapter demonstrates how spatial data and maps are used for monitoring threats and prioritizing locations to safeguard bonobo habitat, including identifying areas of highest conservation value to bonobos and collaboratively mapping community-based natural resource management (CBNRM) zones for reducing deforestation in key corridor areas. We also highlight the development of a range-wide model that analysed a variety of biotic and abiotic variables in conjunction with bonobo nest data to map suitable habitat. Approximately 28 per cent of the range was predicted suitable; of that, about 27.5 per cent was located in official protected areas. These examples highlight the importance of employing spatial data and models to support the development of dynamic conservation strategies that will help strengthen bonobo protection.

Le bonobo en voie de disparition (*Pan paniscus*), endémique à la République Démocratique du Congo (DRC), est menacé par la chasse et la perte de l'habitat. Deux guerres récentes et les conflits en cours dans le DRC menacent les efforts de conservation. Ici, nous montrons comment les données spatiales et les cartes sont utilisées pour surveiller les menaces et prioriser les espaces pour protéger l'habitat bonobo, inclut identifier les zones de plus haute valeur de conservation aux bonobos. En plus, la déforestation est réduite par une cartographie collaborative communale de gestion de ressources dans les zones de couloirs essentiels. Nous soulignons le développement d'un modèle de toute la gamme qui a analysé un variété de variables biotiques et abiotiques en conjonction avec les données de nid bonobo pour tracer la carte d'un habitat adéquat. Environ 28 per cent de la gamme est prédit adéquat; de cela, environ 27.5 per cent est dans une zone officiellement protégée. Ces exemples soulignent l'importance d'utiliser les données spatiales et les modèles pour soutenir le développement de stratégies de conservations dynamiques qui aideront à renforcer la protection des bonobos.

Introduction

Planning for the conservation of wildlife requires information about biological and ecological significance and its geographic relationship to human livelihoods and natural resource use. Considering accelerating and irreversible losses of global biodiversity (Butchart et al., 2010; Jenkins, 2003) and increasing threats from expanding human populations that depend on Earth's natural resources, it is critical to identify efficiently the geographic focus for conservation actions. Mapping and spatial analysis within a Geographic Information System (GIS) provide a spatial dimension that can contribute to conservation prioritization and planning. This can include mapping measures of biological and ecological importance (Potapov et al., 2008; Tyukavina et al., 2016), anthropological threat (Hansen et al., 2016; Sanderson et al., 2002) and identification of areas where conservation actions are most needed (Brooks et al., 2006; Rodrigues et al., 2004).

Classified by the IUCN Red List as Endangered since 2007 (Fruth et al., 2008), the bonobo (*Pan paniscus*) remains highly threatened across its range in the Democratic Republic of Congo

Nackoney, J., Hickey, J., Williams, D., Facheux, C., Furuichi, T., and Dupain, J., *Geospatial information informs bonobo conservation efforts*. In: *Bonobos: Unique in Mind, Brain, and Behavior*. Edited by Brian Hare and Shinya Yamamoto: Oxford University Press (2017).
© Oxford University Press. DOI: 10.1093/oso/9780198728511.003.0017

(DRC). Because the bonobo's range stretches across remote areas that are both less accessible to researchers and have a history of political instability, bonobos have been relatively less studied than other great apes, and therefore information about their ecological preferences, habitat use and distribution is still limited to-date. The biggest threat to bonobos is commercial and subsistence-based hunting, followed by habitat loss (IUCN and ICCN, 2012). Since approximately 66 per cent of the DRC's human population is rural (FAO, 2010), it relies heavily on the country's forests for the provision of natural resources and livelihood subsistence (Klaver, 2009), which includes hunting animals for protein (Fa et al., 2002). Subsistence agriculture and fuel-wood collection make up the majority of deforestation and forest degradation in the country (Hansen et al., 2008; Potapov et al., 2012).

The DRC experienced two wars between 1996 and 2003 that collapsed the country's formal economy and resulted in increased social unrest and poverty. Millions of people were displaced and human fatalities exceeded 5 million in the second Congo war alone (Bavier. 2008). DRC's post-war recovery has been slow; in 2014, the country's GDP per capita was the third lowest in the world (CIA, 2014). Many bonobo-related conservation and monitoring projects were stopped during the wars due to surrounding instability and insecurity, and post-war monitoring of bonobos has revealed enormous declines in bonobo populations in certain sites due to increased wartime hunting (Furuichi et al., 2012). Although DRC law officially protects bonobos from hunting, high poverty rates have driven human populations to hunt them illegally, and localized traditional taboos against hunting bonobos have begun to erode in view of human population growth, economic pressure and exposure to bonobo bushmeat consumers through increased mobility, immigration and emigration (Lingomo and Kimura. 2009). There is high uncertainty about the number of bonobos remaining in the wild (Fruth et al., 2008). Bonobos have long life spans and, like most instances in humans, female bonobos give birth to only one offspring at a time. The average interval between births is about 4.8 years (Furuichi et al., 1998). These life-history traits make bonobos especially vulnerable to extinction.

In recent decades, our capability for monitoring bonobos' forest habitats, assessing habitat suitability and monitoring the pressures that threaten bonobos has increased in large part due to availability of geospatial data derived from satellite imagery. Ground-based surveys are used by biologists and researchers to collect data on specific locations of bonobos and bonobo signs but are normally limited to small geographic areas that are accessible by foot. Satellite data, which cover large geographic areas and can span long time periods (but lack detailed information on species' presence on the ground), complement ground survey data to help analyse habitats and habitat conditions at broader time scales. Information on forest cover and locations of human occupancy can be extracted from satellite data and used as input to models that assess bonobo habitat suitability without surveying the entire bonobo range, thereby saving an enormous amount of effort and resources.

The use of spatial data and models for mapping both the likely suitable habitat of bonobos as well as locations of highest human pressure or threat is necessary for developing conservation plans that will promote connected and viable bonobo populations. This chapter first presents a case study using geospatial information for informing bonobo conservation efforts within the Maringa–Lopori–Wamba (MLW) Landscape located in northern DRC. In this sub-section, we highlight how spatial data have been used for long-term monitoring of bonobo habitat and for identification of areas of highest conservation priority for bonobos, including locations of several potential connectivity areas connecting a set of intact forest habitats. In addition, we describe how mapping is being used by local communities living around an important bonobo connectivity area to develop land-use plans that aim to secure land management rights and conserve forest habitats. Finally, we describe a recent effort that developed a model of suitable bonobo nesting habitat and resulted in a range-wide map for bonobo-specific conservation prioritization and planning.

Development of spatial models for bonobo conservation and land use planning in the Maringa–Lopori–Wamba Landscape, DRC

The geographic range of the bonobo spans approximately 500,000 km² across central DRC. About 17 per cent of its range occurs inside the Maringa–Lopori–Wamba (MLW) Landscape (Figure 17.1), one of 12 priority landscapes selected for having high biological value by the Congo Basin Forest Partnership (CBFP) in 2002. In addition to the bonobo, the landscape harbours an array of threatened terrestrial species such as the Congo peafowl (*Afropavo congensis*) and the forest elephant (*Loxodonta cyclotis*), both classified as Vulnerable on the ICUN Red List of Endangered Species. The landscape covers a 72,000 km² swathe of land in DRC's remote Equateur Province and contains a number of land-use and land-cover types, including 68 per cent moist, dense, equatorial evergreen forest, 25 per cent swamp forest and 5 per cent agriculture. Human population density is relatively low, with approximately three to five inhabitants per square kilometre (CBFP, 2006).

Since 2004, the African Wildlife Foundation (AWF), along with several partner institutions, has worked with the DRC government to develop a participatory land-use plan for the landscape (CBFP, 2005). Conservation prioritization and participatory mapping combined with human livelihood improvement strategies have formed the basis of planning activities. The DRC government

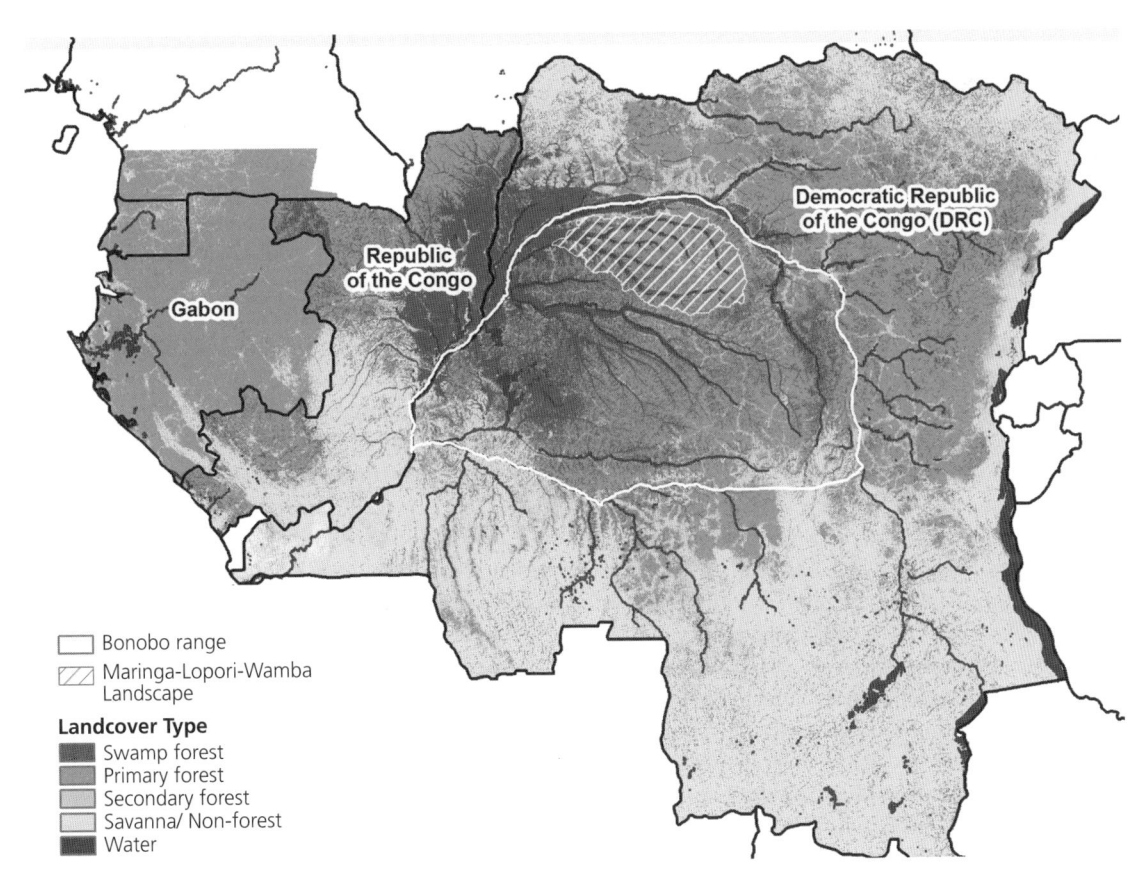

Figure 17.1 The bonobo's large range is outlined in white in northern-central DRC. The MLW Landscape, shown as an interior polygon with hatching, encompasses 17 per cent of the bonobo range.
(La gamme des bonobos est décrit en blanc dans le nord-central du DRC. Le paysage MLW, montré par un polygone jaune avec hachures, inclue 17 per cent de la gamme des bonobos.)

is establishing a framework for national land-use planning that includes the development of a set of national laws and guidelines for forest zoning (MECNT, 2011). The development and initial implementation steps of the landscape's planning activities continue to inform national and regional planning policy frameworks (Dupain et al., 2010).

The role of forest monitoring: forest fragmentation and war

Historically, due to its remote location, the MLW Landscape experienced a relatively low rate of forest loss. Local-scale agricultural activities are primarily responsible for forest loss in this landscape to-date because of a low prevalence of commercial logging and industrial agriculture. The landscape therefore still maintains large tracts of intact forests that sustain bonobo populations and other terrestrial wildlife. However, with demand for bushmeat and agricultural land escalating within the landscape, hunting pressure and habitat degradation are principal threats to these bonobo populations.

Satellite data enable systematic monitoring of forests in the landscape to provide annual assessments of bonobo habitat conditions. Certain threats to bonobo populations, such as forest conversion to agriculture and roads, as well as increased human encroachment into forests, can be detected via satellite monitoring. Forest fragmentation, the process of breaking up previously intact forests into smaller, disconnected patches, can cause isolation of wildlife habitats and prevent genetic exchange between wildlife populations (Botequilha and Ahern 2002). Hickey and colleagues (2012 and 2013) found that bonobo nesting habitat is negatively influenced by forest-edge density (one measure of forest fragmentation). Hickey and colleagues (2012) also found that in relatively intact habitats, forest-edge density can be a strong indicator of hunting accessibility.

Analysis of past and present land-use changes via satellite imagery can elucidate the relative vulnerability of areas of high conservation priority to anthropogenic pressure and habitat fragmentation. Historical analyses depend on the availability of ample satellite data that cover a range of years, do not contain data gaps and are not obscured by clouds. The long-term 30-metre Landsat satellite record spans from 1972 to the present; however, Landsat satellites flown during the 1970s through the 1990s did not collect data systematically and therefore lacked continuous coverage (Goward et al., 2006). Historical analyses are especially relevant in the context of understanding impacts of DRC's wars on its forest habitats. War and civil conflict have been linked to increased wildlife poaching and environmental degradation (Westing, 1992; Yamagiwa, 2003), often as a result of displaced human populations who are forced to rely more heavily on their natural surroundings for food and shelter (Dudley et al., 2002; Hanson et al., 2009). Approximately one-third of the wars' main frontline cut through the centre of the bonobo range (Draulans and Van Krunkelsven, 2002). Vogel (2000), Reinartz and Inogwabini (2001) and Furuichi and colleagues (2012) all reported increased incidences of bonobo poaching during DRC's wars. In addition, soldiers captured bonobos for the pet trade (Draulans and Van Krunkelsven 2002).

In the two protected areas supporting bonobo populations located in south-eastern MLW Landscape, the Luo Scientific Reserve (Luo SR) and the recently created Iyondji Community Bonobo Reserve (ICBR), analysis of satellite imagery collected from 1991 (when civil conflict began in this region) and spanning both wars until the end of 2003 showed an increase in small, isolated clearings in the reserves' primary forests (Figure 17.2). Primary forest loss in the reserves was over twice as high between 1990 and 2000, when conflict and war were predominant in this region, than that occurring between 2000 and 2010, a decade relatively free of conflict in this region (Nackoney et al., 2014). Recorded bonobo losses were especially high in Luo SR during wartime. From 1991 to 2005, the number of bonobos in Luo SR declined by over half, and three groups of bonobos disappeared completely (Hashimoto et al., 2008). Confirming satellite observations analysed by Nackoney and colleagues (2014), bonobo researchers noted in what is now ICBR an increased prevalence of hunting camps and temporary huts, where human populations lived and cleared small areas of surrounding forests to grow cassava crops. Nackoney and colleagues (2014) also mapped increases in forest fragmentation taking place in Luo SR and ICBR during this time period. The dramatic

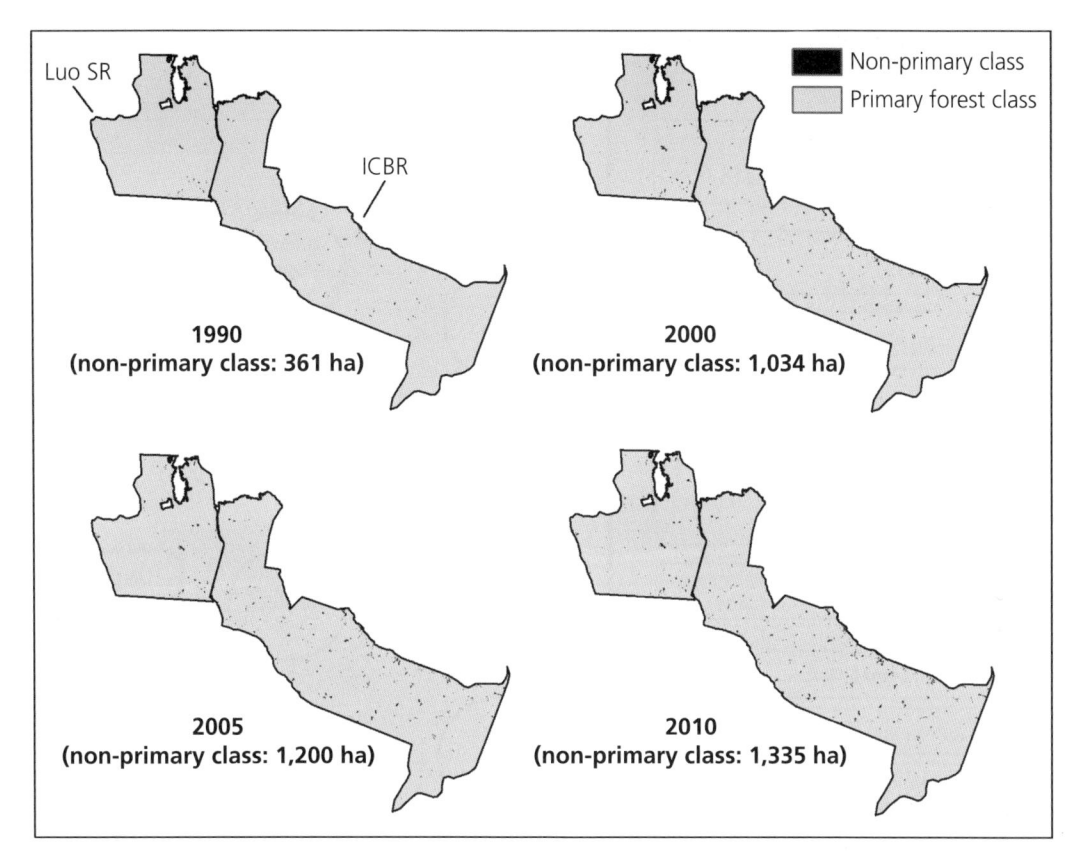

Figure 17.2 Maps show an increase in primary forest loss in Luo SR and ICBR between 1990 and 2010.(Les cartes montrent une augmentation de perte de forêts primaires.
(en rouge) dans Luo SR et ICBR entre 1990 et 2010.)

increases in peripheral (edge) forests caused by human movement patterns during the wars were linked to increased human access in these remote forests that consequently led to the disappearance of the three bonobo populations in Luo SR.

Spatial modelling guides conservation prioritization and identification of connectivity areas

To understand where locations harbouring highest potential for maintaining viable bonobo populations might exist across the MLW Landscape, a spatial model was developed using a simple additive weighting (SAW) process within a multi-criteria decision analysis (MCDA; see Malczewski, 1999). First, a set of evaluation criteria was selected, then each criterion was normalized across multiple map layers so that their scores ranged between 0 and 1. Each criterion was then assigned a weight that defined its relative importance to conservation potential in the landscape, and finally all criterion were added together via an added overlay process. The conceptual diagram of the developed model is shown in Figure 17.3. The model considered 1) potential hunting pressure and 2) habitat degradation. Inputs to the model were mapped at 90-metre resolution.

The model's measure of hunting pressure considered both human accessibility across the landscape as well as relative population demand for bushmeat. It was based on an open-access model of hunting accessibility developed by the Wildlife Conservation Society (WCS) (Didier and LLP, 2006). Human accessibility was determined by evaluating, for each 90-metre pixel in the landscape, relative accessibility based on the distance from the nearest

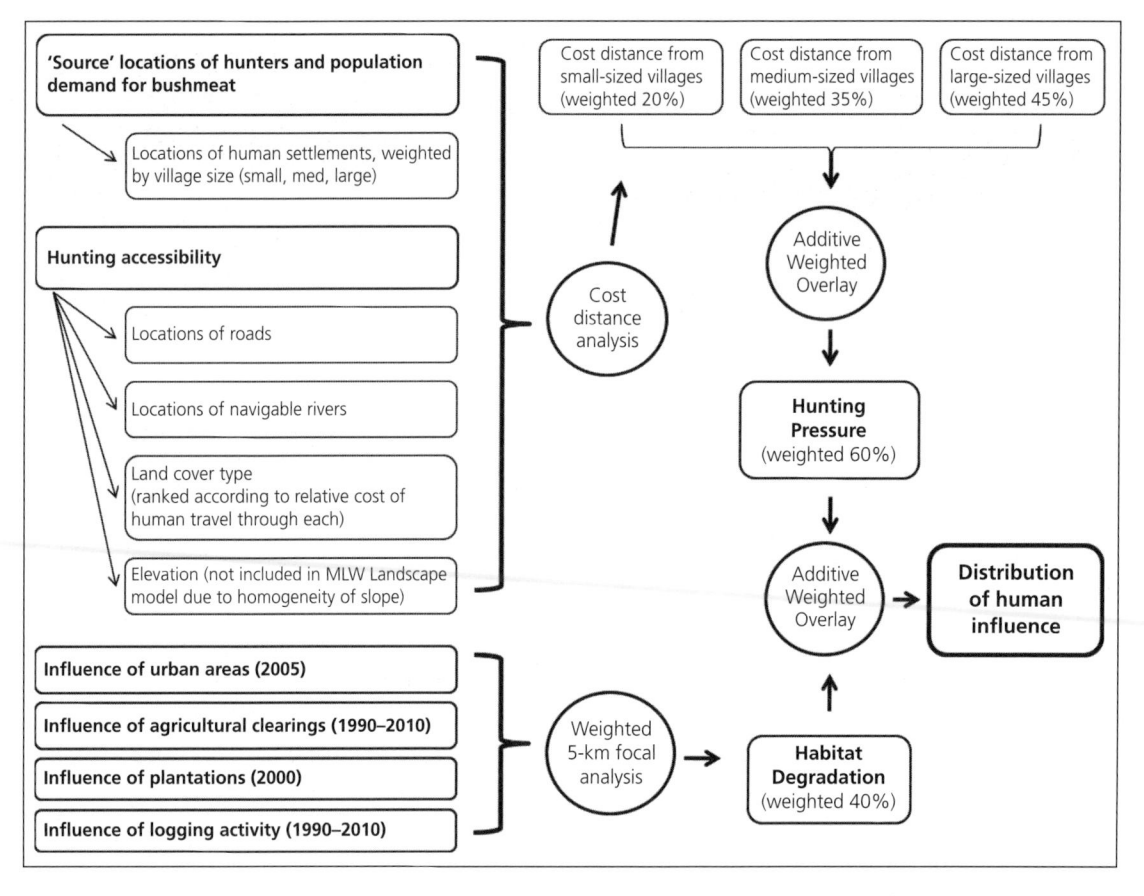

Figure 17.3 A conceptual diagram of the model developed for identifying the spatial distribution of human influence in the MLW Landscape. The major components of the model include factors relating to potential hunting pressure and habitat degradation in the landscape.
(Un diagramme conceptuel du modèle développé pour identifier la distribution spatiale de l'influence humaine dans le paysage du MLW. Les composants majeurs du modèle incluent des facteurs en relation à la pression potentielle de chasse et la dégradation dans le paysage.)

town and the relative 'cost' of accessing it, which considered the influence of land cover type and locations of roads and navigable rivers. The model's measure of habitat degradation considered the relative influence of a variety of factors influencing quality of bonobo habitat, including the influence of agricultural and urban areas, logging roads and industrial plantations. Both sub-models are described in further detail in Nackoney and Williams (2012). The two sub-models were combined in an additive weighting process. Because hunting poses a greater immediate threat to bonobos (Fruth et al., 2008), hunting was assigned a greater weight (60 per cent) than that of habitat degradation (40 per cent). Figure 17.4A presents the mapped result of human influence across the MLW Landscape. Areas of highest human influence (shown in darker shades of grey to black) are clustered around roads where settlements occur.

Next, a set of least-disturbed forest blocks having perhaps the highest potential for bonobo conservation in the landscape were identified systematically. For this, the landscape was divided into a grid of 1 × 1-km planning units and the average human influence value was calculated for each planning unit. Planning units falling below the medium mean threshold of human influence were designated as 'wildland blocks', meaning they are least disturbed by human activity and could be important for conservation

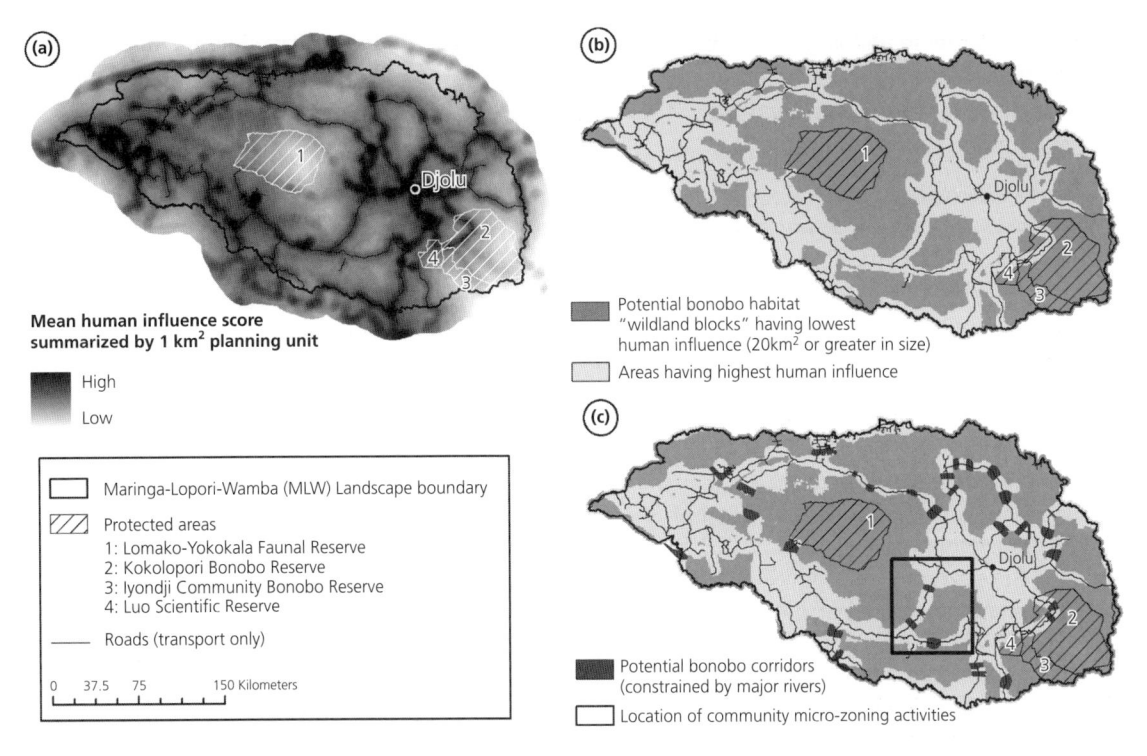

Figure 17.4 A map of intensity of human influence in the MLW Landscape, summarized by 1-km^2 planning units (a); locations of potential bonobo habitat 'wildland blocks' having the lowest human influence (b); and locations of potential bonobo corridors (c).
(Une carte de l'intensité de l'influence humaine dans le paysage MLW, résumée par des unités de 1-km 2 (a); emplacements d'un habitat potentiel pour les bonobos 'wildland blocks' avec le minimum d'influence humaine (b); et les emplacements des couloirs bonobos potentiels (c).)

prioritization. The approximate size of the bonobo home range has been found to vary by population and geography; Hashimoto and colleagues (1998) recorded a home range of 20 km^2. Considering this, 42 wildland blocks, occupying 60 per cent of the landscape, were identified that either met or exceeded this home-range size (Figure 17.4B). The largest identified wildland block encompassed almost 13,000 km^2. Wildland blocks smaller than the 20 km^2 home-range size may hold value for bonobo dispersal and connectivity.

Maintaining connectivity between wildland blocks is especially important in order to facilitate animals' feeding across multiple habitat types (Kozakiewicz, 1995) and to reduce inbreeding (Richards, 2000). A widely used approach for corridor identification, least-cost modelling (Beier et al., 2009; Majka et al., 2007), was used to map potential connectivity areas linking the identified bonobo wildland blocks. As described in more detail in Nackoney and Williams (2012), the output of the human influence model was used to depict the relative conditions promoting or discouraging bonobo movement and to identify the most permeable travel routes between blocks. Because bonobos do not cross major rivers, locations of rivers spanning at least 30 metres across were mapped using Landsat satellite imagery and applied in the model as a prohibitive constraint to connectivity. Thirty-two potential bonobo connectivity areas, occupying 3 per cent of the landscape, were identified and mapped (Figure 17.4C). The connectivity areas generally threaded through agricultural areas occupied by human settlements. Identification of the most vulnerable areas reinforced the importance of both educating nearby communities on the importance of protecting the connectivity areas from further degradation and fragmentation and building capacity for local land-use planning.

Community mapping and engagement in conservation strategies: formalizing land-use plans for forest conservation in a key bonobo connectivity area

DRC lacks a formal system of land tenure; legally, DRC's forests and natural resources belong to the State. In 2002, DRC's Forest Code was established which included Article 22 that allows communities to submit a formal request to manage their own community forest concession (Sidle et al., 2012). This provision has been viewed as perhaps the first real opportunity for communities in DRC to obtain direct authority over the forests and natural resources they depend on. In 2009, a national-level zoning steering committee was formed (USAID, 2010). Through this, a set of guidelines and national laws for forest zoning was developed, including one that mandates extensive public participation (MECNT, 2011). However, DRC's rural villages, which are often located in very remote and impoverished areas, have very little capacity to initiate this process. Lack of community empowerment and access to government services, among other factors, make the establishment of management plans and formal delineation of boundaries for legal recognition extremely challenging in the DRC to-date.

In 2010, the MLW Programme led by AWF initiated participative micro-zoning activities with local communities living within the vicinity of a key bonobo connectivity area (identified from the mapping process introduced earlier) located just south and west of the town of Djolu in the eastern-central part of the landscape (Figure 17.4C, indicated by the black box). This particular connectivity area provides a main linkage between the Lomako Yokokala Faunal Reserve (located in the central part of the landscape) and the Luo–Iyondji–Kokolopori cluster of reserves (located in the eastern part of the landscape). Although a historically well-established agricultural area, agricultural productivity is generally low in this area due to declining seed quality, lack of seed diversity, depletion of soil nutrients and poor agronomic practices such as low planting density and limited weeding capabilities (Nackoney et al., 2013). Slash-and-burn techniques are used to clear forested land to cultivate crops such as cassava, maize and peanuts, grown primarily for subsistence. Poverty is widespread; other challenges besides low agricultural productivity include lack of government extension services and poor road infrastructure that limits farmers' access to markets.

Similar to the rest of the landscape, human settlements in this area occur along roads, and agricultural areas extend outward from the roads into the forest. Recent maps of land-use and land-cover change (OSFAC, 2010) show primary forest loss occurring around the outermost edges of many of the agricultural complexes. From a conservation perspective the outward expansion of agriculture into primary forest contributes to wildlife habitat loss and forest fragmentation that disrupts ecosystem processes. From a human livelihoods perspective, this expansion requires farmers and hunter–gatherers to walk farther to access their fields and collect forest products that provide fuel, food and medicine. Hence, a primary goal of the MLW Programme's engagement with local communities in this area is to help devise planning strategies that will conserve the primary forests surrounding existing agricultural complexes and improve agricultural livelihoods.

Through a voluntary process, 53 village communities belonging to 8 administrative '*Groupements*' (Grpmts) in this region are currently engaging in participative micro-zoning and livelihood improvement programmes instituted by partners in the MLW Programme. Representatives from these village communities have voluntarily signed a Memorandum of Understanding (MOU) with the MLW Programme to define collaboratively the boundaries of two types of respective forest micro-zones according to the 2002 DRC Forest: 1) 'non-permanent' forests (intended for the sustainable expansion of agricultural activities under an approved management plan), and 2) 'permanent' forests (protected for community-based natural resource management or CBNRM). A quid pro quo agreement forms the basis of each MOU whereby the communities agree to respect the defined micro-zone boundaries and engage in collaborative monitoring to ensure that forest loss does not extend into the permanent (CBNRM) forest zone. In exchange, the communities receive agricultural extension services from MLW Programme partners that will help increase agricultural productivity within the defined agricultural micro-zone, which includes receiving a) higher quality and more diverse types of seeds, b) trainings in agro-forestry and product transformation, c) assistance for improving agronomic practices (to increase soil fertility and shorten fallow periods, for example) and d) increased access to markets.

Working directly with local communities in these ways helps provide them with the empowerment and necessary tools to sustainably manage the land they occupy as well as change their perceptions about conservation, thereby helping to decrease pressure on nearby forest habitats and bonobo populations. Efforts for improved law enforcement are ongoing: training for magistrates, prosecutors and paralegals on anti-poaching laws, and training for certain eco-guards as officers of the law, giving them legal authority to arrest people who break wildlife laws. AWF also initiates educational activities with nearby communities focused on building conservation awareness and education. This has helped rehabilitate a community radio station in the main village of Djolu to educate villagers on the importance of conserving wildlife and to explain the hunting laws that protect endangered species such as the bonobo. AWF has also constructed a new 'Conservation School' in the nearby community of Ilima, which teaches primary school children about forest conservation and wildlife, including bonobos. In the nearby ICBR, AWF has trained over 20 eco-guards who are employed by the DRC wildlife authority to protect the Reserve from poachers.

Mapping naturally plays a large role in the forest zoning process; Nackoney and colleagues (2013) provide a thorough description of methods that combine the transfer of local knowledge from participatory mapping to satellite image maps, supplemented by GPS data collection to ensure more accurate delineation of micro-zone boundaries. After the boundaries are defined, the MLW Programme helps to obtain formal government recognition of the zone boundaries with the DRC Ministère de l'Environnement & Développement Durable (formerly the Ministry of Environment, Nature and Tourism, MECNT).

Of the 84 villages with established MOU agreements with the MLW Programme, all have collaboratively defined and mapped the boundaries of their respective agricultural micro-zones, while 31 villages have defined and mapped the boundaries of their respective CBNRM forest zones. Since the agricultural zones are smaller and generally do not extend more than approximately 3 km in each direction from the main roads, they are mapped more quickly and easily, while the CBNRM zones extend much farther from the roads and therefore can take significantly longer to map completely.

Mapping bonobo suitability range-wide

The remote nature of the forest habitats within the bonobo range presents many challenges for long-term monitoring on the ground. Both geographically registered species-occurrence data and data describing their biological habitat are required to study relationships between bonobos and their habitats. Because these data are limited, so is our understanding of a) which environmental conditions most promote bonobo occurrence, and b) where such conditions occur throughout the range. Statistical models that investigate the probability of a species' occurrence in relation to relevant environmental predictor variables (Elith et al., 2006; Guisan and Zimmermann, 2000) can identify and map areas capable of supporting a given species. Such models have been developed by Torres and colleagues (2010) for chimpanzees in Guinea–Bissau and Bergl and colleagues (2010) for Cross River gorillas. Junker and colleagues (2012) also developed a model to map the likelihood of the presence for all great ape species across Africa, including the bonobo. Predictor variables included a range of climatic, vegetation and abiotic factors mapped at 5×5 km resolution.

Hickey and colleagues (2013) developed a bonobo-specific model to map relative suitability across the bonobo range. It differed from Junker and her team's (2012) model in that it a) used only bonobo nest presence data and included additional nest data not captured by Junker and team b) was mapped at a finer 100×100 m resolution, c) used a maximum entropy model (Junker and colleagues used a logistic regression model), d) relied on a slightly different set of predictor variables that focused more on forest metrics and human impact, and eliminated multi-colinearity within the bonobo range and e) dealt with model bias due to the non-random distribution of nest presence data. The remainder of this section describes Hickey and colleagues' (2013) model and results in more detail.

Development of the model began during a collaborative spatial modelling workshop held in Kinshasa in January 2011 and was refined and further developed in the months following. A maximum-entropy method, executed via a habitat-suitability modelling tool called MaxEnt (Elith et al., 2010; Phillips et al., 2006), analysed bonobo nest locations and environmental covariates to predict the spatial distribution of potentially suitable conditions. MaxEnt has been widely applied in the species-distribution modelling and mapping literature and has been found to perform favourably in comparison to other presence-only models (Elith et al., 2006; Hernandez et al., 2006).

Geographic boundaries of the bonobo range were based on IUCN (2010) and expanded using a contiguous boundary that encompassed all locations containing nest presence data, eliminating isolated pockets located in the western portion of the range. Locations of bonobo sleeping nests, collected by a wide variety of bonobo researchers representing multiple project sites, were compiled as part of the IUCN/SSC primate specialist group (PSG) Apes, Populations, Environments, and Surveys (APES) database managed by the Max Planck Institute for Evolutionary Anthropology. The data were checked for accuracy and quality during the Kinshasa workshop. The data included bonobo sleeping nest locations collected between 2003 and 2010. To reduce the effects of spatial autocorrelation caused by clustered nest data, individual nest site locations were aggregated to 100 × 100 m nest blocks, thereby resulting in 2,364 total nest blocks across the range.

Environmental predictor variables included those related to forest metrics (per cent forest, presence of intact forest, and forest-edge density), climate (precipitation), terrain (elevation, soil, wetness and proximity to rivers), and human impact (proximity to roads, agriculture and presence of forest loss). The variables were tested for multi-colinearity; correlated layers having weaker predictive power were excluded. In addition, other layers were excluded because they exhibited too narrow a range of values at the scale of the bonobo range or caused over-fitting.

Models were run using 70 per cent of the nest blocks for training; the remaining 30 per cent were withheld to test model accuracy independently. Classification accuracy was evaluated by the area under the curve (AUC) of a receiver operating characteristic (ROC) plot (an AUC of 0.5 represents a prediction no better than random, whereas a theoretically perfect prediction would approach an AUC of 1). The AUC of the model was 0.82, which confirmed that model performance was significantly greater than random.

Because many of the bonobo nest presence data were collected from specific project sites or from protected areas, the data were distributed non-randomly, and tests were therefore conducted to investigate presence of model bias. For this, iterative models were run while withholding nest-block locations from specific surveyed regions in order to evaluate how well each sub-model performed in the area of withheld data. This procedure accomplished two aims: 1) it evaluated the sensitivity of the models to potential bias from highly sampled sites, and 2) it evaluated the ability of the models to predict into unsampled regions. Results of these sub-models generally supported results from the final model (refer to Hickey et al., 2013, pp. 3096, 3100–1) for more details about sub-model nuances).

Approximately 28 per cent (156,210 km^2) of the bonobo range was predicted suitable; within this area, about 46 per cent had at least some level of survey effort, and 28 per cent (42,980 km^2) was located in official protected areas (Figure 17.5). At least 54 per cent of the area had not yet been surveyed. The majority of Iyondji Community Bonobo Reserve, Lomako–Yokokala Faunal Reserve, Salonga National Park and Kokolopori Bonobo Reserve were predicted suitable (87 per cent, 86 per cent, 82 per cent and 64 per cent, respectively) whereas less than 50 per cent each of Luo Scientific Reserve, Sankuru Reserve and Tumba–Lediima Reserve were predicted as suitable.

Results revealed that distance from agriculture (negative correlation), forest-edge density (negative correlation) and percentage of forest (positive correlation) were the strongest predictors of bonobo nest presence. This suggested that threats associated with human activity (forest fragmentation and proximity to agriculture and roads) were likely to have affected bonobo distribution most, overriding the importance of the biological and climatic conditions modelled. In addition, these threats

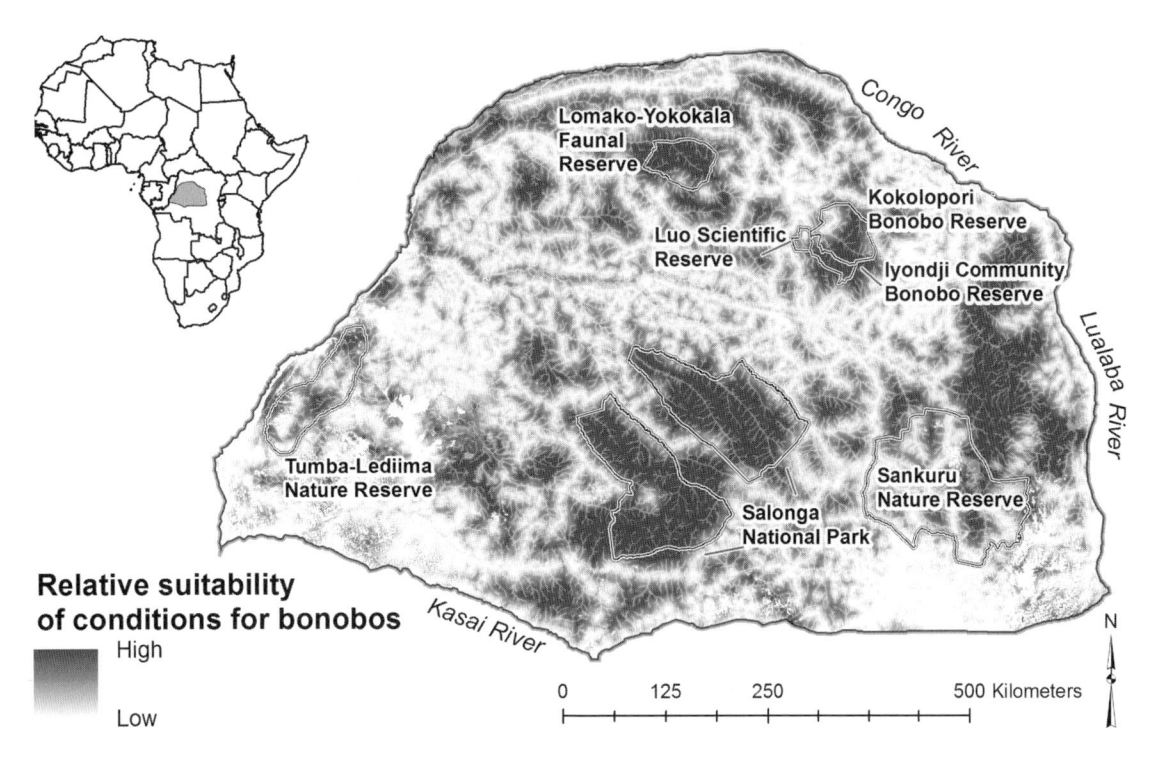

Figure 17.5 Range-wide map of suitable conditions for bonobos as modelled by Hickey et al. 2013, with protected areas. (Carte de toute la gamme de conditions adéquates pour les bonobos modelées par Hickey et al. 2013 avec des zones protégées.)

were probable indicators of hunting impact, as areas located in closer proximity to agriculture and roads were most likely used for hunting by humans who occupied those areas. However, testing a wide range of environmental predictors for the analysis was limited because predictor variables were restricted to those for which spatially explicit data existed across the entire bonobo range.

The resulting suitability map has provided a necessary foundation for developing strategies to help maintain viable bonobo populations. The map has been used to help identify areas of highest conservation priority and target areas not yet surveyed that may be important. These unsurveyed areas either may harbour bonobo populations that we are not yet aware of, or could support a natural expansion of the current bonobo distribution. The resulting suitability map was critical to the identification of 14 future high-priority bonobo survey areas as part of the recent Bonobo Conservation Strategy (IUCN and ICCN, 2012), a collaborative effort undertaken by a large number of active bonobo field researchers following the Kinshasa workshop. Researchers from the Zoological Society of Milwaukee, for example, have used the map to guide future survey efforts in an identified high-priority area located just north of Salonga National Park (G. Reinartz and P. Guislain, personal communication). The map also helped AWF choose the location of a proposed new protected area in northern MLW Landscape. Maps such as these can be very powerful tools to help develop land-use management plans aimed at protecting highly suitable areas, reducing threats to bonobos and promoting conservation and sustainable natural resource management throughout the bonobo range. In addition, they can be updated and revised as more data become available.

Conclusion

This chapter has emphasized the critical role of geospatial data and mapping in prioritizing and planning for bonobo conservation. There is growing awareness about the field of systematic conservation planning and its broad applicability to a range of planning solutions (see Moilanen et al., 2009 for a comprehensive review). Spatial conservation prioritization provides conservation planners and decision makers with location-specific information that helps identify areas where conservation efforts are most needed. Geospatial data and tools have an important role in both determining where to target conservation efforts and monitoring the performance of those efforts. The long-term survival of bonobo populations will depend in part on identifying and protecting critical habitat blocks and supporting healthy forest connectivity areas linking those blocks. Understanding where resources need to be allocated in order to protect these areas efficiently will be of utmost importance, especially considering the bonobo's endangered status.

Monitoring spatial patterns of human impact and subsequent patterns of deforestation are also key to developing conservation strategies that minimize human impact in forest habitats important for the bonobo. Satellite programmes that support long-term, systematic global data collection, such as NASA's Landsat programme, play a crucial role in enabling habitat monitoring. Since 2008, data from Landsat satellites have been distributed free of charge by the United States Geological Survey (USGS). With this development, there has been much progress in automating Landsat data processing in order to produce cloud-free image composites that can be classified and compared across time periods (Hansen et al., 2013; Potapov et al., 2011; Sexton et al., 2013). New web-based tools like Global Forest Watch (GFW, 2015) provide a user-friendly public interface to map and visualize real-time forest loss data (Gunther, 2015) and monitor degradation and human accessibility in interior forests. Such tools require no specialized software or mapping skills and can be extremely valuable for monitoring bonobo habitats and planning more effective planning strategies.

In addition, satellite data play an important role in complementing ground-based data collected on bonobo presence and habitat use. The bonobo range is located in an extremely large and remote area of the DRC and is therefore nearly impossible for biologists to survey completely by foot. Satellite data provide broad-scale information on forest characteristics and human accessibility metrics that cover the entire bonobo range. Paired with finer-resolution data from the ground, both types of data can be used complementarily as input to suitability models that extrapolate to wider areas not yet surveyed within the bonobo range. In addition, the high frequency of satellite data acquisitions can enable suitability models to be systematically updated in order to track patterns of change over time. Enhancing our understanding of multi-scale linkages in the context of bonobo habitats is important for improving our knowledge of the factors influencing the distribution of bonobos and developing maps that locate areas that might be targeted for future survey effort and conservation action.

Acknowledgements

This work was funded by the United States Agency for International Development (USAID) Central Africa Regional Program for the Environment (CARPE). In addition, we thank several other donors that funded various parts of works summarized in this chapter: the Arcus Foundation, Columbus Zoo, Conservation International, European Union, Frankenberg Foundation, IUCN/SSC Primate Specialist Group, Japan Ministry of the Environment, Japan Society of Promotion of Science (JSPS), Margot Marsh Biodiversity Foundation Primate Action Fund, Max Planck Institute (MPI) for Evolutionary Anthropology, United States Fish and Wildlife Service (USFWS) Great Apes Program, United States Forest Service (USFS), University of Georgia, University of Kent, Wildlife Conservation Society (WCS), Woodtiger Foundation and World Wildlife Fund (WWF).

References

Bavier, J. (2008–2001–22). 'Congo war-driven crisis kills 45,000 a month: study'. *Reuters*. Available at: http://www.reuters.com.

Beier P., Majka, D.R., and Newell, S.L. (2009). Uncertainty analysis of least-cost modeling for designing wildland linkages. *Ecological Applications*, 19, 2067–77.

Bergl, R.A., Warren, Y., Nicholas, A., et al. (2010). Remote sensing analysis reveals habitat, dispersal corridors and expanded distribution for the critically endangered Cross River gorilla *Gorilla gorilla diehli*. *Oryx*, 46, 278–9.

Botequilha, L.A. and Ahern, J. (2002). Applying landscape ecological concepts and metrics in sustainable landscape planning. *Landscape and Urban Planning*, 59, 65–93.

Brooks, T.M., Mittermeier, R.A., da Fonseca, G.A.B., Gerlach, J., Hoffmann, M., et al. (2006). Global biodiversity conservation priorities. *Science*, 313, 58–61.

Butchart, S.H., Walpole, M., Collen, B., Van Strien, A., Scharlemann, J.P., Almond, R.E., Baillie, J.E., Bomhard, B., Brown, C., Bruno, J., and Carpenter, K.E. (2010). Global biodiversity: indicators of recent declines. *Science*, 328(5982), 1164–8.

CBFP (Congo Basin Forest Partnership) (2005). *The Forests of the Congo Basin: A Preliminary Assessment*. Washington DC: Congo Basin Forest Partnership.

CBFP (Congo Basin Forest Partnership) (2006). *The Forests of the Congo Basin: State of the Forest 2006*. Congo Basin Forest Partnership.

CIA (Central Intelligence Agency) (2014). The World Factbook 2013–2014.Washington, DC: Central Intelligence Agency. Available at: https://www.cia.gov/library/publications/the-world-factbook/geos/cg.html.

Chen, J.Q., Franklin, J.F., and Spies, T.A. (1992). Vegetation responses to edge environments in old-growth Douglas-fir forests. *Ecological Applications*, 2, 387–96.

Didier, K.D., LLP (Living Landscapes Program) (2006). Building biological and threat landscapes from ecological first principles: a step-by-step approach. Technical Manual 6. Wildlife Conservation Society, New York, NY. Available at: http://programs.wcs.org/DesktopModules/Bring2mind/DMX/Download.aspx?EntryId = 5366&PortalId = 0&DownloadMethod = attachmentPRAQ: Please provide working url in Didier, K.D., LLP (Living Landscapes Program) (2006). Building biological and threat landscapes from ecological first principles: a step-by-step approach. Technical Manual 6. Wildlife Conservation Society, New York, NY. Available at: http://programs.wcs.org/DesktopModules/Bring2mind/DMX/Download.aspx?EntryId = 5366&PortalId = 0&DownloadMethod = attachment.

Draulans, D. and Van Krunkelsven, E. (2002). The impact of war on forest areas in the Democratic Republic of the Congo. *Oryx*, 36, 35–40.

Dudley, J.P., Ginsberg, J.R., Plumptre, A.J., Hart, J.A., and Campos, L.C. (2002). Effects of war and civil strife on wildlife and wildlife habitats. *Conservation Biology*, 16, 319–29.

Dupain, J., Degrande, A., De Marcken, P., Elliott, J., and Nackoney, J. (2010). Landscape land use planning: lessons learned from the Maringa–Lopori–Wamba Landscape. In: Yanggen, D., Angu, K., and Tchamou, N. (eds). *Landscape-Scale Conservation in the Congo Basin: Lessons Learned from the Central Africa Regional Program for the Environment (CARPE)*. Washington DC: IUCN–USAID/CARPE, pp. 46–60.

Elith, J., Graham, C.H., Anderson, R.P., et al. (2006). Novel methods improve prediction of species' distributions from occurrence data. *Ecography*, 29, 129–51.

Elith, J., Phillips, S.J., Hastie, T., Dudík, M., Chee, Y.E., and Yates, C. (2010). A statistical explanation of MaxEnt for ecologists. *Diversity and Distributions*, 17, 43–57.

FAO (Food and Agriculture Organization of the United Nations) (2010). *Global Forest Resources Assessment 2010*. Rome: UNFAO.

Fa, J.E., Peres, C.A., and Meeuwig, J. (2002). Bushmeat exploitation in tropical forests: An intercontinental comparison. *Conservation Biology*, 16, 232–7.

Fruth, B., Benishay, J.M., Bila-Isia, I., et al. (2008). *Pan paniscus*. In: IUCN (ed.). IUCN Red List of Threatened Species, version 2010.4. Available at: http://www.iucnredlist.org/details/15932/0.

Furuichi, T., Idani, G., Ihobe, H., et al. (1998). Population dynamics of wild bonobos (Pan paniscus) at Wamba. *International Journal of Primatology*, 19, 1029–43.

Furuichi, T., Idani, G., Ihobe, H., et al. (2012). Long-term studies on wild bonobos at Wamba, Luo Scientific Reserve, DR Congo: Towards the understanding of female life history in a male-philopatric species. In: Kappeler, P.M. and Watts, D.P. (eds). *Long-term Field Studies of Primates*. Berlin: Springer, pp. 413–33.

GFW (Global Forest Watch) 2015, Monitoring Forests in Near Real Time. Available at: http://www.globalforestwatch.org/.

Goward, S., Arvidson, T., Williams, D., Faundeen, J., Irons, J., and Franks, S. (2006). Historical record of Landsat global coverage. *Photogrammetric Engineering & Remote Sensing*, 72, 1155–69.

Guisan, A. and Zimmermann, N.E. (2000). Predictive habitat distribution models in ecology. *Ecological Modelling*, 135, 147–86.

Gunther, M. (2015). 'Google-powered map helps fight deforestation'. *The Guardian*, 10 March. Available at: http://www.theguardian.com/.

Hansen, M.C., Krylov, A., Tyukavina, A., Potapov, P.V., Turubanova, S., Zutta, B., Ifo, S., Margono, B., Stolle, F., and Moore, R., 2016. Humid tropical forest disturbance alerts using Landsat data. *Environmental Research Letters*, 11(3), 034008.

Hansen, M.C., P.V. Potapov, R. Moore, et al. (2013). High-resolution global maps of 21st-century forest cover change. *Science*, 342, 850–3.

Hansen, M.C., Roy, D.P., Lindquist, E., Adusei, B., Justice, C.O., and Altstatt, A. (2008). A method for integrating MODIS and Landsat data for systematic monitoring of forest cover and change and preliminary results for Central Africa. *Remote Sensing of Environment*, 112, 2495–513.

Hanson, T., Brooks, T.M., Da Fonseca, G.A.B., et al. (2009). Warfare in biodiversity hotspots. *Conservation Biology*, 23, 578–87.

Hashimoto, C., Tashiro, Y., Hibino, E., et al. (2008). Longitudinal structure of a unit-group of bonobos: male philopatry and possible fusion of unit-groups. In: Furuichi, T. and Thompson, J. (eds). *The Bonobos: Behavior, Ecology, and Conservation*. New York, NY: Springer, pp. 107–19.

Hashimoto, C., Tashiro, Y., Kimura, D., et al. (1998). Habitat use and ranging of wild bonobos (*Pan paniscus*) at Wamba. *International Journal of Primatology*, 19, 1045–60.

Hernandez, P.A., Graham, C.H., Master, L.L., and Albert, D.L. (2006). The effect of sample size and species characteristics on performance of different species distribution modeling methods. *Ecography*, 29, 773–85.

Hickey, J., Carroll, J.P., and Nibbelink, N.P. (2012). Applying landscape metrics to characterize potential habitat of bonobos (*Pan paniscus*) in the Maringa–Lopori–Wamba Landscape, Democratic Republic of Congo. *International Journal of Primatology*, 33, 381–400.

Hickey, J.R, Nackoney, J., Nibbelink, N.P., et al. (2013). Human proximity and habitat fragmentation are key drivers of the rangewide bonobo distribution. *Biodiversity and Conservation*, 22, 3085–104.

IUCN and ICCN (2012). Bonobo (Pan paniscus): Conservation Strategy 2012–2022. Gland, Switzerland: IUCN/SSC Primate Specialist Group & Institut Congolais pour la Conservation de la Nature.

IUCN (International Union for Conservation of Nature) (2010). IUCN Red List of Threatened Species, version 2010.4. Available at: http://www.iucnredlist.org.

Jenkins, M. (2003). Prospects for biodiversity. *Science*, 302, 1175–7.

Junker, J., Blake, S., Boesch, C., et al. (2012). Recent decline in suitable environmental conditions of African great apes. *Diversity and Distributions*, 18, 1077–91.

Klaver, D. (2009). *Multi-stakeholder Design of Forest Governance and Accountability Arrangements in Equator Province, Democratic Republic of Congo*. International Union for Conservation of Nature and Natural Resources and Wageningen University & Research Centre, The Netherlands.

Kozakiewicz, M. (1995). Resource tracking in space and time. In: Hanson, L., Fahrig, L., and Merriam, G. (eds). *Mosaic Landscapes and Ecological Processes*. London: Chapman and Hall, pp. 136–48.

Laurance, W.F., Lovejoy, T.E., Vasconcelos, H.L., et al. (2002). Ecosystem decay of Amazonian forest fragments: a 22-year investigation. *Conservation Biology*, 16, 605–18.

Lingomo, B. and Kimura, D. (2006). Taboo of eating bonobo among the Bongando people in the Wamba Region, Democratic Republic of Congo. *African Study Monographs*, 30, 209–25.

Lingomo B. and Kimura, D. (2009). Taboo of eating bonobo among the Bongando people in the Wamba region, Democratic Republic of Congo. *African Study Monographs*, 30, 209–225.

MECNT (Ministry of Environment, Nature Conservation and Tourism) (2011). Normes de macro-zonage forestier: guide opérationnel (Standards of macro-forest zoning: operational guide). Direction des Inventaires et Aménagement Forestiers, Ministère de l'Environnement, Conservation de la Nature et Tourism, République Démocratique du Congo (Directorate of Forest Inventory and Forest Management, Ministry of Environment, Nature Conservation and Tourism, Democratic Republic of Congo), Kinshasa, DRC. [in French].

Majka, D., Jenness, J., and Beier, P. (2007). CorridorDesigner: ArcGIS tools for designing and evaluating corridors. Available at: http://corridordesign.org.

Malczewski, J. (1999). *GIS and Multicriteria Decision Analysis*. New York, NY: John Wiley and Sons Inc.

Moilanen, A., Wilson, K.A., and Possingham, H.P. (2009). *Spatial Conservation Prioritization: Quantitative Methods and Computational Tools*. Oxford: Oxford University Press.

Nackoney, J., Molinario, G., Potapov, P., Turubanova, S., Hansen, M.C., and Furuichi, T. (2014). Impacts of civil conflict on primary forest habitat in northern Democratic Republic of the Congo, 1990–2010. *Biological Conservation*, 170, 321–8.

Nackoney, J., Rybock, D., Dupain, J., and Facheux, C. (2013). Coupling participatory mapping and GIS to inform village-level agricultural zoning in the Democratic Republic of the Congo. *Landscape and Urban Planning*, 110, 164–74.

Nackoney, J., and Williams, D. (2012). Conservation prioritization and planning with limited wildlife data in a Congo Basin forest landscape: assessing human threats and vulnerability to land use change. *Journal of Conservation Planning*, 8, 25–44.

OSFAC (Observatoire Satellital des Forêts d'Afrique Centrale) (2010). Forêts d'Afrique Centrale évaluées par télédétection (FACET): forest cover and forest cover loss in the Democratic Republic of Congo from 2000 to 2010. Brookings: South Dakota State University. Available at: http://osfac.net/data-products/facet/facet-dr-congo.

Phillips, S.J., Anderson, R.P., and Schapire, R.E. (2006). Maximum entropy modeling of species geographic distributions. *Ecological Modelling*, 190, 231–59.

Potapov, P., Turubanova, S., and Hansen, M.C. (2011). Regional-scale boreal forest cover and change mapping using Landsat data composites for European Russia. *Remote Sensing of Environment*, 115, 548–61.

Potapov, P.V., Turubanova, S.A., Hansen, M.C., et al. (2012). Quantifying forest cover loss in Democratic Republic of the Congo, 2000–2010, with Landsat ETM + data. *Remote Sensing of Environment*, 122, 106–16.

Potapov, P., Yaroshenko, A., Turubanova, S., Dubinin, M., Laestadius, L., Thies, C., Aksenov, D., Egorov, A., Yesipova, Y., Glushkov, I., and Karpachevskiy, M., 2008. Mapping the world's intact forest landscapes by remote sensing. *Ecology and Society*, 13(2), 51.

Reinartz, G.E. and Inogwabini, B.I. (2001). Bonobo survival and the wartime mandate. In: *The Great Apes: Challenges for the 21st Century*. Conference proceedings, Brookfield Zoo, Brookfield: pp. 52–6.

Richards, C. (2000). Inbreeding depression and genetic rescue in a plant metapopulation. *American Naturalist*, 155, 383–94.

Rodrigues, A.S., Akcakaya, H.R., Andelman, S.J., Bakarr, M.I., Boitani, L., Brooks, T.M., Chanson, J.S., Fishpool, L.D., Da Fonseca, G.A., Gaston, K.J., and Hoffmann, M. (2004). Global gap analysis: priority regions for expanding the global protected-area network. *BioScience*, 54(12), pp.1092–100.

Sanderson, E.W., Jaiteh, M., Levy, M.A., Redford, K.H., Wannebo, A.V., and Woolmer, G. (2002). The human footprint and the last of the wild. *BioScience*, 52, 891–904.

Sexton, J.O., Urban, D.L., Donohue, M.J., and Song, C. (2013). Long-term land cover dynamics by multitemporal classification across the Landsat-5 record. *Remote Sensing of Environment*, 128, 246–58.

Sidle, J., Dupain, J., Beck, J., et al. (2012). Forest zoning experience in Central Africa. In: de Wasseige, C., P de Marcken, P., Bayol, N., Hiol Hiol, F., Mayaux, Ph., Desclée, B., Nasi, R., Billand, A., Defourny, P., and Eba'a, R. (eds). *Congo Basin Forest—State of Forests 2010*. Luxembourg: EU Publications Office,

Torres, J., Brito, J.C., Vasconcelos, M.J., Catarino, L., Gonçalves, J., and Honrado, J. (2010). Ensemble models of habitat suitability relate chimpanzee (*Pan troglodytes*) conservation to forest and landscape dynamics in Western Africa. *Biological Conservation*, 143, 416–25.

Tyukavina, A., Hansen, M.C., Potapov, P.V., Krylov, A.M., and Goetz, S.J. (2016). Pan-tropical hinterland forests: mapping minimally disturbed forests. *Global Ecology and Biogeography*, 25(2), 151–63.

USAID (United States Agency for International Development) (2010). Democratic Republic of Congo: biodiversity and tropical forestry assessment (118/119) final report. Prosperity, livelihoods and conserving ecosystems indefinite quantity contract (PLACE IQC) Contract Number EPP-I-03-06-00021-00. Available at: http://pdf.usaid.gov/pdf_docs/PNADS946.pdf.

Vogel, G. (2000). Conflict in Congo threatens bonobos and rare gorillas. *Science*, 287, 2386–7.

Westing, A. (1992). Protected natural areas and the military. *Environmental Conservation*, 19, 343–8.

Yamagiwa, J. (2003). Bushmeat poaching and the conservation crisis in Kahuzi–Biega National Park, Democratic Republic of the Congo. *Journal of Sustainable Forestry*, 16, 111–30.

CHAPTER 18

Bonobo population dynamics: Past patterns and future predictions for the Lola ya Bonobo population using demographic modelling

Lisa J. Faust, Claudine André, Raphaël Belais, Fanny Minesi, Zjef Pereboom, Kerri Rodriguez and Brian Hare

Abstract Wildlife sanctuaries rescue, rehabilitate, reintroduce and provide life-long care for orphaned and injured animals. Understanding a sanctuary's population dynamics—patterns in arrival, mortality and projected changes in population size—allows careful planning for future needs. Building on previous work on the population dynamics of chimpanzees (*Pan troglodytes*) in sanctuaries of the Pan African Sanctuary Alliance (PASA; Faust et al. 2011), this chapter extends analyses to the only PASA bonobo sanctuary. Its authors analysed historic demographic patterns and projected future population dynamics using an individual-based demographic model. The population has been growing at 6.7 per cent per year, driven by arrivals of new individuals (mean = 5.5 arrivals per year). Several model scenarios projecting varying arrival rates, releases and breeding scenarios clarify potential future growth trajectories for the sanctuary. This research illustrates how data on historic dynamics can be modelled to inform future sanctuary capacity and management needs.

Les sanctuaires de faune secourent, réhabilitent, réintroduisent, et fournissent des soins pour toute la vie aux animaux orphelins et blessés. Comprendre les dynamiques de la population d'un sanctuaire—les motifs d'arrivée, mortalité, et de changements projetés de la taille de la population—permet une planification prudente pour les nécessités du futur. En se basant sur le travail déjà fait sur les dynamiques de la population chimpanzé (*Pan troglodytes*) dans les sanctuaires du Pan African Sanctuary Alliance (PASA; Faust et al. 2011), nous étendons notre analyse au seul sanctuaire bonobo par PASA. Nous avons analysé les motifs démographiques historiques et avons projeté les futures dynamiques de la population en utilisant un modèle démographique basé sur l'individu. La population augmente de 6.7 per cent par an, poussée par l'arrivée de nouveaux individus (moyenne = 5.5 arrivées par an). Plusieurs scénarios modèles montrent une trajectoire de potentielle croissance pour le sanctuaire. Cette recherche illustre comment modeler les données sur les dynamiques historiques pour informer la capacité future du sanctuaire et les besoins gestionnaires.

Introduction

Wildlife sanctuaries such as the Lola ya Bonobo sanctuary in Kinshasa, Democratic Republic of Congo, serve an important role in providing rescue, rehabilitation and housing for animals which frequently cannot be re-released into the wild; they also often have broader missions of conservation and research (André et al., 2008; Farmer, 2002; Faust et al., 2011). These sanctuaries must balance the need to rescue new animals with the long-term care of their inhabitants. To plan for their long-term sustainability, sanctuaries must understand their past and future population dynamics.

Previous analysis for the chimpanzees housed in 11 sanctuaries within the Pan African Sanctuary Alliance (PASA) illustrated the importance of assessments of population dynamics, especially for long-lived species such as great apes (Faust et al., 2011). As each new arrival or birth represents a

Faust, L. J., André, C., Belais, R., Minesi, F., Pereboom, Z., Rodriguez, K., and Hare, B., *Bonobo population dynamics: past patterns and future predictions for the Lola ya Bonobo population using demographic modelling*. In: *Bonobos: Unique in Mind, Brain, and Behavior*. Edited by Brian Hare and Shinya Yamamoto: Oxford University Press (2017). © Oxford University Press. DOI: 10.1093/oso/9780198728511.003.0018 decades-long commitment to lifelong care, careful consideration of arrival rates, longevity data and management approaches allows sanctuaries to have a more nuanced view of their future. In addition to projections of future population sizes under alternate model scenarios, Faust and colleagues (2011) also highlighted that changes were not just related to total population size, but also described how changes in age structure, including increases in the proportions of adolescent (9–19 years of age) and older (35+) chimpanzees, may necessitate adjustments to management, veterinary care and housing. Here, we build on Faust and colleagues' previous analysis of chimpanzees in African sanctuaries (2011) to investigate population dynamics for bonobos at Lola ya Bonobo, the only sanctuary focused on bonobos.

Study population

Analyses are based on population data collated in the international bonobo studbook (Pereboom, 2015). The studbook is an electronic database which includes data on each individual's birth or capture date, death date, and transfers to different locations throughout its lifespan (Ballou et al., 2010). The studbook includes bonobos who previously or currently live in the globally managed zoo population; for the purposes of these analyses, these include bonobos in institutions in North America, Europe, Japan and Mexico—hereafter 'zoo bonobos' (Reinartz et al., 2014). In addition, the studbook includes individuals at Lola ya Bonobo—hereafter 'Lola bonobos'. For both the zoo and Lola bonobos, data were current up until 31 December 2014.

The majority of analyses are focused on the Lola bonobos in the studbook. The Lola dataset included 123 individuals as of 31 December 2014: 73 living bonobos and 50 individuals who had previously died. For the wild-born Lola individuals, age at arrival at the sanctuary was estimated. Lola has also undertaken two reintroductions to the Ekolo ya Bonobo reserve, and 18 living animals are currently alive at that reintroduction site.

Demographic analysis methods

Historic patterns and current status were analysed using PopLink 1.4 (Faust et al., 2012) and Microsoft Excel 2010. Historical population growth rates (λ) were calculated as the geometric mean of annual λ, calculated as N_{t+1}/N_t, wherein N is the total population size in the sanctuary as of 31 December and t is year (Case, 2000). Annual per-capita death rates were calculated by dividing the annual number of deaths by the total population size as of 31 December of each year. Annual per capita reproductive rates (i.e. the proportion of females reproducing) were calculated by dividing the annual number of births by the number of females aged 8 to 42 during each year, based on the ages of reproduction identified for zoo bonobos. We used Chi-square tests to compare sex ratios of living and dead individuals.

Demographic modelling methods

To project future population dynamics we used ZooRisk 3.8, an individual-based stochastic simulation model that projects demographic and genetic changes in populations over time and is often used with intensively managed populations such as zoo populations (Earnhardt et al., 2008; Faust et al., 2011). Individual-based models track characteristics and fates of individuals in a population over time and provide flexibility in modelling stochasticity, individual-based management decisions, and other complex population dynamics (DeAngelis and Gross, 1992). For this modelling, we initialized ZooRisk with the Lola population's size and structure (e.g. the number of individuals of different ages/sexes) as of 31 December 2014.

The age- and sex-specific mortality rates (Q_x) used in ZooRisk are calculated as the number of deaths in an age class/number of individuals at risk for death for all or a portion of that age class (Caughley, 1977; Faust et al., 2012). Parameterizing Qx values for the modelling was challenging because 1) Lola has a relatively young population for such a long-lived species, with few individuals having aged into older age classes which makes calculation of mortality rates impossible (i.e. no Lola animals have been 30-years old, simply because of the short timespan that the sanctuary has been in operation), and 2) sample sizes were small when only Lola data were analysed. Sample sizes dropped below 20 individuals at risk per age class after ages 11 for males and 9 for females; typically in zoo populations, if an age class has fewer than 30 individuals at risk of events the data are viewed with caution (Ballou et al., 2010). To circumvent these issues, we utilized Lola-only Q_x rates (i.e. calculated from only the Lola ya Bonobo animals, based on a date window of 1 January 1981 to 1 January 2015) until age 11 for males and 9 for females, and then utilized rates from the entire studbook (i.e. all zoo-housed and Lola-housed bonobos, based on a date window of 1 January 1981 to 1 January 2015) thereafter. Although zoo-based mortality rates may not be exactly representative of Lola population dynamics because of potential differences in management and husbandry, both reflect a relatively protected environment, and zoo rates provide the most realistic approximations for the sanctuary (Faust et al., 2011; Lathouwers and Elsacker, 2005).

ZooRisk simulates demographic stochasticity, including random variation in mortality and reproduction in each year (Faust et al., 2008). We did not simulate any environmental stochasticity, as the annual variation in food, predation, weather and other factors that typically contributes to this sort of variability is buffered in a sanctuary where food, shelter and medical care are provided (Faust et al., 2011). To explore the population's stochastic dynamics in full, each model scenario was run for 1000 iterations and final results are reported as the mean values ±1 standard deviation (SD) across all model iterations.

We modelled 12 alternate model scenarios (Table 18.1) to reflect the potential demographic futures that might exist for the population based on changes in arrival rates, releases and other management practices. For scenarios involving arrival of new individuals (scenarios B to L), we modelled these as yearly imports of a specified number of individuals between the ages of 0 and 5, divided equally between males and females. We varied the annual numbers of arrivals depending on the scenario, including the current arrival rate (B, E to L), and higher (D) and lower (C) rates that were ±3 individuals from the current rate. Little data exist to predict trends in future arrival rates, but these values

Table 18.1 Lola ya Bonobo modelling scenarios.
(Les scénarios de modélisation de Lola ya Bonobo.)

Scenario Name	Description
A) no new arrivals	No new individuals arriving
B) current arrival rate (5/year)	5 individuals arrive per year, based on 2010–14 average annual arrival rate
C) lower arrival rate (2/year)	2 individuals arrive per year
D) higher arrival rate (8/year)	8 individuals arrive per year
E) current arrival + 5% p(B)	5 arrivals/year, females have 5% chance of reproduction (20-year interbirth interval)
F) current arrival + 10% p(B)	5 arrivals/year, females have 10% chance of reproduction (10-year interbirth interval)
G) current arrival + 20% p(B)	5 arrivals/year, females have 20% chance of reproduction (5-year interbirth interval)
H) current arrival + each female has one offspring before lifetime contraception	5 arrivals/year, females are allowed one birth and then p(B) = 0% for the rest of their reproductive life
I) current arrivals + single reintro	5 arrivals/year, single reintroduction of 18 individuals (9 males and 9 females between ages 0–25) in year 5
J) current arrivals + single reintro + Totaka	5 arrivals/year, reintroduction of 18 individuals as in scenario I + transfer of 20 individuals to Totaka (10 males and 10 females between ages 0–25) in year 5
K) current arrivals + 20% p(B) + reintros to stabilize growth	5 arrivals/year, females have 20% chance of reproduction, and releases to stabilize population at current size: 5 individuals are released every year (sex ratios even, individuals between ages 0–25)
L) current arrivals + 20% p(B) + Totaka + reintros to stabilize growth	5 arrivals/year, females have 20% chance of reproduction, Totaka transfer occurs in year 5 as described in scenario J, and releases to stabilize population at current size: 4 individuals are released every year (sex ratios even, individuals between ages 0–25)

p(B) = probability of breeding

help roughly characterize two potential alternate scenarios.

In scenarios exploring the impact of births on population dynamics (E to H) we took two approaches to modelling the impact of accidental births. In scenarios E to G we modelled reproduction using ZooRisk's probability of breeding (p(B)), the age-specific probability that a female will have at least one offspring in a year (Faust et al., 2008). For example, p(B) = 0.20 is equivalent to females producing an offspring on average once every 5 years (a 5-year interbirth interval), or 20 per cent of reproductively available females breeding in any given year. For scenarios E to G, we modelled hypothetical levels of p(B) ranging from 5 to 20 per cent, applied to all females in reproductively viable age classes (8 to 42). Finally, scenario H explores a management strategy of allowing each female to have a single offspring and then be restricted from reproducing (in this simplified scenario, modelled via a 100 per cent p(B) in age class 8 and 0 per cent p(B) for the rest of the reproductive lifespan).

We also modelled release strategies: the sanctuary has plans for an additional reintroduction (scenario I), and may also transfer a group of bonobos to Totaka island (scenario J). We also simulated two scenarios (K, L), which included enough releases to stabilize growth. All releases and transfers to Totaka were simulated as exports of individuals aged 0 to 25 in the model years specified in Table 18.1, with exports equally divided between males and females. The number of exports was based on sanctuary plans.

Lola ya Bonobo population dynamics: historical patterns

The Lola ya Bonobo population has grown steadily since the sanctuary's opening in 1994, with an average growth rate over the last decade of 6.7 per cent (λ = 1.067; Figure 18.1a). Population size declined because of two reintroductions into Ekolo Ya Bonobo Reserve: 6 individuals were released in 2009 and 7 in 2011. As of 31 December 2014, Lola ya Bonobo housed 36 male and 37 female bonobos ranging in age from 0 to 28 years old (Figure 18.2a). The median age of the current population is ten years, signifying a relatively young population for such a long-lived species. In comparison, in the zoo bonobo population the median life expectancies from birth are 34.4 for males and 42.8 for females (based on very limited sample sizes/data; these estimates should be treated with caution until more data accumulate), and maximum observed longevity is 63 years for females (based on a wild-caught, living female with an estimated birth date) and 51 years for males (based on a captive-born male with a known birth date; Pereboom, 2015).

Population growth is mainly driven by the arrival of new individuals; the sanctuary has received

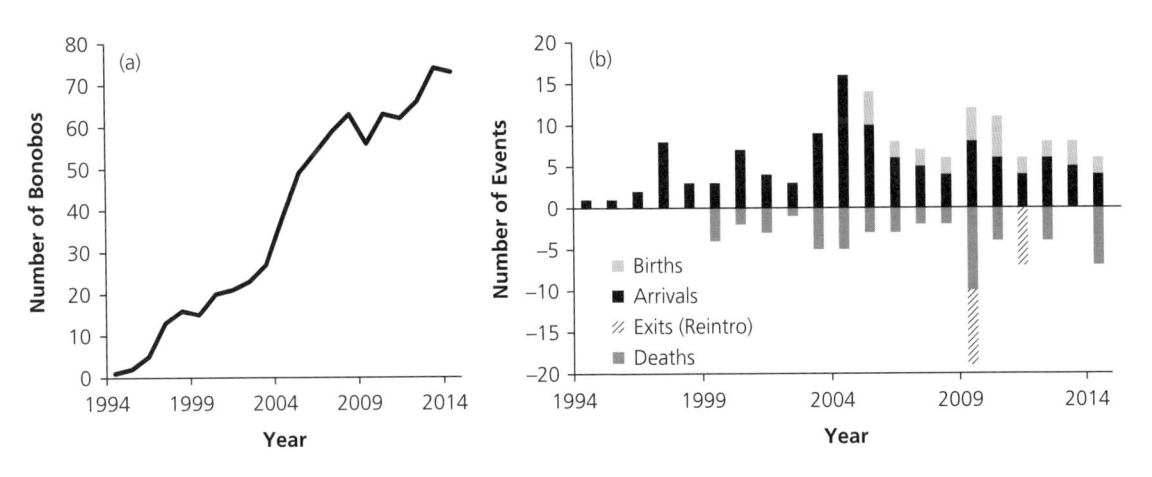

Figure 18.1 Population history of bonobos at Lola ya Bonobo, including (a) total population size over the sanctuary's history, and (b) annual arrival, birth, death and exit (via reintroduction) rates of bonobos at Lola ya Bonobo.
(Histoire de la population bonobo à Lola ya Bonobo, comprenant (a) la taille de la population totale en fonction de l'histoire du sanctuaire, et (b) les taux d'arrivées, de naissances, de mortalités et de sorties (par réintroduction) annuels des bonobos à Lola ya Bonobo.)

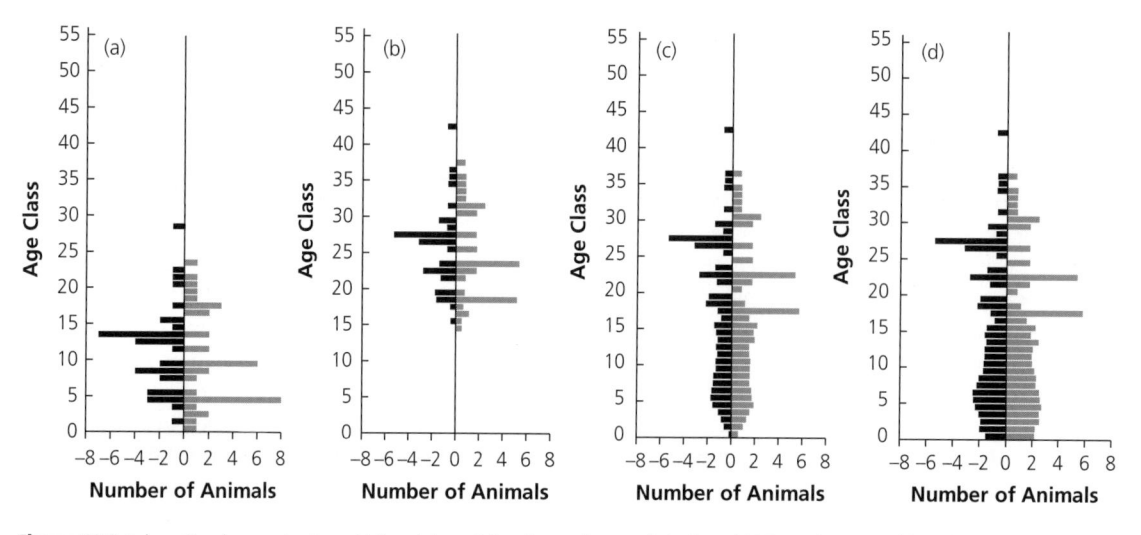

Figure 18.2 Lola ya Bonobo age structure at (a) model year 0 (i.e. the starting population) as of 31 December 2014; (b) model year 15 under scenario A, no new arrivals; (c) model year 15 under scenario B, current arrival rate; (d) model year 15 under scenario G, current arrival rate and 10 per cent p(B). Age structures for b to d are averaged across 1000 iterations. Darker bars are males, lighter bars are females.
(La structure d'âge à Lola ya Bonobo et (a) année modèle 0 (c-a-d la population initiale) dès le 31 Décembre 2014; (b) année modèle 15 sous le scénario A, aucune nouvelle arrivée; (c) année modèle 15 sous le scénario G, taux d'arrivée actuel; (d) année modèle 15 sous le scénario B, taux d'arrivée actuel et 10 per cent p(B). Les structures d'âges pour b–d sont moyennés pour 1000 itérations. Les bandes plus sombres sont mâles, les bandes claires sont femelles.)

108 individuals over its history (Figure 18.1b). Over the entire period, the average number of animals arriving annually was 5.5 ±3.5 (range 1–16). More recently (2010–14), the average number of annual arrivals was only slightly lower at 5.0 ±1.0. The large peak in arrivals in 2004 reflects when the sanctuary expanded holding capacity and took in bonobos from other captive situations. The median age of arriving bonobos was 3.0 ±3.4 years based on 108 arriving individuals with known/estimated ages (range = 0–20), and the sex ratio of arriving individuals was not significantly different than parity (57 males, 51 females, $\chi2 = 0.333$, df = 1, $p = 0.5637$).

The sanctuary has had 28 births (Figure 18.1b). On average across 2002–14 (the years when the sanctuary had reproductive-aged females), the population had 2.2 ±1.6 births/year (range 0–5); over the past 5 years, the sanctuary experienced an average of 2.8 ±1.3 births/year. On average from 2002–14, the annual proportion of reproductive-aged females who reproduced was 0.17 ±0.13 (range 0.0–0.31).

Since 1994, the sanctuary has experienced 50 deaths (Figure 18.1b), with an average of 2.6 ±2.7 deaths per year over the period (range 1–10). More males (29) than females (20) have died, but the difference was not statistically significant ($\chi2 = 1.653$, df = 1, $p = 0.1985$). On average, from 1994–2014 the mean of the annual per capita (total across all age classes) mortality rate was 6.9 per cent ±7.5 (range 0 to 26.7 per cent); more recently over the past 5 years that average has been lower, at 4.4 per cent ±4.2. The sanctuary receives animals in serious medical distress but despite this only 11 per cent ($N = 12$) of all animals arriving at the sanctuary ($N = 108$) die within the first 3 months of arrival. Cause-of-death data were available on a subset of individuals; overall, of the 50 deaths, the most frequent cause was infectious disease (48 per cent, $N = 24$) followed by unknown causes (36 per cent, $N = 18$) and traumatic/accidental causes (12 per cent, $N = 6$; Lola Ya Bonobo, unpublished data).

Lola ya Bonobo population dynamics: future projections

In the model scenario with no new births or arrivals (scenario A), the population was projected to decline from its starting size of 73 slowly, at 2 per cent per year ($\lambda = 0.98$), and still had 41.0 ±4.0 bonobos in 30 years (Table 18.2, Figure 18.3). Although this scenario is unrealistic by design (given past patterns, it is unlikely that there would be *no* arrivals or births), it illustrates the 'minimum' future population for which Lola ya Bonobo managers will need to plan.

In comparison, more realistic alternate scenarios demonstrate an array of potential projected futures for the Lola ya Bonobo population. If bonobos continue to arrive at Lola ya Bonobo under a range of projected rates (scenarios B to D), sanctuary capacity will need to increase to meet their demand (Table 18.2, Figure 18.3). If arrivals continue at the current rate (scenario B), the population would grow at 1.8 per cent annually and in 30 years there would be around 123 bonobos. Lower and higher rates result in population growth rates ranging from 0.1 to 2.9 per cent (Table 18.2).

We evaluated the impact of continued reproduction in scenarios E to H (Table 18.2, Figure 18.3). Scenarios with new arrivals and reproduction grow more quickly (2.4 to 4.1 per cent). New arrivals and births have an impact on the population's future age structure, resulting in many more young animals (Figure 18.2). At model year 15, the proportion of the population under 20 years old was only 24 per cent for scenario A (no new arrivals), but 60 per cent for B (current arrival rate) and 67 per cent for G (current arrivals + 10 per cent p(B)). Scenarios E to G, utilizing different p(B) values, illustrate that as the number of reproductively-aged females increases over time, the accidental birth rate may also be expected to increase (Figure 18.4). Scenario H, giving every female at least one chance at reproducing, is essentially equivalent to maintaining a 5 per cent p(B) (or a 20-year interbirth interval). Model projections can also yield predictions of annual deaths; higher rates are anticipated if arrival and reproduction continues (Figure 18.5).

Finally, we compared four release scenarios (I to L), which all included continued arrival of five bonobos per year but modelled one planned reintroduction, transfers to one island population (Totaka), and/or enough reintroductions to stabilize the population at its current size (Table 18.2, Figure 18.3). Scenarios with one reintroduction or the reintroduction and Totaka transfer (I, J) slowed the population's growth rate slightly: the population increased at 1.8 per cent annually without reintroductions (scenario B) compared to 1.4 per cent and 1.0 per cent for scenarios I and J, respectively. However, given its current arrival and birth rates, Lola ya Bonobo would have to reintroduce four to five individuals annually in perpetuity to remain at its current size (scenarios K and L).

Discussion and conclusions

Our results illustrate that Lola ya Bonobo's population has grown strongly in the past (over the past 5 years at 5.4 per cent annually) and will most likely

Table 18.2 Projected growth rate and population sizes (mean ±1 SD) for Lola ya Bonobo model scenarios.
(Taux de croissance projeté et tailles de population (moyenne ±1 SD) pour les scénarios modèles de Lola ya Bonobo.)

Scenario Name	Annual Lambda (λ)	Population Size at Model Year 0	15	30
A) no new arrivals	0.981	73	53.2 ±3.6	41.0 ±4.0
B) current arrival rate (5/year)	1.018	73	100.3 ±5.5	123.3 ±7.1
C) lower arrival rate (2/year)	1.001	73	72.4 ±4.3	75.2 ±5.7
D) higher arrival rate (8/year)	1.029	73	128.5 ±6.3	173.0 ±8.8
E) current arrival + 5% p(B)	1.024	73	111.5 ±7.1	147.4 ±9.7
F) current arrival + 10% p(B)	1.030	73	123.6 ±8.0	175.2 ±12.7
G) current arrival + 20% p(B)	1.041	73	149.0 ±9.9	244.7 ±18.9
H) current arrival + each female has one offspring before lifetime contraception	1.024	73	147.0 ±7.0	147.0 ±9.3
I) current arrivals + single reintro	1.014	73	86.3 ±5.3	112.2 ±7.0
J) current arrivals + single reintro + Totaka	1.010	73	70.6 ±5.0	99.6 ±6.6
K) current arrivals + 20% p(B) + reintros to stabilize growth	0.994	73	68.2 ±8.4	62.4 ±12.8
L) current arrivals + 20% p(B) + Totaka + reintros to stabilize growth	0.998	73	62.0 ±8.4	70.0 ±13.1

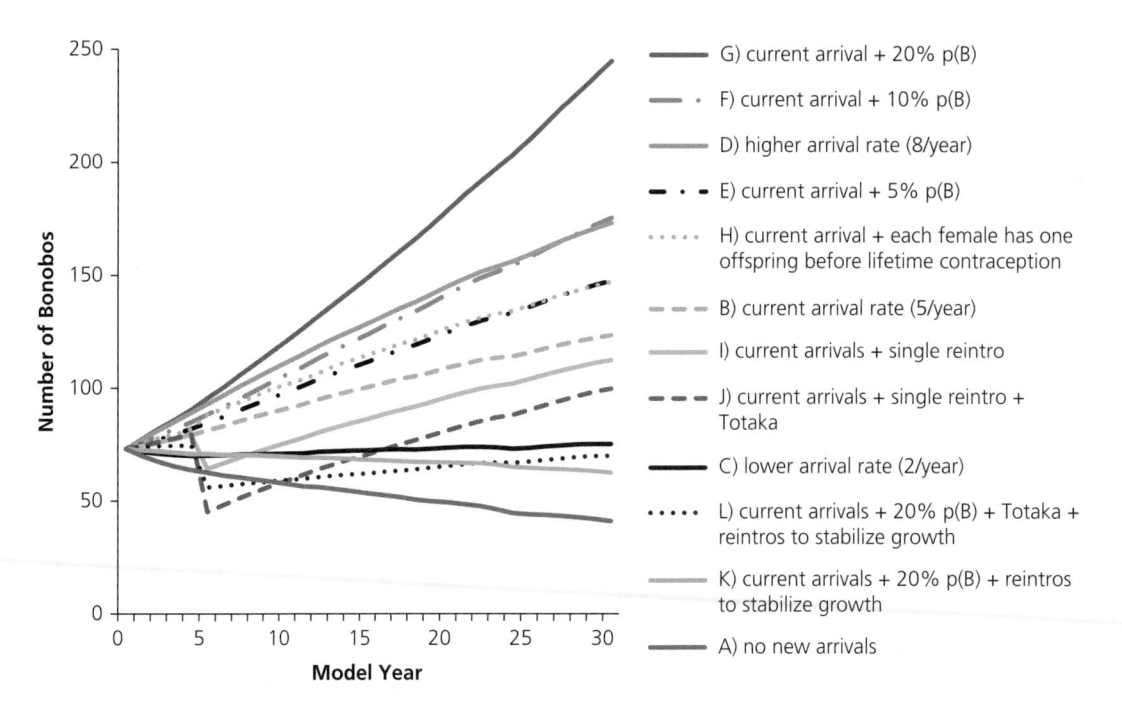

Figure 18.3 Projected population size under alternate model scenarios. Total population size is averaged across 1000 model iterations. Figure legend order matches the order of the final population sizes from largest to smallest.
(Taille de population projetée sous un scénario modèle alternatif. La taille de la population totale est moyennée pour 1000 itérations du modèle. L'ordre des légendes correspond à l'ordre des tailles de la population finale de la plus grande à la plus petite.)

continue to do so; the challenge for sanctuary managers lies in being able to understand and plan for how this population may grow and change in the future. Analyses of historical demographic patterns and modelled future dynamics can help managers understand this future better.

As Lola ya Bonobo dynamics are primarily driven by arrivals (Figure 18.1b), model scenarios projecting varying arrival rates can help inform what to expect in the future. Arrival rates have remained fairly steady at the sanctuary (average 5.5 over the entire period, or 5.0 over the last 5 years); this is in comparison to declining trend in arrival rates in chimpanzee sanctuaries seen in Faust et al. (2011). It is extremely difficult to predict how future bonobo arrival rates might change as these rates are dependent on many factors, such as the size of source populations, the level of hunting pressure, the vigilance of anti-poaching activities, political and economic factors, the success of education campaigns which discourage bonobo ownership and poaching activity and the sanctuary's capacity to add new individuals (Andre et al., 2008). Although scenarios A to D cannot be viewed as definitive predictions of the future, they illustrate a range of potential space needed if only new arrivals were a consideration: at 30 years, the sanctuary should expect to house at least around 41 bonobos (scenario A), but if arrival rates continue at their current pace, up to 123 bonobos (scenario B). Arrival rates that are lower or higher than current arrival rates will require sanctuary space to be adjusted accordingly.

Lola ya Bonobo dynamics also change over time because of mortality. Most bonobos arrive at the sanctuary at a young age (median = 3) and are often in varying states of psychological and physical trauma which require expert care (Farmer, 2002). The sanctuary has experienced multiple outbreaks of respiratory illness (Claudine André, personal communication), and infectious disease was the most frequently listed cause of death (48 per cent). If the sanctuary is successful in reducing disease outbreaks and survival rates improve over time, then that will most likely result in annual deaths lower than those projected in Figure 18.5, and overall the observed population sizes will be higher than the model's prediction.

Lola ya Bonobo also attempts to limit reproduction, recognizing that births place constraints on the ability to sustain the sanctuary and take in new arrivals. Over the past 5 years the sanctuary has experienced 2.8 births/year, and in scenarios E to G we modelled varying per-capita rates to illustrate how numbers of births will change as the number of reproductive-aged females increases as the sanctuary population ages. There are currently 23 reproductive-aged females in the population, but in model year 15 there were about 43 females under scenario B (current arrivals) and around 47 under scenario G (current arrivals + 10 per cent p(B)). As the numbers of reproductive-aged females increase so too do the chances of accidental births occurring (Figure 18.4), with concomitant effects on population size (Table 18.2, Figure 18.3). Although projecting that each female has a 20 per cent chance of giving birth in any given year (scenario G) seems high, historically the population's average rate was 12 per cent, with occasional extreme years (range 0 to 31 per cent). In addition, we modelled the impact of allowing each female to have a single offspring (H), which had final results similar to the 5 per cent p(B) scenario. Managers should continue to strive to control reproduction, as even a few births a year can accumulate into large differences in sanctuary population size in the future. Birth-control methods have been widely and successfully used in zoos in North America and Europe as a management tool for bonobos and great apes with good reversibility and low failure rates (Z. Pereboom, G. Reinartz and C. Asa, personal communication), and Lola ya Bonobo managers could further explore methods used by zoos to limit sanctuary births.

Reintroductions, if they occur in the future, will not have a major impact on numbers of bonobos housed at the sanctuary in the long-term (Figure 18.3, Table 18.2). The population will continue to serve as a robust source for potential future reintroductions, but scenarios K and L illustrates the extreme number (3 to 4/year every year) of individuals that would need to be released to stabilize the population given current arrivals and a moderate birth rate. Although reintroductions are important conservation actions in and of themselves, they do not necessarily have a strong impact on sanctuary population dynamics.

As the Lola ya Bonobo population matures, sanctuary managers can also anticipate changes in social dynamics. Comparing the relatively young current population (Figure 18.2a) to projected populations under current arrival or arrival and birth rates (Figure 18.2c,d), it is clear that sanctuary managers will continue to have to care for young bonobos but will also, increasingly, need to manage the changing social, veterinary and housing needs of adolescent, middle-aged and geriatric bonobos. For zoos, these management challenges have led to revised downward estimates of zoo carrying capacity and

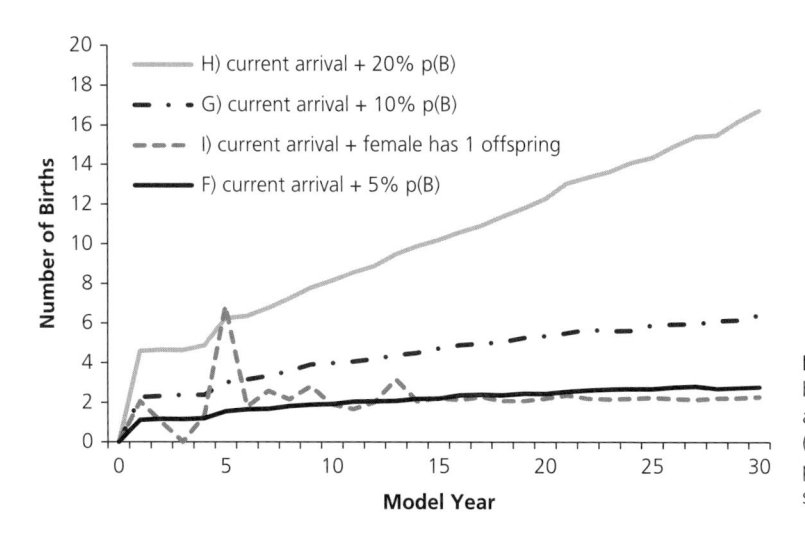

Figure 18.4 Average number of annual births under select model scenarios; births are averaged across 1000 model iterations. (Nombre moyen de naissances annuelles sous plusieurs scénarios modèles; les naissances sont moyennées pour 1000 itérations modèles.)

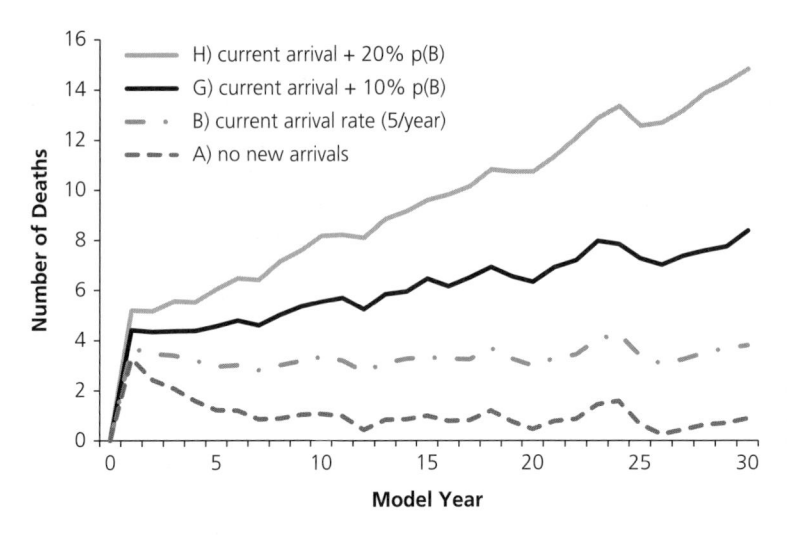

Figure 18.5 Average number of annual deaths under various model scenarios; deaths are averaged across 1000 model iterations. (Nombre moyen de mortalités annuelles sous plusieurs scénarios modèles; les mortalités sont moyennées pour 1000 itérations modèles.)

highlighted the need for flexible housing (G. Reinartz, personal communication).

In conclusion, for Lola ya Bonobo sanctuary to plan strategically for the future management needs of their sanctuary, managers need a full understanding of their population's past dynamics and the potential future trajectories that can be predicted using demographic modelling. This research provides a roadmap for the potential future challenges managers may face, and should help in long-term sustainability of sanctuaries by allowing managers to plan the future better.

Acknowledgements

We thank Kathryn Gamble for providing assistance with categorizing cause of death data. Steve Ross and Gay Reinartz provided valuable comments on this chapter.

References

André, C., Kamate, C., Mbonzo, P., Morel, D., and Hare, B. (2008). The conservation value of Lola ya Bonobo Sanctuary, in: Furuichi, T. and Thompson, J. (eds). *The Bonobos, Developments in Primatology: Progress and Prospects*. New York, NY: Springer, pp. 303–22.

Ballou, J., Lees, C., Faust, L.J., Long, S., Lynch, C., Bingaman-Lackey, L., and Foose, T. (2010). Demographic and Genetic management of captive populations for conservation. In: Kleiman, D.G., Thompson, K.V., and Baer C.K. *Wild Mammals in Captivity*, 2nd edn. pp. 219–52. Chicago, IL: University of Chicago Press.

Case, T. (2000). *An Illustrated Guide to Theoretical Ecology*. New York, NY: Oxford University Press.

Caughley, G. (1977). *Analysis of Vertebrate Populations*. Chichester: John Wiley & Sons.

DeAngelis, D.L. and Gross, L.J. (1992). *Individual-Based Models and Approaches in Ecology: Populations, Communities, and Ecosystems*. New York, NY: Chapman & Hall.

Earnhardt, J.M., Bergstrom, Y.M., Lin, A., Faust, L.J., Schoss, C.A., and Thompson, S.D. (2008). *ZooRisk: A Risk Assessment Tool. Version 3.8*. Chicago, IL: Lincoln Park Zoo.

Farmer, K.H. (2002). Pan-African Sanctuary Alliance: Status and range of activities for great ape conservation. *American Journal of Primatology*, 58, 117–32.

Faust, L.J., Bergstrom, Y.M., Thompson, S.D., and Bier, L. (2012). *PopLink Version 2.4*. Chicago, IL: Lincoln Park Zoo.

Faust, LJ, Cress, D, Farmer, K, Ross, S.R., and Beck, B. (2011). Predicting capacity demand on African chimpanzee sanctuaries. *International Journal of Primatology*, 32, 849–64.

Faust, L.J., Earnhardt, J.M., Schloss, C.A., and Bergstrom, Y.M. (2008). *ZooRisk: A Risk Assessment Tool. Version 3.8 User's Manual*. Chicago, IL: Lincoln Park Zoo.

Furuichi, T., Idani, G., Ihobe, H., Kuroda, S., Kitamura, K., Mori, A., Enomoto, T., Okayasu, N., Hashimoto, C., and Kano, T. (1998). Population Dynamics of Wild Bonobos (*Pan paniscus*) at Wamba. *International Journal of Primatology*, 19, 1029–43.

Lathouwers, M.D. and Elsacker, L.V. (2005). Reproductive Parameters of Female *Pan paniscus* and *P. troglodytes*: Quality versus Quantity. *International Journal of Primatology*, 26, 55–71.

Pereboom, Z. (2015). *International Bonobo Studbook*. Antwerp: Royal Zoological Society of Antwerp.

Reinartz, G.E., Pereboom, Z., and Ray J. (2014). *AZA Species Survival Plan® Yellow Program Population Analysis and Breeding & Transfer Plan*. Chicago, IL: Lincoln Park Zoo.

Afterword

Richard Wrangham

The naturalist Pliny the Elder (23–79 AD) could have been referring to bonobos in the twenty-first century when he wrote: 'There is always something new coming out of Africa.' Hare and Yamamoto designed *Bonobos: Unique in Mind, Brain and Behavior* to show that comparing humans with bonobos is fully as informative as comparing humans with chimpanzees. In meeting their goal they and their colleagues have produced a realm of new material.

No conspiracy theories are needed to explain why bonobos have received little attention until now. To begin with, chimpanzees have been known in Europe since the seventeenth century whereas bonobos were discovered only in 1929. In addition, bonobos are elusive. According to the *World Atlas of Great Apes and their Conservation* (Caldecott and Miles, 2005), between 172,000 and 301,000 chimpanzees live in 21 countries in the wild. By contrast, an unknown number of bonobos, probably fewer than 100,000, occupy a relatively small area of a single country (albeit still half a million square kilometres).

Many of the countries occupied by chimpanzees are relatively easy to visit and work in, thanks to relatively stable governance. Unfortunately, the bonobo homeland of the Democratic Republic of Congo has been riddled with violence and corruption and contains minimal infrastructure, unpredictable warlords and numerous other obstacles to health, safety and budgets. So it is not surprising that the first long-term study of wild bonobos, by Takayoshi Kano and colleagues, was not published until the 1980s (Kano, 1992). Even for that pioneering Japanese team, logistical problems stopped them from maintaining a continuous presence at their Wamba fieldsite, but they were still able to confirm the indications from captive work that bonobos represented a dramatically different species to chimpanzees.

Bonobos, in short, are a recent discovery that is very difficult to study in the wild, but research in the wild is invaluable. As Nackoney and colleagues observe in Chapter 17, even the data from satellite technology must be repeatedly ground-truthed. Patterns of behavior described in captivity must similarly be validated to find out whether they are representative of the species in the wild and how they make sense in an adaptively relevant environment. So it does not matter whether researchers are concerned with the critical efforts to identify areas of habitat destruction and illegal activity, or with behavioral data from habituated individuals in sites such as Wamba and LuiKotale. Either way, commitment to rugged fieldwork remains a vital part of bonobo evolutionary biology. If this book encourages the establishment of even a single new field site, let alone a new generation of adventurous scholars, it will have more than done its job. As experience has repeatedly shown, ape research sites have the added benefit of making major contributions to long-term conservation (Wrangham and Ross, 2008).

Meanwhile, for decades, thousands of chimpanzees have been kept in medical centres, zoos and other institutions throughout Europe and the United States, and in Africa the bushmeat trade has consigned to sanctuaries almost a thousand further chimpanzees. Throughout the world there are only a few hundred bonobos in captivity including, as Faust and colleagues describe in Chapter 18, the sole African bonobo sanctuary, Lola ya Bonobo, which contains the most natural captive groups. In short, even in captivity bonobos provide less chance for research than chimpanzees do. Yet as Hare and Yamamoto noted in Chapter 1, there are definitely more opportunities waiting for captive studies to be conducted, such as in zoos and sanctuaries in the United States and Japan.

The comparative ease with which chimpanzees can be studied is not the only reason why they have tended to arouse more research interest than bonobos. Although the two apes are equally closely related to humans, much evidence indicates that bonobos have diverged more than chimpanzees from their common ancestor (Pilbeam and Lieberman, 2017). To the extent that the last common ancestor (LCA) of humans, chimpanzees and bonobos, is the focus of concern, the chimpanzee is more relevant.

However as this book resoundingly demonstrates, the comparative study of bonobos and chimpanzees has merits far beyond mere reconstruction of the LCA. Where bonobos are more similar to humans than chimpanzees, the resulting questions are particularly provocative since they raise the possibility of bonobos and humans having been subject to similar selection pressures that did not apply to chimpanzees.

Yamamoto and Furuichi's new account of 'courtesy food-sharing' in Chapter 9 provides a fascinating example. The researchers describe bonobos begging from each other for pieces of junglesop fruit even when portions are available that are apparently equally high quality and easily obtained without begging. Those kinds of request make no sense in terms of optimal foraging. Yamamoto and Furuichi therefore suggest that bonobos are using begging for purely social purposes, such as to assess a social relationship or, as the researchers prefer, to make friends. Either way the implication is very different from chimpanzees employing a superficially similar behavior, since among chimpanzees food-begging seems to be almost entirely a way to get much-desired food. As Yamamoto and Furuichi note, the bonobo style of food-begging seems to be more in line with human than chimpanzee behavior.

'Courtesy food-sharing' is still not richly studied so alternative explanations have yet to be firmly ruled out, but it is tempting to interpret the behavior in terms of the increasingly abundant evidence reviewed by Krupenye and colleagues in Chapter 6 that bonobos have greater theory-of-mind skills than chimpanzees. The pervasive accounts of empathic behavior in both wild and captive bonobos are as stimulating as they are inherently surprising. A few years ago probably no-one would have predicted that juveniles would be dominant to adults in feeding competition, or that individuals would be routinely affiliative towards strangers.

Explanation of these and many similar observations is an exciting prospect not only for the fun of solving puzzles related to a curious ape, nor even just for what it will teach us about human evolution. Bonobo research may go further by helping to understand evolutionary dynamics relevant to many other species. The example I like best is of bonobos as candidates for self-domestication. Hare and Woods (Chapter 15) show the plausibility of this putative heterochronic process. The idea seems to explain much about bonobos, to offer many ideas for a comparable process in humans and in addition to be a prime example showing that the adaptationist paradigm is incomplete. In its most extreme form the adaptationist program sees all traits as adaptive. The concept of self-domestication is that selection against the propensity for reactive aggression leads to diverse incidental consequences that were not directly selected for, including changes in anatomy, physiology, behavior and cognition. Bonobos are the first wild species to which the self-domestication hypothesis has been applied. Many others could follow, providing a major example of a perspective that importantly complements the adaptationist paradigm.

Bonobos contains further ideas that are similarly provocative for their applicability both to bonobos and to other systems. If bonobos generally score higher on tasks of social cognition and worse on tasks of physical cognition, can we relate those differences to the anatomy, physiology and genetics of the brain, as Hopkins and colleagues propose in Chapter 14? Is it helpful to see bonobos and chimpanzees as being on opposing ends of a spectrum of empathizing and systemizing, as Clay and Genty suggest in Chapter 8? To what extent will we find social cognition explicable by foraging cognition, as Rosati suggests in Chapter 11? Do differences in oxytocin biology underlie the flexible communication and ready gazing of bonobos, as Hare and Woods speculate in Chapter 15? The questions, rich in implications beyond the apes, tumble out of the new studies and make us all the more aware of the importance of finding ways to nurture this surprising and generous species. So the excitement that this book represents should be cause for optimism. Bonobos are too fascinating, and too like us, for them to pass away without a fight. Let us hope that bonobos will continue to make Pliny look good by showing, with intensified research in future field sites as well as in captivity, that new things will still be coming out of Africa for a long time to come.

References

Caldecott J. and Miles L. (eds). (2005). *World Atlas of Great Apes and Their Conservation*. Berkeley, CA: University of California Press.

Kano, T. (1992). *The Last Ape: Pygmy Chimpanzee Behavior and Ecology*. Stanford, CA: Stanford University Press.

Pilbeam, D.R. and Lieberman, D.E. (2017). Reconstructing the Last Common Ancestor of chimpanzees and humans. In: Muller, M.N., Pilbeam, D.D., and Wrangham, R.W. (eds). *Chimpanzees and Human Evolution*. Cambridge, MA: Harvard University Press.

Wrangham, R.W. and Ross, E.A. (eds). (2008). *Science and Conservation in African Forests: The Benefits of Longterm Research*. Cambridge: Cambridge University Press.

Index

Notes

As the subject of this book concerns bonobos *(Pan paniscus)*, all entries in the index refer to this species unless otherwise indicated.

Tables and figures are indicated by an italic *t*, and *f* following the page number.

vs. indicates a comparison

A

ACP (arginine vasopressin) 206–7, 207*f*, 208*t*
adults
  interactions 5
  rough-and-tumble play 69–70
African Wildlife Foundation (AWF) 253–4, 257*f*, 258
Afroparvo congensis (Congo peafowl) 253
age
  behavioural variables 189, 190–1, 191–2*t*
  personality studies 189, 190, 191*f*
age-specific mortality rates (Q_x) 267–8
aggression 5
  baby dominance study analysis 55
  male–male relationships 42
  vocal communication 111–12
agonistic alliances, male–male relationships 41, 42–3
Agreeableness (HPQ) 195
  chimpanzee *vs.* 185*t*
  developmental changes 186
Allen's swamp monkey *(Allenopithecus nigroviridis)* 242
American Sign Language (ASL) 96, 113
Amici, F 147
amygdala 207–8, 209–10
Angolan colobus *(Colobus angolensis)* 242
animal phylogeographical studies, Congo River 243–4
Anomalurus 126, 127*t*
Anonidium mannii 127, 127*f*, 127*t*
arginine vasopressin (AVP) 206–7, 207*f*, 208*t*

ASL (American Sign Language) 96, 113
Assertiveness (HPQ) 193–4
  developmental changes 186
association
  behaviour linkage in male–male relationships study 39
  juveniles 56, 56*f*
  male–male relationships 39, 42
  social associations, female roles 21–2, 36
Ateles 70
attentional state
  chimpanzee 83
  gestures 69
  Primate Cognition Test Battery 202, 203*t*
  theory of mind 90*t*
attention-getters 98
  gestures 95
Attentiveness (HPQ) 195
Austin 97–8
avoidant relationships 39
AVPR1A 200, 209
  polymorphisms 206–7
AWF (African Wildlife Foundation) 253–4, 257*f*, 258
aye-aye *(Daubentonia)* 70

B

baby dominance study 49–63
  analysis 55–6
  limits of 60–1
  methods 53–6
  results 56–7
Baldwin, DA 98–9
beckoning gesture 117–18, 117*f*
begging for social bonds 129–32

abundant food 129–30
courtesy food sharing 129–32
behaviour 4*t*
  age 189
  personality 191, 193*t*
  personality observations 184
  sex differences 189
behavioural experiments
  male–male relationships study 37
  personality studies 188–9, 188*f*
beliefs, false *see* false beliefs
Belyaev, Dmitry 215–16
Bermejo, M 106
births, population dynamics 271*t* 273, 273*f*
black-crested macaque *(Macaca nigra)* 186
blue monkey *(Cercopithecus mitis)* 242
boldness 176
Bompusa community, male–male relationships study 37
bonobo-like hypothesis of human evolution 217
Brachystegia laurentii 127*t*
brain organization studies 203–8
  microstructural studies 204–8
  Voxel-Based Morphology 203–4, 205*f*, 206*t*, 210
Broca's area 208
Brodmann's area 208

C

Cacajao (Uakari monkeys) 70
Calithrix jacchus (common marmosets) 160
call combinations 112–13
Call, J 82, 173

captivity studies 6
  female dominance 25–6
  foraging cognition 161
  tool use 174, 174f
case markers, language studies 101
catarrhine monkeys 73
causal reasoning, chimpanzee 90
CBFP (Congo Basin Forest Partnership) 253
CBNRM (community-based natural resource management) 258–9
CDH (co-dominance hypothesis) 50, 51
Cercopithecus ascanius (red-tailed monkey) 242
Cercopithecus mitis (blue monkey) 242
Cercopithecus neglectus (De braza's monkey) 242
ceropthecids (monkeys) 70
Chantek 96
chimpanzee (*Pan troglodytes*)
  American Sign Language 96
  behaviours 4t
  causal reasoning 90
  courtship play fighting 70
  estrus periods of 22–5, 23f
  evolution of 3–4
  food sharing 134–5
  food types 160–1
  foraging 160–1
  gaze following 201–2, 201f
  gender roles 17–18
  general characteristics 3
  Hominoid Personality Questionnaire 184–5, 185t
  knowledge/ignorance/false belief attribution to others 84–5
  male–male relationships 36
  manipulation of others' perception 83–4
  patience 161
  perception of others' hearing 84
  Play Face/Full Play Face use. 67–8
  point following 201–2, 201f
  prosociality 140–1, 141t
  relative party size 18, 19–20
  sexual relations 24–5, 24f
  spatial memory 161, 200
  territoriality 144
  theory of mind *see* theory of mind in chimpanzee
  tool use 161, 172–3, 200
  vocal communication 107, 108f
chimpanzee-like hypothesis, human evolution 217
China 7

Chlorocebus cynosures (Mallbrouck monkey) 242
cichlid fish 243–4
citations in Google scholar 2f
Clay, Z 111–12, 112–13
co-dominance hypothesis (CDH) 50, 51
cognition 9
  chimpanzee *vs.* 214–32
  ecological hypothesis *see* ecological cognition evolution hypothesis; foraging cognition
  foraging in *see* foraging cognition
  non-kin prosociality 142
  studies *see* cognitive studies
cognitive development/evolution 1–2, 86–8, 214
  ecological hypothesis *see* ecological cognition evolution hypothesis
  Empathizing–Systemizing hypothesis 86f, 88
  self-domestication hypothesis 86–8
  tool use 176–7
cognitive studies 201–3
  gaze following 201–2
  point following 201–2, 201f
  Primate Cognition Test Battery 202–3, 203t
Cola chlamydantha 127t
Colobus angolensis (Angolan colobus) 242
combined pairwise affinity value (CPAV), male–male relationships study 38–9
common marmosets (*Calithrix jacchus*) 160
common squirrel monkey (*Saimiri sciureus*) 186
communication 8, 105–22
  aspect of 95–6
  facial expressions 67–9
  Play Face *see* Play Face (PF)
  Primate Cognition Test Battery 202, 203t
  studies *see* language studies
  see also gestures; vocal communication
community-based natural resource management (CBNRM) 258–9
community mapping 258–9
comparative studies/research 3f
Congo Basin Forest Partnership (CBFP) 253
Congo peafowl (*Afroparvo congensis*) 253
Congo River 235–48

animal phylogeographical studies 243–4
biogeographical implications 242–3, 243f
crossings of 136, 241–2
evolution in 3–4
formation in origin/evolution hypothesis 235–6, 237t, 238f, 240
historical discharge reduction 241
history of 9–10
connectivity area identification 257, 257f
Conscientiousness (HPQ) 195
  chimpanzee 184–5, 185t
conservation 251–65
  spatial model development 253–9
  tools of 10
constructions, language studies 101
construct validity, personality studies 189–90
contest hoots 107, 109f
cooperative nature 8–9
  non-kin prosociality 142
copulation
  aggression reduction and increased rates of 60
  baby dominance study analysis 55–6
  group encounters 28–9, 29f
  rates and, offspring social status association 57–8, 58f
  see also sexual relations; socio-sexual behaviour
copulation calls 110
corridor hypothesis, origin/evolution 235–6, 237t, 238f, 240–1
cortisol 218
cotton-top tamarin (*Saguinus oedipus*) 160
courtesy food sharing 129–32
courtship play fighting 69–70
CPAV (combined pairwise affinity value), male–male relationships study 38–9
cranio-morphology 106
Crockford, C 85

D

Dacroydes edulis 127t
data analysis, male–male relationships study 37
Daubentonia (aye-aye) 70
De braza's monkey (*Cercopithecus neglectus*) 242
decision-making, foraging cognition 160, 163, 163f, 166

Demidoff's dwarf galago (*Galagoides demidoff*) 242
Democratic Republic of Congo (DRC) 2, 251–2, 253*f*
  conflicts in 252, 254–5
  land tenure lack 258
  demographic modelling 266–77
    methods in population dynamics 267–9
development
  chimpanzee *vs.* 5
  food begging 132, 132*f*
de Waal, F B M 106–7
Dialium pachyphyllum (eimilimi) 127*t*
dietic gestures (humans) 116
distribution 2–3, 5–6, 6*f*, 200, 236*f*, 253*f*
  geographic boundaries 260
  range-wide mapping 259–61, 261*f*
dogs, self-domestication hypothesis 1216
Dominance (HPQ) 184, 185*t*
dominance rank 187–8
DRC *see* Democratic Republic of Congo (DRC)

E

ecological cognition evolution hypothesis 158–9
  see also foraging cognition
ecology, species divergence 222
eimilimi (*Dialium pachyphyllum*) 127*t*
emotional responsiveness 175–6
Empathizing–Systemizing hypothesis 86*f*, 88
empathy 89
endangered species classification 251–2
environment
  cognitive effects 209
  food sharing, evolution of 136
  predictor variables 260
  tool use 177
estrus
  periods of 22–5, 23*f*
  sex ratio 24*t*
ethics in personality studies 190
event-participant structure, language studies 101
evolution/origins 3–4, 244
  chimpanzee *vs.* 3–4
  cognition of 1–2
  Congo River crossing 241–2
  genetic information 237
  geological/geographic information 238–40, 239*f*
  hypotheses of 235–6, 237*t*, 238*f*, 240–1

Extroversion (HPQ) 193–4, 194–5
  chimpanzee *vs.* 185*t*
eye-tracking, prosociality 219

F

facial expressions 67–9
  open mouth display 67
  Play Face *see* Play Face (PF)
factor analysis studies, personality 184
false beliefs
  attribution to others by chimpanzee 85
  theory of mind 90*t*
female(s) 17–34, 30*f*
  chimpanzee in 17
  group encounters 28–9
  grouping patterns 18–21
  ranging 18–21, 30
  same-sex interactions 110
  social associations 21–2, 36, 61, 135–6
  social status 25–6, 30
  social status of male offspring 26–8, 26*f*, 27*f*, 52, 53–4*t*, 57–8, 60
  see also offspring dominance hypothesis (ODH)
  sons, bonds with 36, 43
female dominance 5, 25–6, 58
  masculinization 49–50
  non-masculinization 51
  other primate species *vs.* 51
female dominance through coalitions (FDCH) 50
fights, maternal support 26
5-HTTLPR gene 210
food
  adult tolerance around 72
  personality studies 189
  tolerance in non-kin prosociality 142–4, 143*f*
  transfers 128, 128*f*
  vocal communication 112
  see also food begging; food competition; food sharing; food type
food begging
  abundance and 129–30
  sex difference & developmental change 132, 132*f*
food competition
  chimpanzee 83
  female dominance 26
food sharing 125–39, 218–19
  courtesy food sharing 129–32
  environment and evolution 136
  food types 127*t*
  high-cost prosociality test 149–50

juvenilization 220*t*
male–male relationships 41, 42–3
meat 128
prosociality 151
reciprocity 126
seasonal changes 133, 133*t*
social relationships 129, 129*t*
food type
  foraging cognition 164, 164*f*
  Wamba field site 126–8
  see also terrestrial herbaceous vegetation (THV)
foraging
  call combinations 112–13
  chimpanzee 160–1
  chimpanzee *vs.* 222
  effort required 160–1
  preferences 5
  problems encountered 160–1
  seasonal variation 158
  vocal communication 20–1, 106
foraging cognition 158–9
  captivity studies 161
  decision-making 160, 163, 163*f*, 166
  definition 159–60
  divergence of, chimpanzee *vs.* 161–4
  food type 164, 164*f*
  foraging problem characteristics 160–1
  human evolution implications 165–6
  predictions of 160–1
  risk-averse character 163–4
  spatial memory 159–60, 161–3, 161*f*
  targeted divergence evidence 164–5
forest elephants (*Loxodonta africanus*) 243, 253
forest fragmentation 254–5
forest losss 254–5, 255*f*
forest monitoring 254–5
forest refugia hypothesis 235–6, 237*t*, 238*f*, 240
formalized greeting signals 107
FPF *see* Full Play Face (FPF)
Full Play Face (FPF) 67, 68*f*
  audience effect 68
  chimpanzee 67–8
functional flexibility, vocal communication 109–10

G

Galagoides demidoff (Demidoff's dwarf galago) 242
Galagoides thomasi (Thomas' dwarf galago) 242
gaze direction, theory of mind 90*t*

gaze following 173–4
  chimpanzee 83, 201–2, 201f
  cognitive studies 201–2, 201f
  Primate Cognition Test Battery 202, 203t
gender differences, personality 184
gender ratios
  food begging 132, 132f
  relative party size 19, 20f
gender roles, chimpanzee 17–18
general characteristics 3
genetics
  chimpanzee vs. 30
  male–male relationships study 39
genomic comparisons
  chimpanzee vs. 3
  human evolution 217
  human vs. 4
Genty, Emile 115, 116, 118
Geographic Information System (GIS) 251
geography
  boundaries in distribution 260
  evolution/origins 238–40, 239f
gestures 68–9, 113–18
  adjustment for goal 115
  attention-getters 95
  beckoning gesture 117–18, 117f
  definition 114
  flexibility of 115–16
  hand-begging 114
  language evolution 113–14
  meanings in 116–18
  sexual solicitation 114, 115, 116, 117f
  shared knowledge 115–16
  stretch-over gesture 114
  vocalization combination 118
GFW (Global Forest Watch) 262
giant mouse lemurs (Mirza) 70
GIS (Geographic Information System) 251
Global Forest Watch (GFW) 262
goals
  recognition in chimpanzee 85
  theory of mind 90t
Google scholar citations 2f
gorillas
  language teaching 96
  Play Face 68
grammar, language studies 99–102
Greenberg, J R 147
Greenfield, P M 99, 102
greetings 135–6
grooming 36
grooming density 187
grooming diversity 187

Groos, Karl 69
group encounters
  female roles 28–9
  vocalizations 28
group feeding studies, non-kin prosociality 143–4
grouping patterns, female roles 18–21
Guarea laurentii 127t

H
hamadryas baboon (Papio hamadryas) 70
hand-begging 114
Hare, B 82–3, 84–5, 216, 217–18
Hayes, C 96
Hayes, K J 96
hearing
  chimpanzee sensitivity to others perception 84
  theory of mind 90t
Herman, L 173
Heyes, C M 82
Hickey, Jena 259–60
high-cost prosociality, food sharing 149–50
high hoots 107, 108f, 113
Hirata, S 173
hominids, evolution of 244–5
Hominoid Personality Questionnaire (HPQ) 9, 184
  Agreeableness see Agreeableness (HPQ)
  Assertiveness see Assertiveness (HPQ)
  Attentiveness 195
  chimpanzee vs. 185t
  Conscientiousness see Conscientiousness (HPQ)
  Dominance 184, 185t
  Extroversion see Extroversion (HPQ)
  Neuroticism 185t
  Openness see Openness (HPQ)
hoots 107
Hopkins, W D 203, 223
hormones, prosociality 218
HPQ see Hominoid Personality Questionnaire (HPQ)
human evolution
  cognition 22
  foraging cognition implications 165–6
  theories of 216–17
humans, play and tolerance 73
hunter-gatherers 166
hunting pressure, conservation 255–6
hypothalamus 206–7, 207f

I
ICBR see Iyondji Community Bonobo Reserve (ICBR)
ignorance
  attribution to others by chimpanzee 84–5
  theory of mind 90t
immigration 144–5, 145f
  population dynamics 269–70, 272
innovation, theories of 177–8
instrumental helping task, low-cost prosociality 147–9
insular cortex 204–5, 210
intentions
  recognition by chimpanzee 85
  theory of mind 90t
Isolona congolana 127t
IUCN Red List 251–2, 253
Iyondji Community Bonobo Reserve (ICBR) 254
  boundary mapping 260

J
juveniles/juvenilization 220–1, 220t
  adults, displacement of 56–7
  interactions with adults 59
  see also baby dominance study

K
Kano, Takayoshi 5, 114, 173
Kanzi 100–1
  language studies 97–9, 97f
Karg, K 84
Kellogg, L A 96
Kellogg, W N 96
knowledge
  attribution to others by chimpanzee 84–5
  theory of mind 90t
Koko 96

L
Landsat programme 262
land tenure, lack of 258
land-use changes 254
language studies 95–104
  American Sign Language 96
  Austin 97–8
  grammar 99–102
  history of 96
  lexigrams 97–8
  Sherman 97–8
  symbol combinations 99–102
  symbols 97–9
  syntax 99–102
  Washoe 96

see also gestures; vocal communication
leaf sponging 173
Lemur catta (ring-tailed lemur) 73
lexigrams, language studies 97–8
Lola ya Bonobo 6
  baby dominance study methods 53
  population dynamics *see* population dynamics
Lomako Yokokala Faunal Reserve
  boundary mapping 260
  connectivity areas 258
  food sharing 134
long-distance vocal communication 113
low-cost prosociality
  instrumental helping task 147–9
  standard prosocial choice task 145–7, 146*t*
Loxodonta africanus (forest elephants) 243, 253
Lualaba River detour hypothesis 235–6, 237*t*, 238*f*, 240–1
Lui Katole (Salonga National Park) 5
  food sharing 134
  male–male relationships study 37
Luo–Iyondji–Kokolopori reserves, connectivity areas 258
Luo Scientific Reserve (LuoSR) 5, 254

M

macaques (*Macaca* spp)
  black-crested macaque (*Macaca nigra*) 186
  5-HTTLPR gene 210
  play and tolerance 73–4
MacLean, E L 84–5
male(s)
  coercion intolerance 5
  dominance in chimpanzee 5
  juveniles, displacement by 56–7
  maternal bonds 36
  non-dominance 50
  patrolling behaviour 28
  social status and maternal influence 26–8, 26*f*, 27*f*, 44, 57, 60
male–male relationships 35–46, 61, 223
  aggression 42
  agonistic alliances 41, 42–3
  diversity of 39–41, 40*f*, 42*f*
  food sharing 41, 42–3
  maternal kinship 41, 43
  philopatric groups 21, 36
  results 39–41
  social-sexual behaviour 42, 42*f*, 43

study methods 37–9
Mallbrouck monkey (*Chlorocebus cynosures*) 242
Maringa–Lopori–Wamba (MLW) Landscape 252, 257*f*, 258
masculinization
  decrease in 89
  female dominance 49–50
maternal dependance, juvenilization 220*t*
maternal kinship, male–male relationships 41, 43
Mauss, M 129
MCDA (multi-criteria decision analysis) 255
McGrew, W C 178
meat, food sharing 128
Meliponinae 127*t*
Melis, A P 147
Memorandum of Understanding (MOU) 258
memory tasks 222
microstructural studies, brain organization 204–8
mind 1–13
Mirza (giant mouse lemurs) 70
mitochondrial DNA (mtDNA) 237
MLW (Maringa–Lopori–Wamba) Landscape 252, 257*f*, 258
modelling scenarios, population dynamics 268–9, 268*t*, 270*f*
model predators, personality studies 188–9
monophylectic origins 238
mortality, population dynamics 272–3
mosaic hypothesis of human evolution 216–18
motivation
  non-kin prosociality 142
  theory of mind 91
MOU (Memorandum of Understanding) 258
multi-criteria decision analysis (MCDA) 255
Musanga cercoploides 127*t*

N

National Science Foundation (USA) 6, 7*f*
naturalistic observation, personality studies 187–8
nearest neighbour 55
neural crest hypothesis 221
neuropeptides, social behaviour 209
neuropil fraction, Broca's area 208, 208*f*
Neuroticism (HPQ) 185*t*

nightfall, vocalizations 21
non-food items, prosociality 151
non-kin prosociality 140–54
  basis of 142
  cooperation with 142
  food tolerance 142–4, 143*f*
  social tolerance 144–5
nuclear DNA 237
number of neighbours 187
nut-cracking 178

O

object choice tasks 89
offspring dominance hypothesis (ODH) 50, 51–3, 52–3
  feeding behaviour 52
  maternal response to male interference 52
  social context of infant 52
  sub-adult individuals 52–3
olive baboon (*Papio anubis*) 70
Omedes, A 106
open mouth display 67
Openness (HPQ) 193–4
  chimpanzee *vs.* 185*t*
  developmental changes 186
orangutan (*Pongo*) 96
OXTR polymorphisms 206–7
oxytocin 206–7, 207*f*, 208*t*
  prosociality 219–20

P

pairwise affinity value (PAV), male–male relationships 37–8, 39–40
Pakia bicolor 127*t*
Pan African Sanctuary Alliance (PASA) 266–7
Pancovia laurentii 127*t*
pant grunt, chimpanzee 107
pant hoot, chimpanzee 107, 108*f*
Pan troglodytes see chimpanzee (*Pan troglodytes*)
Papio anubis (olive baboon) 70
Papio hamadryas (hamadryas baboon) 70
paraventricular nucleus (PVN) 206
Parr, L 223
party affinity, male–male relationships 38
patience, chimpanzee 161
PCTB (Primate Cognition Test Battery) 202–3, 203*t*
perception, understanding of 83–4
Perodicticus (potto) 70
personality 183–98
  behavioural variables 191, 193*t*

definition 183–4
developmental changes 185–6
gender differences 184
measures of 184
personality studies 187–96
  analyses 189–90
  methods 187–9
  results 190–2
PF *see* Play Face (PF)
phenotypes 4
  chimpanzee *vs.* 4–5
  human evolution 217
physical cognition in juvenilization 220t, 221
pitch, vocal communication 106
play 65–77
  adult tolerance around food and 72
  definition 65, 66
  development of 70–1, 71f
  importance of 69–72
  juvenilization 220t
  networks 72, 72f
  roles of 66
  social tolerance 70
  symmetrical play 70–1
  tolerance effects 72–4
Play Face (PF) 67, 67f
  audience effect 68
  chimpanzee 67–8
  gorillas 68
play-fighting *see* rough-and-tumble play
Play of Animals (Groos) 69
point following
  chimpanzee 201–2, 201f
  cognitive studies 201–2, 201f
Pollick, A S 118
Pongo (orangutan) 96
population dynamics 266–77
  births 269
  demographic analysis methods 267
  demographic modelling methods 267–9, 268
  future projections 270–1, 270f, 271t
  historical patterns 269–70, 269f
  modelling scenarios 268–9, 268t, 270f
  study population 267
potto *(Perodicticus)* 70
Povinelli, D J 82
Premack, D 81–2, 96
Primate Cognition (Tomasello & Call) 82
Primate Cognition Test Battery (PCTB) 202–3, 203t
proactive nature of prosociality 151
prolonged pseudo-estrus 22–5
pronouns 98–9
prosociality 141t
  chimpanzee 140–1, 141t, 151

definition 140
high-cost, food sharing 149–50
juvenilization 220t
low-cost *see* low-cost prosociality
non-food items 151
non-kin *see* non-kin prosociality
proactive nature 151
social cognition 219–20
stranger preference 218–19
studies of 145–50
proximity
  baby dominance study methods 54
  male–male relationships study 38–9
psychological bias 177–8
puzzle feeder experiments 189
PVN (paraventricular nucleus) 206

Q

questionnaire ratings, personality studies 187

R

range-wide mapping 259–61, 261f
ranging, female roles 18–21, 30
rank
  baby dominance study analysis 55
  copulation calls 110, 111f
  personality studies 189, 190
Raphia 127t
raspberrys 109
recipient initiation, food transfers 128, 128f
reciprocity, food sharing 126
red-tailed monkey *(Cercopithecus ascanius)* 242
reduced scramble competition 222
reintroduction, population dynamics 271t, 272f, 273
relative party size 18–20, 19f
  chimpanzee 18, 19–20
  gender ratios 19, 20f
release scenarios 271
reproduction
  limitation 273
  population dynamics 271, 272f, 273f
reproductive isolation, evolution in 4
research problems 6–7, 7f
Rilling, J K 203
ring-tailed lemur *(Lemur catta)* 73
risk-averse character 163–4
rough-and-tumble play 66, 69–70
  adults 69–70
  gestures 69

S

Saba florida 127t
Saginus oedipus (tamarins) 73

Saguinus oedipus (cotton-top tamarin) 160
Saimiri sciureus (common squirrel monkey) 186
Savage-Rumbaugh, E S 96, 97–8, 99, 102, 114
SAW (simple additive weighting) 255
SDH *see* self-domestication hypothesis (SDH)
seasonal variation
  food sharing 133, 133t
  foraging 158
seeing, theory of mind 90t
self-domestication hypothesis (SDH) 90, 214–32
  cognitive evolution 86–8
  developmental predictions 221
  development of 215–16, 217–18
  neural crest hypothesis 221
  psychology of 216–18
  study support for 226
  theory of mind 88–9
  see also juveniles/juvenilization
serotonin (5-HT) receptors in amygdala 207–8, 209–10
sex differences
  behavioural variables 189, 190–1, 191–2t
  personality studies 189, 190, 190f, 194
sex-specific mortality rates (Q_X) 267–8
sexual relations 24–5, 24f
  see also copulation
sexual solicitation, gestures 114, 115, 116, 117f
Shannon–Wiener index 187
shared goals, theory of mind 90t
shared knowledge, gestures 115–16
sharing-under-pressure hypothesis 126
Sherman 97–8
signal comprehension 110
simple additive weighting (SAW) 255
sit alone, personality studies 187
snake alerting studies, chimpanzee 85
social assessment, play 71
social associations, female roles 21–2, 36
social awareness 105–22
  vocal communication 108–12
social behaviour 8
  conflict and vocal communication 111–12
  neuropeptides 209
  population dynamics 270f, 273–4
social bonds
  begging 125–39
  begging for *see* begging for social bonds
social cognition, prosociality 219–20

social information sensitivity 89
social inhibition
  developmental delay 71, 71f
  juvenilization 220t
social intelligence, technical
    intelligence vs. 173–4
social relationships, food sharing
    129, 129t
social status, female roles 25–6, 30
social tolerance 65–77
  non-kin prosociality 144–5
  play 70
  strangers 144–5
socio-sexual behaviour 36, 70, 110
  baby dominance study methods 54
  juvenilization 220t, 221
  male–male relationships 42, 42f, 43
solitary play 67
spatial memory
  chimpanzee 161, 200
  foraging cognition 159–60, 161–3
  juvenilization 220t
spatial modelling 255–7, 256f, 257f
  hunting pressure 255–6
speciation in drier habitat hypothesis
    235–6, 237t, 238f, 240
specific superior females 21–2
spoken English, understanding of 100
standard prosocial choice task 145–7,
    146t
stranger call studies 223–5, 225f
stranger preference *see* xenophilia
    (stranger preference)
strangers, reactions to 223–5
  see also xenophilia (stranger
    preference); xenophobia
stretch-over gesture 114
study population, population
    dynamics 267
study sites
  baby dominance study methods 53
  male–male relationships study 37
subgroups (parties) 18
subjects
  baby dominance study methods
    53–4, 53t, 54t
  male–male relationships study 37,
    37t
submissive grunt 107
sunset calls 21
supraoptic nucleus (SON) 206
symmetrical play 70–1
syntax, language studies 99–102

T

tamarins *(Saginus oedipus)* 73
technical intelligence, social
    intelligence vs. 173–4

temperament
  bonobos *vs.* chimpanzee 174–5
  non-kin prosociality 142
Terrace, H S 96
terrestrial herbaceous vegetation
    (THV) 161
  chimpanzee, lack of 222
territoriality, chimpanzee 144
theories of innovation 177–8
theory of mind (ToM) 81–94, 82f,
    90t
  chimpanzees *see* theory of mind in
    chimpanzee
  developmental delays and 87
  motivation 91
  Primate Cognition Test Battery
    (PCTB) 202, 203t
  self-domestication hypothesis
    88–9
  study lack 91–2
  study replication 91
theory of mind in chimpanzee 82–6,
    90t
  historical perspective 82–3
  understanding of perception
    83–4
Thomas' dwarf galago *(Galagoides
    thomasi)* 242
THV *see* terrestrial herbaceous
    vegetation (THV)
tolerance 135–6
ToM *see* theory of mind (ToM)
Tomasello, M 82, 83, 100
tool use 171–80, 174f
  captivity studies 174,
    174f
  chimpanzee 200
  cognitive development 176–7
  environment 177
  food gathering 161
  leaf sponging 173
  nut-cracking 178
  temperament 174–5
  wild in 172–3
Treculia africana 127, 127t

U

Uakari monkeys *(Cacajao)*
    70
United States Geological Survey
    (USGS) 262
unit groups (communities) 18

V

vocal communication 106–13,
    200
  aggression 111–12

call combinations 112–13
chimpanzee 107, 108f
contest hoots 107, 109f
copulation calls 110
feeding events 112
flexibility in 108–9
foraging in 20–1
functional flexibility 109–10
gesture combination 118
group encounters 28
high hoots 107, 108f,
    113
hoots 107
long-distance 113
at nightfall 21
nightfall at 21
pitch 106
raspberrys 109
signal comprehension 110
social awareness with 108–12
social conflict 111–12
stranger call studies 223–5,
    225f
submissive grunt 107
sunset calls 21
vocal flexibility 113
Voxel-Based Morphology
    (VBM) 201
  brain organization studies 203–4,
    205f, 206t, 210

W

Wamba field site 5
  food sharing 126–9, 133–4
Washoe 96
whistles 113
wider habitat hypothesis,
    origin/evolution 235–6, 237t,
    238f, 240
wildland blocks 256–7
Wobber, V 52
Woodruff, D S 81–2
word-order consistency, language
    studies 100

X

xenophilia (stranger preference) 9,
    144–5, 223
  juvenilization 220t
  prosociality 218–19
xenophobia 223

Y

Yerkes, Robert 3

Z

Zuberbühler, K 112–13, 116